Advances in Thermodynamics

Advances in Thermodynamics

Editor   **G. Ali Mansoori**
Department of Chemical Engineering
University of Illinois
P.O. Box 4348, Chicago, IL 60680

Editorial Board

Larry G. Chorn
Mobil Research & Development Corp.
Dallas Research Laboratory
P.O. Box 819047, Dallas, TX 75381

Enrico Matteoli
Instituto di Chimica Quantistica ed
Energetica Molecolare del C.N.R.
Via Risorgimento 35, 56100 Pisa, Italy

Peter Salamon
San Diego State University
San Diego, CA 92182

Stanislaw Sieniutycz
Warsaw Technical University
Warsaw, Poland

Finite-Time Thermodynamics and Thermoeconomics

EDITED BY

Stanislaw Sieniutycz and Peter Salamon

Taylor & Francis

New York • Bristol, PA • Washington D.C. • London

USA	Publishing Office:	Taylor & Francis New York Inc.
		79 Madison Ave., New York, NY 10016-7892
	Sales Office:	Taylor & Francis Inc.
		1900 Frost Road, Bristol, PA 19007-1598
UK		Taylor & Francis Ltd.
		4 John St., London WC1N 2ET

Finite-Time Thermodynamics and Thermoeconomics

First published 1990
Printed in the United States of America

Library of Congress Cataloging in Publication Data

Finite-time thermodynamics and thermoeconomics / edited by Stanislaw
    Sieniutycz and Peter Salamon.
          p.   cm. -- (Advances in thermodynamics ; v. 4)
    Includes bibliographical references and index.
    ISBN 0-8448-1668-X
    1. Thermodynamics. 2. Thermodynamics--Economic aspects.
I. Sieniutycz, Stanislaw.   II. Salamon, Peter, 1950-
III. Series.
TJ260.F56   1990
621.402'1--dc20   90-11259
                                                           CIP

Contents

v

Contributors

B. Andresen
Physics Laboratory, University of Copenhagen, Denmark

P. Le Goff
Laboratoire des Sciences du Genie Chimique, Nancy, France

J.M. Gordon
Department of Mechanical Engineering, Ben Gurion University of the Negev Beersheva, Israel

K. H. Hoffmann
Institut fur Theoretische Physik, Universitat Heidelberg, FRG

K. Lund
Department of Mechanical Engineering, San Diego State University, San Diego, CA

B. Å. Månsson
Theoretical Ecology, Research Center Jülich, FRG, and Institute of Physical Resource Theory, Göteborg, Sweden

S. de Oliveira
Laboratoire des Sciences du Genie Chimique, Nancy, France

M. Ondrechen
Department of Chemistry, Northeastern University, Boston, MA

R. Rivero
Laboratoire des Sciences du Genie Chimique, Nancy, France

P. Salamon
Department of Mathematics, San Diego State University, CA

B. Schwarzer
Laboratoire des Sciences du Genie Chimique, Nancy, France

S. Sieniutycz
Institute of Chemical Engineering, Warsaw Technical University, Poland

J. Szargut
Institute of Thermal Engineering, Technical University of Silesia, Gliwice, Poland

Z. Szwast
Institute of Chemical Engineering, Warsaw Technical University, Poland

D. Tondeur
Laboratoire des Sciences du Genie Chimique, Nancy, France

Foreword

Finite-time thermodynamics, at least under that name, had its origin in the context of improving the efficiency of energy use. Comparisons had been made between actual consumption of free energy in industrial processes and the ideal thermodynamic limits for the same processes. People questioned the value of these comparisons because the ideal limits were based, of course, on ideal, reversible processes, not on processes operating at real, nonzero rates. And, while reversible models are sometimes good for describing real processes, many real processes operate at too far from reversibility to be considered "infinitely" slow What would be the realistic limits on the performance of a process if the specification of the process itself contained constraints on the time rate of the process? Would it be possible to find limits of performance of processes with such constraints? If so, would it also be possible to find the pathways that would yield those limits? What criteria could be used to evaluate performance? Addressing these questions was the origin of the subject of this volume.

Thermodynamics has a special role in science because of its remarkable generality and economy of structure. It sometimes gives people the impression of being a closed, completed subject. Yet by asking questions a little different from previous ones, we occasionally discover whole new areas that extend the subject in useful directions, like finding new chambers in a cavern whose old rooms are well known. Onsager's linear model for irreversible processes and the reciprocal relations that from it is one such discovery. In a sense, this is *the* prototypical thermodynamic representation of irreversibility, done in local, microscopic terms of differential equations, like the differential equations of motion mechanical system.

Finite-time thermodynamics, stemming from a global question about optimization, is a kind of integral complement to the differential viewpoint of Onsager. Its equations are integral, at least to begin with, rather than differential, although the Lagrangian or Hamiltonian formalism immediately brings one back to the differentials. Fortunately, it was possible to prove that for a large class of irreversible systems, one can define qualities that act as potentials, in a sense that their changes give the limiting values of work or heat exchanged in a finite-time process. This proof was enough to justify pursuing the task of using a thermodynamic line of approach to optimize energy use for processes constrained to operate in finite time.

But there is one other very important way in which finite-time thermodynamics extends thermodynamics. The focus on optimization, on finding the pathway that yields optimum performance, introduces characteristics that are normally not part of a traditional thermodynamic approach. One is the freedom of choice of objective function, the function is to be optimized. In traditional thermodynamics, one has considerable choice of the constraints but little choice of what is optimized--work, heat efficiency of coefficient of performance (CEP), and these are linked to one another. When one moves to a system operating in finite time, other rich criteria such as power and economic performance (net return) can be introduced. The flexibility in turn raises questions of comparison of the performance of systems optimized for different criteria, a problem outside classical thermodynamics.

The very concern with optimization is itself a shift from the attitude of traditional thermodynamics or of physical science generally. In a conventional scientific approach, the observer is passive, recording and describing what the system does. By adopting a course of optimization, the scientist becomes an active participant who interacts with the system. This is nothing new to engineers, who operate this way regularly, but it is novel to other scientists. And it is a style which seems to be spreading to other parts of physical science, outside of thermodynamics. It is an approach that implies we want to do *something* with our scientific insights besides observe and describe systems.

As this volume shows, finite-time thermodynamics and its close relative, thermoeconomics, encompass a full scale from basic questions to very practical problems. The subject, like thermodynamics itself, illustrates a fascinating historical aspect of science which was sometimes overlooked during the twentieth century. It has been applied, in much of what contemporary scientists do and say, that the applied follows from the basic: that science and technology progress by a sequence beginning with a basic discovery, such as the discovery of quantum physics, and from this come insights that lead to applications and new technology. What thermodynamics showed in the 18th Century and finite-time thermodynamics demonstrated in our own time is that the flow may go equally well in the opposite direction. Asking a very practical question, for example about how to improve the efficiency of energy use, may lead us to addressing very basic issues indeed. That science flows both ways, between basic and applied, should be one of the overarching lessons of this volume.

R. Stephen Berry
The University of Chicago
Chicago, Illinois 60637

Introduction

This multiauthored volume deals with applications of nonequilibrium thermodynamics and availability theory to (mostly lumped) systems where a certain external control can be applied in order to achieve improved performance. Approaches from nonequilibrium thermodynamics are capable of providing quite realistic performance criteria and bounds for real processes occurring in a finite time. Model systems have been developed which incorporate friction, heat loss, inertial effects and finite heat conductance for real energy conversion processes. Finite-time thermodynamics, exergy analysis and thermoeconomics seek the best adjustable parameters of various engines, (thermal, solar, combustion, acoustic, convection cells, etc.), unit operations and unit processes (distillation, evaporation, chemical reactions, etc.) and systems of these operations or processes in chemical plants working under definite operational constraints. Optimal paths (or a set of optimal steady-state parameters) and optimal controls (for instance temperatures maximizing chemical efficiency), have been found. The role of optimization approaches and in particular optimal control theory is essential when solving these problems. The optimization results for a few important industrial and ecological systems are given.

Thermodynamics and Optimization

Stanislaw Sieniutycz
Warsaw Technical Institute,Warsaw, Poland

Peter Salamon
San Diego State University, San Diego, CA 92812

ABSTRACT
The role of thermodynamic optimization approaches is defined
within the family of various optimization problems. Static and
dynamic methods of optimization are briefly characterized. Optimal
control theory is central to formulating and solving problems of
optimal trajectories and optimal decisions required by availability
analysis, entropy source minimization, finite-time thermodynamics,
thermoeconomics, and variational formulations of the irreversible
equations of motion. We find that the difficulties in many variational
and control theoretic approaches to irreversible processes can be
overcome by assuming that the (generalized) kinetic potentials or
Lagrangians contain an action variable explicitly. The relative merits
of these approaches are discussed. The contents of the papers on
thermodynamic optimization, finite-time thermodynamics and
thermoeconomics appearing in the later part of this volume are
briefly reviewed.

1. INTRODUCTION
In the last two decades more and more scientists have devoted their
research to the role of optimal control theory in macroscopic physical
systems. This has been true in the context of availability analysis
(exergy optimization) and the closely related optimizations of
thermodynamic processes in finite-time (Andresen 1983; Andresen
et al. 1984; Bejan 1988; Sieniutycz and Szwast 1982; Szargut and
Petela 1965), as well as in the related field called the
thermoeconomics (Berry et al. 1978; Mansson 1985; Sieniutycz 1978)

which links economic analyses with thermodynamics. The results of these analyses have explored in-principle limits to the efficiency of energy conversion in finite time (Salamon et al. 1980; Salamon and Berry 1983; Nulton et al. 1985), have found ways to improve the power that can be extracted from various engines (Rubin 1980; Salamon et al. 1982; Salamon and Nitzan 1981; Band et al. 1980, 1981, 1982a, 1982b; Hoffmann et al. 1985; Mozurkewich and Berry 1981, 1982) and have provided methods to improve important industrial processes (Mansson and Andresen 1985). Associated directions have also appeared in the context of "cooling" such quasi-physical systems as simulated annealing problems in which one exploits a physical analogy to solve optimization problems plagued by many local minima (Nulton and Salamon 1988; Harland and Salamon 1988).

Finding the strategy which optimally drives a physical system is always a nontrivial task. Accordingly, the mathematical theory of optimal control has become a standard tool for solving these problems. The importance of thermodynamics as a tool in many disciplines endows these problems with an interdisciplinary character which touches the theory of heat engines, solar energy engineering, mechanical and chemical engineering, economics of industrial processes, ecology, etc.

The structure of this paper is as follows. Section 2 discusses the role of optimization theory in physical analyses. Then, after brief review of optimum seeking methods (sections 3A and 3B), the application of optimization theory to the variational formulations of irreversible dynamics is analyzed in section 4. Section 5 discusses the papers appearing in the later part of this volume.

2. OPTIMIZATION IN THERMODYNAMIC SYSTEMS
Roughly speaking there are the two broad categories of approaches where the methods of the contemporary mathematical theory of optimization can be used for analyzing thermodynamic systems. In the first approach, when one wants to *optimize* these systems by changing some external parameters (decisions or controls), both extremal and non-extremal solutions pertain to real physical situations. In this case, the governing equations describing the changes of the internal state are known and a control is imposed through the system boundaries. This first approach includes studies which seek in-principle limits to the operation of thermodynamic processes and studies which seek to improve existing engineering

systems. In the second approach, one wants to predict the system behavior (usually irreversible) under prescribed external conditions and therefore seeks to derive its governing equations from certain *variational or extremum principles*. In these systems, the extremal control is set by nature rather than by man.

In this section we discuss the first approach, i.e. the role of thermodynamics for optimization. This is done rather briefly and only in general terms since detailed analyses are the subject of a number of subsequent papers in this volume.

While thermodynamics (mostly irreversible) is the dominant ingredient in availability analysis, exergy optimization and the minimization of entropy production, it plays only a secondary role in thermoeconomics where the overall problem deals with the balance of value (economic balance). Since the economic performance index is always formulated in economic terms (prices), the solution of any thermoeconomic problem is not equivalent to that of the corresponding thermodynamic problem. It does however reduce to thermodynamic optimizations in the limit when the price of certain thermodynamic quantities such as the availability used or the power produced becomes much larger than the prices of other participating quantities (Andresen et al. 1983). This limit represents an energy theory of value, i.e. a value system in which one considers energy as the single valuable commodity (Berry et al. 1978).

A second way for economic optima to coincide with thermodynamic optima is if the economic value of the exergy unit is the same for all forms of matter and energy taking part in the process. In this case the thermodynamic problem of minimum availability loss (due to the internal irreversibilities in the system) is equivalent to minimum economic costs. This case is however rather special, i.e. the prices of the exergy units generally differ (Sieniutycz 1978).

Hence, the general (thermo)economic optimization problem cannot be, as a rule, broken down to the problem of minimum irreversibility. Even the thermodynamic problem of minimizing the driving exergy (available energy of the input flows) is not equivalent to minimizing irreversibility since since it does not take into account the exergy of the outgoing flows. The problem of minimum cumulative exergy consumption (a generalization of the previous problem), which is of interest for some complex industrial and ecological systems with many units and interconnections, is also not equivalent to the problem of minimum dissipation for the same reasons. These problems belongs rather to energy management. Additional unconventional terms in the performance index appear in ecological analyses where a cost-type optimization criterion takes into account not only the cumulative consumption of exergy but also the exergy losses resulting from the deleterious impact of the wastes on human activity and health, quality of agricultural crops, forestry, natural resources etc.

Nevertheless, a number of complex economic and ecological problems were born as extensions and generalizations of irreversibility analysis and their connection with thermodynamics is still strong enough to justify the common thermodynamic context. Although various performance criteria can be formulated for various problems, the physical constraints (balance equations, equations of state, thermal equilibria, and kinetic relationships) are the product of thermodynamic analyses. Even if the thermodynamic criteria are replaced by the more natural economic ones, the set of process constraints frequently remains unchanged and the same optimum seeking method can be used in both thermodynamic and economic cases. Thus most of the methodological experience in the formulation of the performance index and even frequently the majority of the constraints can be preserved when passing from the thermodynamic to the economic optimization.

3. MATHEMATICS OF OPTIMIZATION

Optimization is seeking of the best solution under given conditions (constraints). Such a general definition of optimization includes most conscious activity pursued by mankind during its long history. Its mathematical pursuit also has a long history, but several major strides have been made during the last three decades. We pause to review several of the most important recent milestones for the development of the discipline. These are: Dantzig's methods for linear programming (Dantzig 1968; Llevellyn 1963), the Karush-Kuhn-Tucker conditions for nonlinear programming, (Kuhn and Tucker 1951; Zangwill 1969), Bellman's dynamic programming (Bellman 1958, 1964; Bellman and Kalaba 1965; Aris 1964), and Pontryagin's maximum principle (Pontryagin at al 1962, Leitman 1981, Fan 1964a; Fan and Wang 1964; Sieniutycz 1978). These methods are complemented by the classical methods of the differential calculus and Lagrange multipliers as well as the calculus of variations (functional optima; Gelfand and Fomin 1963). Together they make a powerful collection of optimization methods which can solve many difficult practical problems (Beveridge and Schechter 1970). For this development, progress in computational techniques has been essential.

The mathematical nature of the optimization problem lends itself naturally to a division between finite and infinite dimensional problems. In the first, called the mathematical programming problem, the optimization criterion (performance index) is the *function* of the decisions (and possibly also other additional variables like constants, parameters etc., called collectively the uncontrolled variables). In the second, called the dynamic optimization problem or the variational problem, the optimization criterion is the *functional* of the decisions (as well as state variables and constant parameters). In the first case, one is searching for an optimum in the form of an n-tuple *of numbers*. In the second case the optimizer's task is to determine the *functions* describing the dynamic behavior of the optimal control as a function of time or as a function of any time-like independent variable characterizing the evolution of the process (length, residence time etc.). These functions constitute the optimal control which characterizes the best external action on the system in time. The optimal control is obtained as the basic ingredient of the optimization solution simultaneously with another vector function which characterizes the best time behavior of the process state, called the optimal trajectory. The stable asymptotic solution of the optimal control problem (if it exists), is associated with a steady-state situation when the process time tends to infinity and the external controls are time independent. These asymptotic solutions are often the solutions to a corresponding static problem. However, this fact is seldom exploited directly, i.e. we seldom to seek solutions to the static problems as the special (steady-state) solutions of dynamic problems. The static optimization problems (mathematical programming problems) have their own methods which are simpler than those of the dynamic problems. Hence tractable static problems may involve many more components of the control vector than their dynamic counterparts.

The complexity of the optimization problems is as a rule connected with the complexity and variety of the constraints. The constraint equations and inequalities may be algebraic, difference, differential, integral and integro-differential. For mathematical programming problems (static optimization problems) the constraints are algebraic. For variational problems (dynamic optimization) differential equations are the natural constraints although any other type of constraints specified above can additionally appear in more involved variational problems. This may result in an extremely complex structure for dynamic optimization problems. This is not to say that static problems of mathematical programming cannot be complex and difficult to solve. These difficulties are due to the nonlinearities that may appear in the constraining equations and/or the performance index.

A. Static methods

When both the constraining equations and the performance index are linear, the optimization problem is a linear programming problem (Dantzig 1968; Llevellyn 1963). The linear programming (LP) problems include transportation problems, distribution from sources to sinks, traveling salesman problems, allocation of resources among activities, management decisions, etc. The LP problem is usually too restrictive for thermodynamic applications where at least the objective function is nonlinear by nature.

When one or more of the constraints, or the objective function are nonlinear, the statics problem is a nonlinear programming problem. It can be formulated as follows. Determine values for n variables $y = (y_1, y_2, ..., y_n)$ that optimize the scalar objective function

$$A(y) = A(y_1, y_2, ...,y_n) \qquad (1)$$

subject to l equality constraints

$$g_i(y) = g_i(y_1, y_2, ...,y_n) = 0, \qquad i = 1,....., l \qquad (2)$$

and to m-l inequality constraints

$$g_i(y) = g_i(y_1, y_2, ...,y_n) \geq 0, \qquad i = l+1,....., m \qquad (3)$$

Nonlinear programming problems may have widely varying properties, and certain limitations on the forms of the functions appearing therein are necessary if the problems are to be solved. The Karush-Kuhn-Tucker conditions, generalizing the classical Lagrange multipliers to the case involving inequality constraints (Greig 1980; Varaiya 1972; Kuhn and Tucker 1951; Zangwill 1969), is the basic theoretical tool for solving nonlinear programming problems.

$$\nabla A(\mathbf{y}) = \sum_{i=1}^{m} \lambda_i g_i$$

$$\lambda_i \geq 0, \qquad g_i(\mathbf{y}) \geq 0, \qquad i = l+1,\ldots, m \qquad\qquad (4)$$

$$\lambda_i g_i = 0, \qquad i = 1,\ldots, m$$

Typical algorithms for its solution proceed by absorbing the constraints, eqs. (2) and (3), into an augmented optimization criterion. We refer the reader to the many books available on the subject (Greig 1980; Varaiya 1972; Bracken and McCormick 1968; Zangwill 1969; Beveridge and Schechter 1970). We mention an important special case: the equilibrium state of a thermodynamic system is in a natural fashion a nonlinear programming problem where the free energy of a system is minimized. For the example of a chemical mixture, the free energy is minimized subject to the linear constraints resulting from the conservation of the atoms of the elements; the minimization determines the concentrations at chemical equilibrium (White, Johnson and Dantzig 1958).

The Kuhn-Tucker method is very general and hence not always the most effective. Sometimes, when only simple equality constraints are present, a number of variables (equal to the number of the equality constraints) can be eliminated and the problem can be reduced to the problem of extremizing an unconstrained function of the remaining variables. For this case (unconstrained optimization) of a multivariable function) a variety of iterative non-gradient as well as gradient techniques can be used to find the optimum. Most frequently they are based on iterative searches for optima along certain directions in the decision space. These searches start from an arbitrary point and terminate close to the optimum. Many reviews of these methods and their applications are available (Beveridge and Schechter 1970; Sieniutycz and Szwast 1980) along with the associated proofs of convergence (Zangwill 1969). The constraints can be taken into account not only by the Lagrange multiplier approach, as in the Kuhn-Tucker method, but also by introducing various penalty terms. These add terms to the objective function which penalize violations of the constraints by increasing the objective if any constraint is violated. This forces any search procedure to leave this region quickly (Bracken and McCormick 1968).

<u>B. Dynamic methods</u>
These methods can be further subdivided into discrete and continuous. We begin by addressing methods for discrete problems.

Such problems are also refered to as multistage decision processes. Perhaps the widest used method for such processes is the discrete version of the dynamic programming method (Bellman 1957, 1967). This is based on the Bellman principle which asserts that at each stage of the process we need compare only the costs of the current decision plus the minimal cost from the state to which this decision moves us. Unfortunately, the effectiveness of numerical algorithms implementing this method work effectively only when the dimensionality of the state vector is low. Hence, for larger dimensional problems, other methods are needed. The other popular method is a discrete version of the Pontryagin maximum principle (Fan and Wang 1964; Sieniutycz 1978). The discrete problem is defined by an objective function in the form of a sum over the costs of single-stages, and constraints in the form of the difference equations (some local algebraic constraints for the decisions and/or state variables at each stage can also appear). The corresponding optimization solution comprises the sequence of the optimal controls and the sequence of the optimal state variables (the discrete trajectory).

When the number of the discrete intervals (number of process stages) as well as the initial and final states and times of the process remain the same, the discrete problem tends to a limiting continuous problem. This is our primary tool in the present volume and so we present the formalism in some detail. The optimal objective function A^o (the cumulative value of the optimum cost corresponding to a final state $\mathbf{X}^f$ and the final time t^f of the process) can be viewed as a function of all the components of the current enlarged state vector, $\mathbf{X} = (x_0, x_1, ..., x_n) = (x_0, \mathbf{x})$, and the current time t. This is central to Bellman's principle (Bellman 1957, 1967).

In the Mayer formulation of the continuous control problem, considered here, the objective function A is the zero-th component of the enlarged state vector, $\mathbf{X} = (x_0, x_1, ..., x_n) = (x_0, \mathbf{x})$, at some final time t^f, i.e. $A = x_0(t^f)$. The optimization is subject to the following constraints

$$\frac{d\mathbf{X}}{dt} = \mathbf{F}(\mathbf{X}, \mathbf{u}, t)$$

$$x_k(t^i) = x_k^i, \qquad k \in K$$

$$x_j(t^f) = x_j^i, \qquad j \in J \tag{5}$$

$$\mathbf{u} \in \mathbf{U}$$

with the n+1 dimensional enlarged state vector $\mathbf{X}$, enlarged rate vector $\mathbf{F} = (f_0, f_1, ...f_n)$, r- dimensional control vector $\mathbf{u}$, and the time t. The continuous problem, described above, pertains to a time evolution of a continuous system with an external control, under the specified boundary conditions. These conditions usually preassign the initial time and some of the coordinates x_k of the state vector at the initial time and another part of the state coordinates x_j at its end. The zero-th component must of course be unspecified as this is the objective to be maximized or minimized. This is the classical problem of optimal control as formulated by Pontryagin. It is best attacked by the very general and powerful optimization method called the maximum principle (Pontryagin at al 1962; Leitman 1981; Fan 1964a). The name of the method is associated with the following basic property of optimal solutions: in order to make a performance index x_0 maximum (minimum) over the whole path, the Hamiltonian function, $H= \mathbf{Z \cdot F(X, u}, t)$, must attain the global maximum (minimum) with respect to the admissible control vector $\mathbf{u}$ at each time instant on the optimal path.

The Hamiltonian H is defined as the scalar product of all the rates $\mathbf{F}$ and the so-called adjoint variables $\mathbf{Z}$. Note that the definition of H includes the zeroth component, i.e., the rate of change of the cost and its adjoint. The latter can be taken equal to 1 along the whole path when the rates $\mathbf{F}$ do not contain the variable x_0 explicitly. The origin of the adjoint variables is seen in the Lagrange multiplier representation of the control problem considered; the adjoint variables are just the Lagrange multipliers associated with the constraints resulting from the state equations. For specific problems, other suitable interpretations of the adjoint variables are possible, e.g., as the partial derivatives of the optimal cost function or as counterparts of the momenta in classical mechanics.

The properties of the optimal solution in terms of the Hamiltonian are described in terms of the canonical equations for the state and the adjoint variables:

$$\frac{d\mathbf{Z}}{dt} = -\frac{\partial H}{\partial \mathbf{X}} \qquad \text{and} \qquad \frac{d\mathbf{X}}{dt} = \frac{\partial H}{\partial \mathbf{Z}} \equiv \mathbf{F(X,u},t) \qquad (6)$$

and the extremum condition on $H(\mathbf{X}, \mathbf{Z}, \mathbf{u}, t)$ with respect to the controls $\mathbf{u}$. Along the extremal path, $H^e(\mathbf{X}, \mathbf{Z}, t) = \max_{\mathbf{u}} H(\mathbf{X}, \mathbf{Z}, \mathbf{u}, t)$. For an interior maximum of H, this condition has the form:

$$\frac{\partial H}{\partial \mathbf{u}} = 0 \tag{7}$$

The boundary conditions for the canonical set (5) and (6) have usually a complex two-point form depending on the specific formulation of the problem; part of the conditions is specified at the beginning and part at the end of the process. For instance, if the initial state of the system is completely specified, the initial conditions specify $\mathbf{X}(t^i)$. The right-end (final) conditions have the simplest form when the final process state is free (unconstrained). This state should then be found as a component of the optimal solution. In this case, the final conditions deal only with the components of the adjoint vector and have the form $z_0(t^f) = 1$, as well as $z_k(t^f) = 0$ for $k=1, 2, \ldots, n$. However, more complex transversality conditions appear when the final state coordinates are constrained to lie on some final submanifold in the state space (Pontryagin at al 1962; Leitman 1981; Sieniutycz 1978).

One consequence of the canonical equations of the continuous problem is an equation describing the time dependence of the optimal Hamiltonian function

$$\frac{dH}{dt} = \frac{\partial H}{\partial t} \tag{8}$$

This equation shows that the Hamiltonian has the constancy property along any optimal path when the system does not involve the time explicitly. This represents, of course, the conservation of energy for conservative systems of analytical mechanics, i.e., the existence of a definite first integral for any autonomous physical system. If the final time is left unspecified (free-end time problem), then the final value of the Hamiltonian must vanish, $H(t^f) = 0$.

A discrete principle can be formulated which is strongly analogous to Pontryagin's principle provided that the time increments $\theta^n = t^n - t^{n-1}$ at the stages of the discrete process are unconstrained except for a possible constraint on their sum, i.e. on the total time (Sieniutycz 1978; Sieniutycz and Szwast 1982). The freedom of choosing θ^n

allows one to formulate the discrete optimal control problem in the canonical form governed by the Hamiltonian $H^{n-1} = Z^{n-1} \cdot F^n(X^n, u^n, t^n)$ obeying the following canonical equations (note the presence of both continuous time t and discrete time n in the equations)

$$\frac{Z^n - Z^{n-1}}{t^n - t^{n-1}} = -\frac{\partial H^{n-1}}{\partial x^n} \qquad \text{and} \qquad \frac{X^n - X^{n-1}}{t^n - t^{n-1}} = \frac{\partial H^{n-1}}{\partial Z^{n-1}} \equiv F^n(X^n, u^n, t^n) \qquad (9)$$

$$\frac{\partial H^{n-1}}{\partial u^n} = 0 \qquad (10)$$

$$\frac{H^n - H^{n-1}}{t^n - t^{n-1}} = \frac{\partial H^{n-1}}{\partial t^n} \qquad (11)$$

The boundary conditions for this discrete set have a form which is similar to the continuous version, discussed above. As distinguished from the continuous version the relationship characterizing the optimal Hamiltonian, eq. (11), does not result from the canonical equations (9) but it constitutes an additional optimality condition associated with the optimal choice of the (unconstrained) time increments $\theta^n = t^n - t^{n-1}$. Equation (11) is here the optimality condition coming from the additional freedom of choosing the time increments, θ^n. The use of this discrete formalism is illustrated in this volume in the article by Szwast treating the example of a multistage fluidized drying process.

The maximum principle reduces the global problem of the maximizing the functional to the problem of the maximizing the function H at each instant. In the most frequent case, when an analytical solution is impossible or very difficult, these methods allow one to organize an approximating numerical scheme using a large but finite number of the time instants which give a discrete approximation for the optimal path and the corresponding optimal control.

Changes in the boundary conditions do not influence the form of the Hamiltonian function and the canonical equations. Nonetheless, they may drastically simplify or complicate the associated numerical solution. The two-point boundary conditions appearing as a rule in problems of this kind are quite involved from the computational view point, and require the use various iterative approaches to gain an acceptable match between the computed and the prescribed boundary values. These methods are discussed comprehensively in the textbooks on optimal control (Lee and Marcus 1967; Leitman 1981; Sieniutycz 1978; Beveridge and Schechter 1970).

Other optimization conditions may be derived from the maximum principle. The continuous version of the dynamic programming results in the form of a partial differential equation of the Hamilton-Jacobi type with an additional condition of the Erdmann-Weierstrass type which expresses the maximality of H with respect to the control vector $\mathbf{u}$ (Bellman 1967; Sieniutycz 1978). The Euler equations of variational calculus can be recovered by investigating an extremum of the objective $x_0(t^f)$. This is a simple and yet illustrative example. Writing the objective in the form of the integral

$$A \equiv x_0(t^f) = x_0(t) + \int_t^{t^f} f_0(\mathbf{X}, \mathbf{u}, t)dt \tag{12}$$

we consider the necessary optimality conditions for the maximum of $x_0(t^f)$ under the simple differential constraints

$$d\mathbf{x}/dt = \mathbf{u} \tag{13}$$

for the n-component vector $\mathbf{x} = (x_1, x_2,..., x_n)$, the subset of the original $n + 1$ dimensional vector $\mathbf{X} = (x_0, x_1, x_2,....., x_n)$. Here dim $\mathbf{u} =$ dim $\mathbf{x} = n$. From eq. (12) one has in terms of $\mathbf{x}$ and x_0 rather that $\mathbf{X}$

$$dx_0/dt = f_0(\mathbf{X}, x_0, \mathbf{u}, t) \tag{14}$$

For the problems of analytical mechanics, the negative of f_0, usually designated by $L(\mathbf{x}, \mathbf{u}, t)$, is the so-called Hamilton's kinetic potential equal (nonrelativistically) to the difference between the kinetic and potential energies. Classically, it depends on all the coordinates of the original state vector $\mathbf{X} = (x_0, \mathbf{x}) = (x_0, x_1,...., x_n)$ except x_0. This results in the constancy of $z_0 = 1$ along the path. The variable x_0 is, of course, the negative action variable in this case. The Hamiltonian has the structure $H = f_0 + z_1 u_1 + ...z_n u_n$, and the stationarity condition for the Hamiltonian, eq. (7), lead to the relations $z_k = -\partial f_0/\partial u_k = \partial L/\partial u_k$, k=

1,...., n, defining the momenta ($p_k = z_k$). Hence the extremal Hamiltonian H is the energy of the process in agreement with the standard definition E = $\mathbf{p.u}$ - L. For conservative systems, where L=f_0 does not contain the time explicitly, H is a constant along the optimal arc. We associate the motion of such a system with "conservative" or "reversible " behavior. The canonical equations (6) for the momenta yield

$$d/dt\,(\partial L/\partial \dot{\mathbf{x}}) = \partial L/\partial \mathbf{x} \qquad (15)$$

which are the well known Euler equations of the variational calculus written in vector form. The are usually suitable for describing the reversible microscopic motion of elementary particles or the frictionless motion of macroscopic bodies in inviscid media or in a vacuum.

4. A SIMPLEST VARIATIONAL FORMULATION OF DISSIPATIVE DYNAMICS

Consider a macroscopic body moving in a (resting) viscous fluid where friction phenomena result in certain additional frictional forces. Can we describe the system in terms of a variational principle taking into account dissipation? Various kinetic potentials generalizing the classical L have been advanced. These potentials have had very limited success even for simple systems (Sieniutycz 1976, 1978; Kobe at al 1986). One such method uses $f_0 = - $ L exp(t/τ) where τ is a time constant describing an interaction of the body with the fluid and L is the classical Hamilton's kinetic potential. The reader can verify that for L = $m\mathbf{u}^2/2$ - V($\mathbf{x}$), the integrand $f_0 = - $ L exp(t/τ) leads to the additional ("frictional") force $\tau^{-1}\,\partial L/\partial u) = mu/\tau$ in the equation of motion corresponding to laminar friction. However the physical interpretation of the related Hamiltonian in energy terms is awkward since H is not a constant of the motion. This poses a serious problem since the autonomous character of the dynamical system involved dictates that the Hamiltonian should be a constant of motion.

We explore a different modification by introducing an extra additive term into f_0 of the form $\tau^{-1}x_0$ so that the action variable x_0 will explicitly appear in f_0. Thus,

$$f_0 = - \{m\mathbf{u}^2/2 - V(\mathbf{x}) + \tau^{-1}x_0 \} \qquad (16)$$

The Hamiltonian is $H = \mathbf{z}.\mathbf{u} + z_0 f_0$. The canonical eqs. (6), yield eqs. (13) and (14) and the dynamical equations

$$\frac{d\mathbf{z}}{dt} = -\frac{\partial H}{\partial \mathbf{x}} = -z_0\frac{\partial V}{\partial \mathbf{x}} \qquad \text{and} \qquad \frac{dz_0}{dt} = -\frac{\partial H}{\partial x_0} = z_0\tau^{-1} \qquad (17)$$

whereas the condition of the stationarity of the Hamiltonian with respect to the control $\mathbf{u}$, eq. (7), yields $\mathbf{z} = -z_0\partial f_0/\partial\mathbf{u} = z_0\partial L/\partial\mathbf{u}$ which shows that the generalized momenta of the process are $z_0\partial L/\partial\mathbf{u}$ rather than $\partial L/\partial\mathbf{u}$. Using the transversality condition $z_0(t^f) = 1$, we can evaluate the time behavior of z_0. On the basis of the second of eqs. (17), we see that z_0 is exponential in time, i.e., $z_0(t) = \exp(t/\tau)$. Thus $\mathbf{z} = \exp[(t - t^f)/\tau]\partial L/\partial\mathbf{u}$ which explains the limited success of the modified kinetic potential $L\exp(t/\tau)$, mentioned earlier. Incidentally, both modifications lead to the same, or proportional, generalized momenta $\mathbf{z}$. Substituting $\mathbf{z} = \exp[(t - t^f)/\tau]\partial L/\partial\mathbf{u}$ into the first of eqs. (17) yields the simplest possible equation of motion with a friction term

$$d/dt\,(\partial L/\partial\dot{\mathbf{x}}) + \tau^{-1}(\partial L/\partial\dot{\mathbf{x}}) = \partial L/\partial\mathbf{x} \qquad (18)$$

In terms of the classical momenta $md\mathbf{x}/dt = m\mathbf{u}$ and the potential energy V

$$d/dt(m\dot{\mathbf{x}}) + \tau^{-1}m\dot{\mathbf{x}} = -\partial V/\partial\mathbf{x} \qquad (19)$$

The case $\tau^{-1} = 3\pi\nu\delta/m$ corresponds with the Stokes frictional force exerted by a spherical body of diameter δ and (effective) mass m, moving in a fluid with dynamic viscosity ν.

Consider the Hamiltonian $H = \mathbf{z}.\mathbf{u} + z_0 f_0$ along the extremal trajectory. It is defined on the space $(\mathbf{X}, \mathbf{Z})$ or $(\mathbf{x}, x_0, \mathbf{z}, z_0)$. The extremum Hamiltonian $H^e(\mathbf{x}, x_0, \mathbf{z}, z_0)$ is obtained by substituting $\mathbf{u}$ expressed in terms of the $(\mathbf{x}, x_0, \mathbf{z}, z_0)$ into H. For our model $\mathbf{z} = -z_0\partial f_0/\partial\mathbf{u} = z_0\partial L/\partial\mathbf{u} = z_0 m\mathbf{u}$ and hence $\mathbf{u} = \mathbf{z}/mz_0$ leading to

$$H^e\,(\mathbf{x}, x_0, \mathbf{z}, z_0) = \frac{\mathbf{z}^2}{2mz_0} + z_0 V(\mathbf{x}) - \frac{z_0 x_0}{\tau} = \exp(\frac{t - t^f}{\tau})(\frac{m\mathbf{u}^2}{2} + V(x) - \frac{x_0}{\tau}) \qquad (20)$$

This Hamiltonian is the constant of an "irreversible" (damped) motion governed by the canonical equations written for H^e in the space of the natural variables $(\mathbf{x}, x_0, \mathbf{z}, z_0)$. These are dynamical equations (17) for $H = H^e$ describing the time evolution of the generalized momenta, and the equations defining the velocities in terms of these variables

$$\frac{d\mathbf{x}}{dt} = \frac{\mathbf{z}}{mz_0} \qquad \text{and} \qquad \frac{dx_0}{dt} = -\frac{\mathbf{z}^2}{2mz_0^2} + V(\mathbf{x}) - \frac{x_0}{\tau} \qquad (21)$$

Thus we have found a variational formulation for our irreversible system. However the price for taking the irreversibility into account is not low. First, the usual space of the mechanical variables $(\mathbf{x}, \mathbf{p})$ had to be replaced by the enlarged space $(\mathbf{x}, x_0, \mathbf{z}, z_0)$. Second, the extremum Hamiltonian is not the classical energy.

However, H^e is a constant of motion of the irreversible canonical equations (17) and (21). Also, any generalized momentum z_i is the constant of motion when the external field $V(\mathbf{x})$ vanishes or is independent of the coordinate x_i. This proves that the (autonomous) irreversible process is in fact conservative if the macroscopic energy and momenta are properly defined. It only behalves like nonconservative when we are using the standard definitions of the energy and momenta formulated for reversible processes. Furthermore, in the limiting case of infinite τ, the Hamiltonian becomes the classical energy of the body. This limit corresponds to a vanishing role of the frequency τ^{-1} of collisions between the body and the molecules of the bath. In this limit, the adjoint variables $\mathbf{z}$ become the usual momenta $\mathbf{p}$. Hence the correspondence of the formalism with the frictionless case is assured. In conclusion, one may expect that a variational treatment exists for more general irreversible dynamics. These treatments should require extended definitions of the energy and momenta and give an explicit role to the individual actions of the bodies involved. These expectations are confirmed by other results obtained for distributed parameter systems which require a vector of actions (Sieniutycz 1990).

Variational and extremum principles for continua have a weaker connection with optimal control theory than do the corresponding lumped systems discussed here. The former are investigated in several papers in Vols. 3 and 6 of this series. The description of the thermo-hydrodynamic systems there correspond to the field representation of motion (Rund 1970) rather than to the Lagrangian representation.

5. DISCUSSION

We return now to the first category of applications of extremal control outlined in section 2. Recall that this category did not try to characterize natural evolution but rather asked for the optimal way to control a process. This volume is the collection of works by various investigators who have been involved with the problems of thermodynamic optimization, finite-time thermodynamics (FTT) and thermoeconomics.

In the paper by K.H. Hoffman the aims and methods of the finite-time thermodynamics are discussed from the viewpoint of finite-time bounds and optima for irreversible engines and energy conversion systems. These bounds are a significant improvement on the classical (infinite-time) bounds from reversible thermodynamics. Applications of FTT to heat engines and light-driven engines are discussed in the context of maximum work and minimum entropy production. Extending the Salamon and Berry (1983) formula for the bound on the dissipated availability, it is shown how work deficiency may be bounded by using a geometric quantity: the thermodynamic length. Application of FTT methods to simulated annealing is discussed.

B. Andresen article analyses FTT as a specific branch of irreversible thermodynamics which uses aggregated macroscopic characteristics (e.g. friction coefficients, heat conductances, reaction rates etc.) rather than their microscopic counterparts. Treating processes which have explicit time or rate dependences, he discusses finite-time thermodynamic potentials and the generalization of the traditional availability to the finite-time case involving entropy production and constraints. Paths for endoreversible engines are analyzed from the viewpoint of various performance criteria and a comparison is made of their properties with those of traditional engines (e.g. Carnot engines).

J.M. Gordon reviews the applications of FTT to problems involving applications of solar energy. By treating the Earth's atmosphere as the working fluid and investigating the solar-driven convection known as the Earth's winds, the natural convection process is modelled as an irreversible heat engine. He also analyses solar-driven heat engines for conventional power generation. The functional form of the heat transfer is shown to have an essential effect on the performance of heat engines operating at maximum power with the most pronounced effects occurring for radiative heat transfer.

K.O. Lund analyses the thermal conversion of solar power using a heat engine working between receiver and sink temperatures. FTT is applied to determine the upper limit to power output for the case of terrestrial and space-based solar-thermal engines. The differences between the optimal strategies for the receiver temperature obtained for terrestrial and space-based systems are discussed. The current design parameters are discussed for both cases considered leading to the conclusion that FTT brings the theoretical limits closer to achievable practice.

M.J. Ondrechen investigates FTT of engines driven by chemical processes. Attention is focused on the constraints governing the source of heat driving the process. Using an exotermic chemical reaction as the heat source makes two features essential for the operation of the processes in the finite time: the finite size (and hence heat capacity) of the heat sources and sinks and the finite rate of heat generation. It is shown that these features impose certain limits on the efficiency and the power of energy conversion devices.

B.A. Mansson discusses some general aspects of the significance of thermodynamics for economic analyses. The former is to be used to identify and quantify basic constraints resulting from conservation laws as well as losses due to irreversibility. Also, thermodynamics provides clear and quantitative definitions of efficiency. The similarity of the value concept in thermodynamics (exergy) and economics is illustrated along with the conclusions about the inappropriateness of efforts to reduce all aspects of value to a single physical standard. An example of the maximum outgoing exergy obtained from given set of sources is given. A simple macroeconomic growth model is presented to show the role of the thermodynamic limitations.

D. Tondeur asks a basic and practical question: in what sense can one say that the distribution of the driving forces in one industrial configuration is better than in another. Several realistic chemical engineering examples, involving heat exchangers, plate distillation columns, chromatographic separators, etc, are analyzed. The duty (useful effect) for each process is defined and the following equipartition of the entropy production is advanced: "for a given duty, the best configuration of the process is that in which the entropy production rate is most uniformly distributed". A generalization of the principle is also proposed for systems with a distributed design variable: "the optimal distribution of an investment is such that the investment in each element is equal to the cost of the energy degradation in this element". It follows that a uniform distribution of the ratio of these two quantities should be preserved. These results show the importance of irreversibility analysis for rational design.

Z. Szwast develops a rigorous mathematical theory of discrete optimal control for multistage crosscurrent drying processes with granular fluidized solids. The sum of the operational costs, based essentially on the exergy tariff, and the investment costs, is accepted as a performance criterion. The discrete maximum principle (see section 3 of the present paper) is extended to processes with product recycling. The parameters of the drying agent (process controls) are varied with the stage number. The performance of controlled drying processes are compared with those obtained for conventional processes in which the drying agent is constrained to be in the same state at each process stage and where only the problem of the optimal level of the decisions arises. Countercurrent drying with interstage heating of the gas as well as simple crosscurrent solid heating are also investigated. The superiority of controlled processes as compared to the conventional ones is shown for expensive drying equipment. Increases in exergy consumption are balanced by increases in equipment costs in agreement with Tondeur's principles, discussed above.

P. Le Goff, R. Riviero, S. de Oliveira and B. Schwarzer present a mature comparison of exergy-based optimization and economic optimization of industrial processes. Energy, exergy and value balancing are discussed with various definitions of the related profits, costs and efficiencies. Scales of value based on enthalpy, exergy, ecology and economics are compared. Detailed examples of economic and energy accounting are presented for industrial systems treated as energy converters with various definitions of the energy equivalent periods and cost equivalent periods. Operating cost optimizations according to energy, exergy and monetary scales are given for a heat exchanger, a beet pulp dryer and some heat transformers which upgrade industrial waste heat. This paper should be certainly of interest to workers who have to organize a realistic industrial optimization with its characteristic many-compromise property.

J. Szargut analyzes a complex technological network as regards its cumulative exergy consumption, outgoing exergy and cumulative exergy losses. He defines constituent exergy losses, related to fabrication of particular partially finished products, and partial exergy losses, connected with particular links of the technological network. His analysis provides information about possible improvement of the network. An illustrative example is presented concerning heat delivery from a complex heat-power station. Exhaustion of restorable natural resources is determined in terms of cumulative exergy. By taking into account the deleterious impact of waste products on human activity, health, agriculture forestry etc., ecological economics is introduced aiming at a minimization of the consumption of unrestorable natural products. This aim can be accomplished using exergy exclusively. His arguments show exergy to be an important ecological quantity.

ACKNOWLEDGEMENT

This research was supported in part by the Office of Naval Research, grant number N00014-88-K-0491.

REFERENCES

Andresen, B. 1983. <u>Finite-Time Thermodynamics</u>, Copenhagen: University of Copenhagen.

Andresen, B., P. Salamon, and R.S. Berry. 1984. Thermodynamics in finite time, <u>Physics Today</u>.

Aris, R. 1964. <u>Discrete Dynamic Programming</u>. New York: Blaisdell.

Band, Y. B., O. Kafri, and P. Salamon. 1980. Maximum work production from a heated gas in a cylinder equipped with a piston, <u>Chem. Phys. Lett.</u>, 72, 127.

Band, Y. B., O. Kafri, and P. Salamon. 1981. Optimal heating of a working fluid in a cylinder with a moving piston, <u>J. Appl. Phys.</u> 52, 3745.

Band, Y. B., O. Kafri, and P. Salamon. 1982a. Optimal expansion of a heated working fluid, <u>J. Appl. Phys.</u>, 53, 8.

Band, Y. B., O. Kafri, and P. Salamon. 1982b. Optimization of a model external combustion engine, <u>J. Appl. Phys.</u>, 53, 29, 1982.

Bejan, A. 1988. <u>Advanced Engineering Thermodynamics.</u> New York : J.Wiley.

Bellman, R.E. 1957. <u>Dynamic Programming</u>. Princeton: Princeton University Press.

Bellman, R.E. 1967. <u>Introduction to Mathematical Theory of Control Processes</u>. New York: Academic Press.

Bellman, R.E., and R. Kalaba. 1965. Dynamic Programming and Modern Control Theory. New York: Academic Press.

Berry, R. S., P. Salamon, and G. Heal, On a relation between thermodynamic and economic optima, Resources and Energy, 1, 125, 1978.

Beveridge, G.S., and S. Schechter. 1970. Optimization: Theory and Practice. New York: Mc Graw Hill.

Boltyanski, V.G. 1971. Mathematical Methods of Optimal Control. New York: Holt, Reihart and Wilson.

Bracken, J., and G.P. McCormick. Selected Applications of Nonlinear Programming. New York: J.Wiley.

Dantzig, G.B. 1968. Linear Programming and Extensions. Princeton: Princeton University Press.

Fan, L.T., and C.S. Wang. 1964. The Discrete Maximum Principle: A Study of Multistage System Optimization. New York:Wiley.

Fan, L.T. 1964a. The Continuous Maximum Principle: A Study of Complex System Optimization. New York. Wiley.

Gelfand, I.M., and S.V. Fomin. 1963. Calculus of Variations. Englewood Cliffs: Prentice-Hall.

Greig, D. M. 1980. Optimisation, New York: Longman.

Harland, J.R., and P. Salamon. 1988. Simulated annealing: a review of the thermodynamic approach Nuclear Physics B (proc. Suppl) 5A, 109.

Hoffmann, K. H., S. J. Watowich, and R. S. Berry. 1985. Optimal paths for thermodynamical systems: J. Appl. Phys. 58, 2125, see also ibid 58, 2893.

Kobe, D.H., G. Reali, and S. Sieniutycz. 1986. Lagrangians for dissipative systems. Amer. J. Phys. 54: 997-999.

Kuhn, H.W., and A.W. Tucker. 1951. Nonlinear programming. In Proceedings of the Second Berkeley Symposium on Mathematical Statistics and Probability, ed. J. Neyman, 481-493. Berkeley: Univ. of California Press.

Lee, E. B. and L. Marcus. 1967. Foundations of Optimal Control Theory, New York: John Wiley.

Leitman, G. 1981. The Calculus of Variations and Optimal Control. New York: Plenum Press.

Llewellyn, R.W. 1963. Linear Programming. New York: Holt, Rinehalt and Wilson.

Mansson, B. A. G. 1985. Contributions to Physical Resource Theory, Chalmers University of Technology, Göteborg.

Mansson, B. A. G. and B. Andresen. 1985. Optimal Temperature Profile for an Ammonia Reactor, in Mansson, B. A. G.

Contributions to Physical Resource Theory, Chalmers University of Technology, Göteborg.

Mozurkewich, M. and R. S. Berry. 1981. FTT: Engine performance by optimized piston motion. _Proc. Natl. Acad. Sci. U.S.A. 78_:1986.

Mozurkewich, M. and R. S. Berry. 1982. Optimal path for thermodynamic systems: the ideal Otto cycle. _J. Appl. Phys. 53_: 34 (see also _J. Appl. Phys. 54_: 3651)

Nulton, J., P. Salamon, B. Andresen, and Q. Anmin. 1985. Quasistatic processes as step equilibrations, _J. Chem. Phys._ 83, 334.

Nulton, J., and P. Salamon. 1988. Statistical mechanics of combinatorial optimization, _Phys. Rev. A_ 37, 1351.

Pontryagin, L.S., V.G. Boltyanski, R.V. Gamkrelidze, and E.F. Mishchenko. 1962. _The Mathematical Theory of Optimal Processes_. New York: Interscience Publishers.

Rubin, M. 1980. Optimal configuration of an irreversible heat engine with fixed compression ratio _Phys. Rev. A_ 22, 1741.

Rund, H. 1970. _The Hamilton-Jacobi Formalism in the Calculus of Variations_. London: Van Nostrand.

Salamon, P., A. Nitzan, B. Andresen, and R.S. Berry. 1980. Thermodynamics in finite time IV: minimum entropy production in heat engines, _Phys. Rev. A_, 21, 2115.

Salamon, P. and A. Nitzan. 1981. Optimizations of a finite time Carnot cycle, _J. Chem. Phys. 74_: 3546.

Salamon, P., Y. B. Band, and O. Kafri. 1982. Maximum power from a cycling working fluid, _J. Appl. Phys._, 53, 197.

Salamon, P., and R.S. Berry. 1983. Thermodynamic length and the dissipated availability. _Phys. Rev. Letters 51_: 1127-1130.

Sieniutycz, S. 1978. _Optimization in Process Engineering_. Warsaw: WNT.

Sieniutycz, S. 1976. A stationary action principle for a laminar motion in a viscous fluid. _Inz Chem 6_: 673-691.

Sieniutycz, S. 1990. Thermal momentum, heat inertia and a macroscopic extension of de Broglie microthermodynamics I. The multicomponent fluids with the sourceless continuity constraints. _Adv. Thermodyn. 3_.

Sieniutycz, S., and Z. Szwast. 1982. _Practice of Optimization_ Warsaw: WNT.

Szargut, J., and R. Petela. 1965. _Exergy_. Warsaw: WNT.

Varaiya, P. P. 1972. _Optimization_, New York: Van Nostrand Rheinhold.

White, W.B., S.H. Johnson and G.B. Dantzig. 1958. Chemical equilibrium in complex mixtures. _J. Chem. Phys. 28_: 751-755.

Zangwill, W.I. 1969. _Nonlinear Programming; a Unified Approach_. New York: Prentice Hall.

Optima and Bounds for Irreversible
Thermodynamic Processes

Karl Heinz Hoffmann
Institut für Theoretische Physik, Universität Heidelberg
Heidelberg, Federal Republic of Germany

Bounds and optima for irreversible thermodynamic processes and their application in different fields are discussed. The tools of finite time thermodynamics are presented and especially optimal control theory is introduced. These methods are applied to heat engines, including models of the Diesel engine and a light–driven engine. Further bounds for irreversible processes are introduced, discussing work deficiency and its relation to thermodynamic length. Moreover the problem of dissipation in systems composed of several subsystems is studied. Finally the methods of finite time thermodynamics are applied to thermodynamic processes described on a more microscopic level. The process used as an example is simulated annealing. It is shown how optimal control theory is applied to find the optimal cooling schedule for this important stochastic optimization method.

22

1. INTRODUCTION AND OVERVIEW

In 1826 Sadi Carnot, a French engineer, published his famous article 'Réflexions sur la puissance motrice du feu' (On the motive power of heat), in which he reported the results of the first systematic study of the physical processes governing steam engines. This was the start of a new science: thermodynamics and among its earliest successes was the formulation of bounds and optima for thermodynamic processes. Carnot showed that any engine, which produces work taking heat from a heat reservoir at temperature T_H has to deposite part of that heat in a colder reservoir at temperature T_L. Moreover, no engine can convert more than the fraction

$$\eta_C = 1 - \frac{T_L}{T_H} \tag{1.1}$$

of the heat taken from the heat reservoir to work. η_C is the well–known Carnot efficiency and it is an upper bound for the efficiency of any heat engine running between two heat baths.

In the early development of thermodynamics the emphasis remained on finding limits for process variables like work, heat transfer or efficiency. But that changed with Gibbs: the introduction of entropy and internal energy shifted the interest away from the process variables and towards those new state variables. Nonetheless in the subsequent development of thermodynamics also criteria for the performance of processes emerged, of which we mention just two:

- even though being state functions the thermodynamic potentials such as the free energy H introduced by Hermann von Helmholtz for isothermal, isochoric processes or the Gibbs' free energy G for isothermal, isobaric processes provided bounds for the work done by or required during the respective process.

- Gibbs introduced the 'availability' as a measure for the maximum work that can be extracted by relaxing a system from an internally equilibrated but constrained state to a state in equilibrium with its surroundings.

The oil crisis in the early seventies brought the process variables back into the focal point of interest: the scarcity of organic fuels and other natural resources raised again the question of what are the bounds and the optima for engines and energy conversion systems and, more broadly, for thermodynamic processes in general. The bounds provided so far by thermodynamics shared all one characteristic: they compared the real processes with reversible processes. But reversible processes take an infinite amount of time to complete and are thus usually not close enough to real processes to be useful – for instance in guiding the future improvement of a real process.

So the question was whether one could find better and more realistic bounds for processes which had to be completed in a finite amount of time.

In 1975 R. Stephen Berry, Bjarne Andresen and Peter Salamon started a research effort into that direction, out of which has grown a whole new field: Finite Time Thermodynamics. It evolved around two central questions:

Under which conditions and how can realistic bounds for process variables of thermodynamic processes, which are performed in finite time, be determined

and

what are the optimal process paths to achieve the optimal process values?

Finite time thermodynamics tried to attack these questions in a fashion very common to traditional thermodynamics: the processes and systems under consideration were not supposed to be extremely detailed descriptions of real processes or systems with all their specific properties, they should be rather generic for a broad class of systems concentrating on the typical properties of the real processes – just as Carnot had idealized a real steam engine to a working fluid operating between two heat baths.

The Curzon Ahlborn engine (Curzon and Ahlborn 1975) has become a paradigm to illustrate this approach:
This engine consists of a Carnot engine operating between two heat baths with temperatures T_H and T_L, to which it is coupled through finite heat conductancies. When operated to produce maximum power the efficiency of this engine is given as

$$\eta = 1 - \sqrt{\frac{T_L}{T_H}} \tag{1.2}$$

independent of the values of the heat conductancies used for the coupling to either of the heat baths.

In 1982 I became interested in this field and chapters 3 – 5 of this paper are a review of the work I have been involved with.

Finite time thermodynamics developed different tools to achieve its goals. Finite time potentials, an extension of the above mentioned thermodynamic potentials, and the tricycle formalism were used to determine bounds and optima for those processes with finite rates; and control theory became the most important tool to determine optimal process paths. These by now classical parts of finite time thermodynamics are reviewed in chapter 2.

Finite time thermodynamics was soon applied in various fields, ranging from the optimization of destillation and exothermic reactions to that of almost realistic heat engines. That the latter have always been at the very heart of finite time thermodynamics is certainly not surprising for this field born out of the oil crisis. These engines are used in chapter 3 to exemplify how bounds and optimal pathways are determined for important irreversible thermodynamic processes. But finite time thermodynamics did not only evolve in applied areas, there was also progress made on its fundamental questions. Two selected advances are reviewed in chapter 4:

– It was shown that the loss of availability and – for processes in which the dissipated energy remains partly inside the system – the work deficiency can be bounded by using a geometrical quantity of the process: its thermodynamic length. This connection to another field in thermodynamics, namely its geometrical formulation, was quite influential on the further development of finite time thermodynamics.

- Progress was also made on the question of whether bounds for process variables could be calculated for systems which consist of a large number of subsystems interacting irreversibly. While the early models of finite time thermodynamics had been quite simple, its application to more and more realistic systems had stressed this question again.

Recently the ideas and methods of finite time thermodynamics have been applied to a new field: Simulated Annealing. It was introduced in 1983 as a new technique to find near optimal solutions to complex optimization problems which were too large to be enumerated. Simulated annealing works by an analogy to a real physical annealing process: just as for a real physical system its ground state can be found by suitably lowering its temperature so can be the minimum state of an optimization problem. As simulated annealing uses a large amount of costly computer time and moreover as it should provide a solution to the optimization in a finite time the natural questions are:
What are the performance bounds for this simulated annealing procedure?
and even more important
How does the annealing proceed in an optimal fashion?

These are the well–known questions from finite time thermodynamics and due to the analogy to a real annealing process the methods of finite time thermodynamics have been used to address these questions. Eventhough being highly interesting in itself in chapter 6 simulated annealing serves as an example to show that optimal paths and extrema for process values can also be determined for irreversible processes described on a statistical level.

2. THE CLASSICAL TOOLS OF FINITE TIME THERMODYNAMICS

In this chapter three of the by now classical tools of finite time thermodynamics (Andresen 1983) are introduced. They provide the answers to the basic questions already raised in the introduction:

- under which conditions and how can bounds for the process variables of thermodynamic processes be determined and
- what are optimal paths to achieve the optimal process values.

The first question is addressed by the topics of the first section in this chapter, while the second question is addressed by the second section.

2.1 Finite time potentials and the tricycle formalism

As pointed out above thermodynamics started out as a practial science dealing with energy conversion, and throughout its development one basic question remained of interest at least from a practical point of view: how much work or heat can one get out of a system undergoing a thermodynamic process from one state to another. The answer to this kind of question was given for certain reversible processes by thermodynamic potentials (Callen 1960, Tisca 1966). For irreversible processes these could provide only bounds on the process variables which might be far from being realistic. The usually known potentials for reversible processes are characterized by the constancy of some state variables like volume, pressure or temperature. This is, however, not necessary, and Salamon, Andresen and Berry presented an extension of the Legendre transformations used in classical thermodynamics which allows to handle a much wider class of constraints including those which depend on time. These extended potentials are called finite time potentials. They can handle finite rate and finite time constraints and provide a bound for process variables in the form of the difference of their value between final and initial states. For certain processes they can explicitly depend on time. The bounds provided by finite time potentials contain the losses by the irreversibilities and are thus more realistic than those for ordinary potentials. The same applies to bounds derived from finite-time availability, which is an extension of the usual availability to irreversible processes (Andresen 1983, Andresen et al. 1983).

The tricycle formalism (Andresen et al. 1977b) is a convenient way to find bounds and optima for thermodynamic processes for which the irreversibilities can be determined as the function of the flows between different heat or work reservoirs. It does so by making use of the conservation laws for the process in question. The formalism allows to handle different loss mechanisms simultaneously and provides the additional advantage of giving the costs or losses by an off–optimal operation.

A system which is in contact with 3 heat reservoirs or 2 heat reservoirs and a work reservoir, e.g. a flywheel, is graphically depicted as a triangle, with heat flows q_1, q_2, and q_3 into the reservoirs at temperatures T_1, T_2, and T_3. In the case of a work reservoir the corresponding temperature is infinite and the energy flow is not accompanied by an entropy flow. The idea behind the tricycle is that it represents a cyclic or continuously operating energy conversion system, which does not produce energy itself, thus

$$q_1 + q_2 + q_3 = 0 \qquad (2.1.1)$$

where for a cyclic operating system the q's are cycle averages. Using $\tau_i = T_i^{-1}$ the entropy production s is given by

$$s = q_1\tau_1 + q_2\tau_2 + q_3\tau_3 \geq 0. \qquad (2.1.2)$$

One now decomposes the tricycle process into a reversible part and into a totally irreversible part as shown in figure 1.

To make the decomposition unique, q_e is set to zero. Such a decomposition becomes useful if by specifying the loss mechanisms the dependencies of the irreversibilities on the reversible flows or rates are determined.

Once that has been achieved the objective (like power, efficiency, minimum entropy production etc.) for the optimization must be chosen and then the optimal flows and thus the corresponding size of the irreversibilities can be determined.

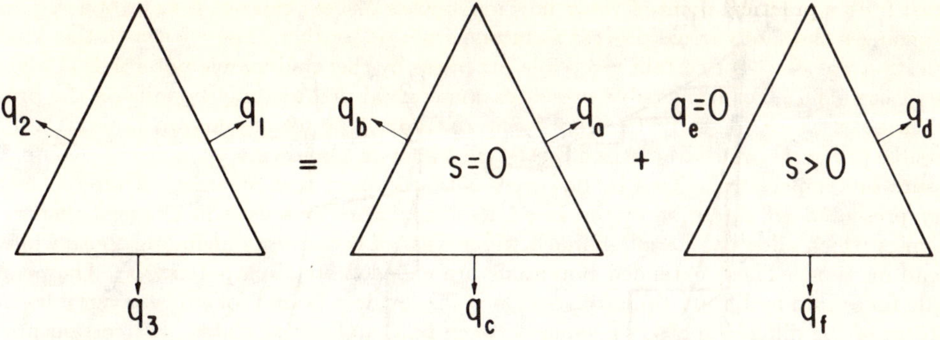

Figure 1. Tricycle decomposition of the general heat flows q_1, q_2, q_3 of a process into reservoirs with temperatures T_1, T_2, T_3 into a reversible part and a totally irreversible part consisting of a heat flow between reservoirs 1 and 3 uncoupled from reservoir 2.

2.2 Optimal paths

Up to now the focus has been on what is the optimal value for a process variable in a given thermodynamic process, or what are bounds for it. But there are situations where one is not only interested in that optimal value – one number – but where one also wants to know how this value can be achieved. In other words, the question is what is the optimal path along which the system has to proceed to attain the optimal value for the process variable in question.

An early attempt in that direction was a paper (Andresen et al. 1977a) by B. Andresen, R.S. Berry, A. Nitzan and P. Salamon which investigated the 'step–Carnot' cycle. The model is a working fluid in a cylinder with a piston. On this acts an external force changing discontinuously for instance through an unlubricated stick–slip motion, as shown in figure 2. The system now strives to follow a Carnot cycle but as the external pressure varies discontinuously the system comes into equilibrium with the surroundings only in a finite number of points along the cycle. The objective then is to distribute these points in such a way along the cycle that either the efficiency, the effectness or the work produced is maximized. As the number of parameters (i.e. the position of the points along the cycle) influencing the objectives is only finite this optimization problem could be handled by ordinary calculus.

The results showed already two of the features which are so common to finite time processes (Andresen 1983): the optimal trajectory is sensitive to what it optimized, i.e. the objective, and generally the performance degrades with decreasing time.

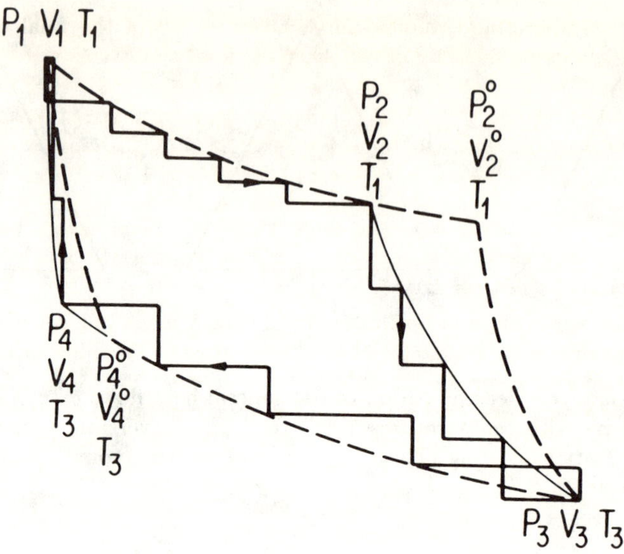

Figure 2. PV diagram of the step-Carnot cycle (heavy line) with its envelope (light line) and the reversible Carnot cycle (dashed) operating between the same extreme states. The indicated pressure for the step-Carnot cycle is the external pressure, not the internal pressure which may be undefined except at the points touching the envelope.

For continuous time problems where the system can be influenced not only at a finite number of points or through a finite number of parameters but throughout the entire process, <u>the</u> tool for obtaining the optimal path is optimal control theory (Boltyanskii 1971, Bryson and Ho 1975, Hsu and Meyer 1968, Tolle 1975). In my opinion, this is certainly the one mathematical theory which influenced finite time thermodynamics and especially its application much more than any other theory.

In order to apply optimal control theory to given physical problems one has to specify

- the objective function, i.e. the quantity which is to be optimized; usually given as the time integral of a function along the process and/or the value of another function at the end of the process (example: work $\int p\dot{V}dt$)

- whether the duration of the process is fixed or part of the optimization.

- constraints for the system; either in integral or differential form (example: conservation law for energy). To these constraints also belong the equations governing the time evolution of the system, usually given in the form of differential equations (example: heat transfer: $\dot{q} = \kappa(T_{ex} - T)$)

- the initial conditions and maybe none, a few or all of the final conditions for the
  state of the system depending on whether the final conditions are given or part of the
  optimization

- the controls, i.e. those quantities which can be varied along the process to achieve
  optimal performance (example: the piston velocity in an engine)

  and finally

- the limits on the controls and the state variables. These may be needed to avoid
  unphysical or undesired situations like negative temperatures or piston accelerations
  beyond technical abilities.

Typically the application of optimal control theory leads to a set of coupled nonlinear
differential equations with given initial and final states. To illustrate the technique of
optimal control we give a simple example:

Let $\underline{u}$ be the control variables, and U the set of allowed controls. Let $\underline{x}$ be the n–dimensional
vector of the state variables, which evolve according to the dynamics

$$\underline{\dot{x}} = \underline{f}(\underline{x}, \underline{u}). \tag{2.2.1}$$

Let $\underline{x}_0$ be the initial condition for the state of the system. Let

$$\int_0^\tau I(\underline{x}, \underline{u}) dt \tag{2.2.2}$$

be the objective which has to be minimized for a process of given finite time τ by the
choice of a suitable $\underline{u}$ from U.

Then a so–called Hamiltonian H is formed with the help of adjoint (or costate) variables $\underline{\lambda}$, one for each state variable:

$$H(\underline{x}, \underline{\lambda}, \underline{u}) = I(\underline{x}, \underline{u}) + \underline{\lambda}^{tr} \underline{f}(\underline{x}, \underline{u}). \tag{2.2.3}$$

The optimal path has to satisfy three conditions:

- The necessary conditions for an extremal path give the canonical equations of motion:

$$\underline{\dot{\lambda}} = -\partial_{\underline{x}} H(\underline{x}, \underline{\lambda}, \underline{u}) \tag{2.2.4}$$
$$\underline{\dot{x}} = \partial_{\underline{\lambda}} H(\underline{x}, \underline{\lambda}, \underline{u}). \tag{2.2.5}$$

- Then the optimal control $\underline{u}^*$ is determined which minimizes H for fixed $\underline{x}$ and $\underline{\lambda}$:

$$H(\underline{x}, \underline{\lambda}, \underline{u}^*) \leq H(\underline{x}, \underline{\lambda}, \underline{u}) \quad \text{for all } \underline{u} \in U. \tag{2.2.6}$$

Note that this will usually depend on $\underline{x}$ and $\underline{\lambda}$. $u^*(\underline{x}, \underline{\lambda})$ is then inserted into the
dynamics, which thus become a closed set of differential equations for $\underline{x}$ and $\underline{\lambda}$.

– For each state variable which has not a final value prescribed the final value $\lambda(\tau)$ of the corresponding adjoint variable is required to be zero.

Together with the initial conditions $\underline{x}_0$ and those final conditions for the state variables which are given, this totals $2n$ boundary conditions. Thus it remains to solve $2n$ coupled first–order differential equations with $2n$ boundary conditions. These are usually too hard to be solved analytically. Nonetheless they are open to either a numerical treatment or a qualitative analysis. For both cases we will have examples in the following chapters.

Summarizing the above we note that the determination of an optimal path for an irreversible thermodynamic process requires much more effort than the determination of the optimal value itself. But once the price has been paid all quantities of interest for this optimal process can be determined as the process is known completely.

3. THE APPLICATION OF FINITE TIME THERMODYNAMICS TO IRREVERSIBLE HEAT ENGINES

Soon after the first tools of finite time thermodynamics had been developed, the field widened as other researchers became interested. Different systems and models were investigated in the spirit of those basic questions raised in the introduction: what are the performance bounds and how can they be achieved?

The list of topics ranged from the efficiency of solar cell converters (Ross and Collins 1980, Ross and Nozik 1982) to the optimization of exothermic reactions (Ondrechen et al. 1980a, Ondrechen et al. 1980b), from optimal distillation (Mullins and Berry 1984, Su and Engel 1980) to systems continuously distributed in space (d'Isep and Sertorio 1982, d'Isep and Sertorio 1983), and from the minimization of dissipation by oscillatory instead of steady state operation (Richter et al. 1981, Richter and Ross 1978, Termonia and Ross 1981) to the problems centered around heat engines (Band et al. 1981, Gutkowics-Krusin et al. 1978, Hoffmann et al. 1985, Mozurkewich and Berry 1981, Mozurkewich and Berry 1982, Rubin 1979a, Rubin 1979b, Rubin 1980, Salamon et al. 1982, Salamon and Nitzan 1981). The latter have always been of central interest to thermodynamics, and they will serve in this chapter to exemplify how bounds and especially optimal paths can be determined.

The first full fledged application of optimal control theory to a heat engine is due to M. Rubin (Rubin 1979a, Rubin 1979b). He investigated a class of engines called endoreversible. An endoreversible heat engine is defined to be an engine such that during its operation its working fluid undergoes reversible transformations.

The same techniques were then used in the work on internal combustion engines (Hoffmann et al. 1985, Mozurkewich and Berry 1982), which is reviewed in section 3.1. These models are probably still the most complex ones for which optimal paths and performance bounds have been determined so far. The dissipation mechanisms incorporated include the major loss terms of real combustion engines.

In section 3.2 light engines (Mozurkewich and Berry 1983b, Nitzan and Ross 1973, Watowich et al. 1985, Watowich et al. 1989, Zimmermann and Ross 1984) are reviewed.

They are prototypical for a class of heat engines called 'dissipative' engines (Wheatley et al. 1983). Those are constantly in contact with a hot and a cold heat reservoir, with the size of the heat flows being controlled by the state of the working fluid. As for those engines a reversible limit does not exist, only with the use of finite time thermodynamics sensible performance bounds and optima can be determined.

Finally in section 3.3 we discuss the ways in which the optimized thermodynamic processes discussed in the previous chapters can be realized in real engines.

3.1 Internal combustion engines

As pointed out in the introduction finite time thermodynamics started out as a reaction to the oil crisis in the early seventies. So it was very natural that after the necessary tools had become available the attention focused on the application of finite time thermodynamics to combustion engines. Very much in the tradition of early thermodynamics it was not the idea to repeat the nearly 100 years of careful engineering and optimization but to abstract the engines enough to make them treatable yet to include at the same time all major loss terms (Taylor 1966) so that the results of the analysis would be useful in guiding which loss terms could be most easily reduced. So the focus was not on optimizing the technical realization of the engines but on optimizing the thermodynamic process itself and on finding its inherent limits when performed in a finite time.

First the Otto engine was investigated by M. Mozurkewich and R.S. Berry (Mozurkewich and Berry 1982) and later the Diesel engine by S. Watowich, R.S. Berry and myself (Hoffmann et al. 1985). Both engines have a four–stroke cycle of successive intake, compression, power and exhaust stroke. They operate at constant period and with a fixed fuel consumption per cycle. While keeping constant those engine parameters determining friction and heat leakage we optimized the time path of the piston to yield the maximum work per cycle.

Both models are endoreversible in that the working fluid is in internal equilibrium. The working fluid consists of N mols of perfect gas with constant volume heat capacity C per mol. In addition both models incorporate the for real engines dominating loss mechanisms:

– Friction

For the well lubricated motion of the piston in the cylinder a frictional force proportional to the piston velocity v is used, such that the frictional loss for a stroke of time t is expressed as

$$W_f = \int_0^t \alpha v^2 \, dt. \qquad (3.1.1)$$

Owing to the greater pressure on the piston during the power stroke, the mechanical friction constant α is roughly twice as large in the power stroke as it is in the other strokes. On all strokes the friction coefficient is assumed constant. Heat generated by frictional work is removed by the engine's cooling system. Possible enhancements of performance based on using this energy as a low–grade heat source are not discussed.

– Pressure drop

The pressure drop is due to the viscosity of the air while being pumped into the cylinder. It is incorporated into the models as a friction–like term by setting the friction coefficient in eq. (3.1.1) to 3α for the intake stroke.

– Heat leak

The heat leak loss is linear in the temperature difference between working fluid and cylinder walls and proportional to their area. With the piston at position x the heat leak is given by

$$q = \kappa \pi b \left(\frac{b}{2} + x\right)(T_w - T) \tag{3.1.2}$$

where κ is the heat conductance, b is the bore, and T_w is the time–independent temperature of the cylinder walls. Since the average value $T_w - T$ is small along the nonpower strokes, the heat leak losses along those strokes are assumed to be negligible.

For the intake, compression, and exhaust strokes the one dominating loss term is friction, so their dissipation is minimized when the piston is operated at constant velocity for as long as possible. If the acceleration is limited, then at the beginning and at the end of each stroke there is a phase with the highest possible acceleration till the desired velocity is reached.

The most interesting part of the cycle is the power stroke. Here the effects of the heat leak have to be balanced against those from friction. This results in the following optimal control problem for the power stroke of the Otto engine:
the objective which is to be maximized is the net work delivered by the piston:

$$W_p = \int_0^{t_p} (NRTv/x - \alpha_p v^2)dt. \tag{3.1.3}$$

The constraints governing the time evolution of the working fluid are

$$\dot{T} = \frac{1}{NC}\left(q - \frac{NRTv}{x}\right) \tag{3.1.4}$$

$$\dot{x} = v \tag{3.1.5}$$

$$\dot{v} = a \tag{3.1.6}$$

where the acceleration a is the control. This is restricted to be

$$\mid a \mid \le a_{max}. \tag{3.1.7}$$

For the case of unrestricted acceleration, eq. (3.1.6) is dropped and v becomes the control. For the Otto engine the combustion of the fuel is almost instantaneous on the time scale of the whole cycle resulting in a step–like temperature increase at the beginning of the power stroke. This is taken into account in the model by choosing an appropriately increased initial temperature T_0:

$$T(0) = T_0. \tag{3.1.8}$$

As the end temperature for the power stroke is not prescribed but open to optimization, the corresponding adjoint variable λ_T is required to fulfill

$$\lambda_T(t_p) = 0. \qquad (3.1.9)$$

These conditions are supplemented by

$$x(0) = x_0 \qquad (3.1.10)$$
$$x(t_p) = x_f \qquad (3.1.11)$$
$$v(0) = 0 \qquad (3.1.12)$$
$$v(t_p) = 0 \qquad (3.1.13)$$

to complete the optimization problem. The optimal path was determined numerically.

To avoid the heat leak losses the optimal path requires that the piston accelerates to a high velocity in order to cool down the working fluid very quickly by extracting work. If the acceleration could be infinite, the initial velocity should be about twice as large as the maximum velocity for the corresponding conventional cycle.

Figure 3 shows the optimized piston velocity and position for the whole cycle. For a range of engine parameters the net work delivered from the piston is increased by 8–15% compared to an engine with conventional linkage to the crankshaft.

For the Diesel engine the optimization is even more complex:
One has to take into account the finite rate with which the heat is transferred into the working fluid during the power stroke. Usually part of the fuel is combusted immediately after injection (after a very short delay time) while the rest is burned slowly during most or all of the time of the power stroke. The finite rate of the combustion process is approximated by the following time–dependent function which describes the extent of the reaction

$$Rn(t) = F + (1 - F)[1 - \exp(-t/t_b)]. \qquad (3.1.14)$$

The explosion fraction F is the fraction of the fuel charge consumed in the initial instantaneous burn, and t_b is the relaxation or burn time during which most of the combustion occurs. For the corresponding heating function $h(t)$, we take

$$h(t) = Q_c\, \dot{Rn}(t) \qquad (3.1.15)$$

where Q_c is the heat of combustion per molar fuel–air mixture charge. We assume that Q_c is temperature independent.

This heating function $h(t)$ modifies eq. (3.1.4) to

$$\dot{T} = \frac{1}{NC}(q + h(t) - \frac{NRTv}{x}). \qquad (3.1.16)$$

Also the changing composition of the working fluid due to the ongoing combustion had to be included, as it is accompanied by changes in the mole number N and in the heat capacity C:

$$N = N(t) = N_i + (N_f - N_i)Rn(t) \qquad (3.1.17)$$

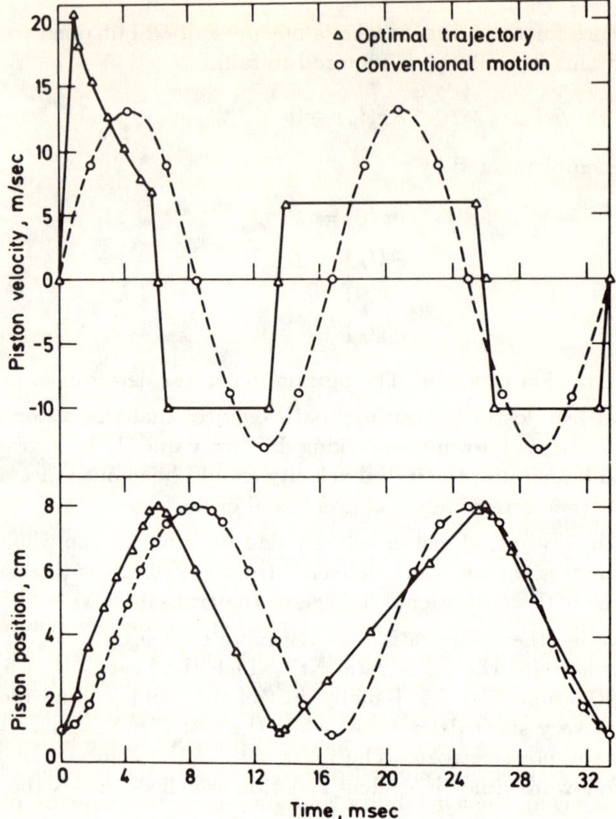

Figure 3. The time path of the Otto cycle has been optimized for maximum work output with all parameters fixed at values representative of a real car engine. By extracting work rapidly before the hot combustion gases cool too much on the cylinder wall, and by reducing friction in the remaining three strokes through constant piston speed, the power output can be increased by 8 to 15%.

and

$$C = C(t) = C_i + (C_f - C_i)Rn(t). \qquad (3.1.18)$$

Finally the finite combustion rate results in an additional loss term due to incomplete combustion at the end of the power stroke, which was incorporated into the model.

The Hamiltonian for this optimization problem thus becomes

$$H = NRT\, v/x - av^2 - (\lambda_T/NC)$$
$$\times\, [NRT\, v/x + k\pi b(b/2 + x)(T - T_w) - h(t)] + \lambda_x v + \lambda_v a. \qquad (3.1.19)$$

The set of coupled differential equations which describe the optimized path are given by eqs. (3.1.4, 3.1.5, 3.1.6) and

$$\dot{\lambda}_T = -\frac{\partial H}{\partial T} = NR\, v/x[\lambda_T(NC)^{-1} - 1] + (\lambda_T/NC)k\pi b(b/2 + x) \qquad (3.1.20)$$

$$\dot{\lambda}_x = -\frac{\partial H}{\partial x} = NRT\, v/x^2[1 - \lambda_T(NC)^{-1}] + \lambda_T(NC)^{-1}k\pi b(T - T_w) \qquad (3.1.21)$$

$$\dot{\lambda}_v = -\frac{\partial H}{\partial v} = 2av - \lambda_x - (NRT/x)(1 - \lambda_T/NC). \qquad (3.1.22)$$

From eq. (3.1.19) the optimal acceleration a has to be determined. This results in a singular control problem, the details of which are beyond the scope of this paper. The optimal path for the limited acceleration case has two switchings. The switching points as well as the complete optimal trajectory were determined numerically.

The results for the Diesel engine differ markedly from those for the Otto engine, with the optimal path for the piston showing an even more intriguing behavior:
Initially the piston is not allowed to move at all. This seemingly bizarre behavior increases losses from heat leakage because the delay increases the maximum temperature of the working fluid. The delayed motion also increases the frictional losses because the piston has less time left to move across the given distance. However, these losses are outweighed by gains in efficiency: the energy of combustion is delivered to the systems at a much higher temperature thus increasing the availability. In figure 4 the temperature profile along the optimal power stroke path is contrasted with the lower temperatures reached along the conventional power stroke. It turns out that the optimized system would start its power stroke by increasing the compression further to achieve even higher internal temperatures if it would not be hindered by the constraint of having a fixed compression ratio. This behavior is reminiscent of the work of Band, Kafri, and Salamon (Band et al. 1981) who found that the optimal expansion stroke for their particular heating function was one which also at first compressed the gas. For our model the optimized position and velocity of the piston as well as the temperature of the working fluid as a function of time is shown in figure 5.

The unbounded arcs of the power stroke show a large initial acceleration. The rapid volume increase limits the temperature rise, even though combustion within the cylinder continues. In a plot of temperature versus time, as in figure 5, one clearly sees the temperature declining at its greatest rate immediately following the motion delay time. In this way the heat conduction losses which dominate the loss mechanisms in the high–temperature regions are minimized. As the temperature of the working fluid decreases, frictional losses become relatively more important. The piston velocity changes slowly, mimicking the optimal constant velocity trajectory characteristic of the nonpower strokes. However, instead

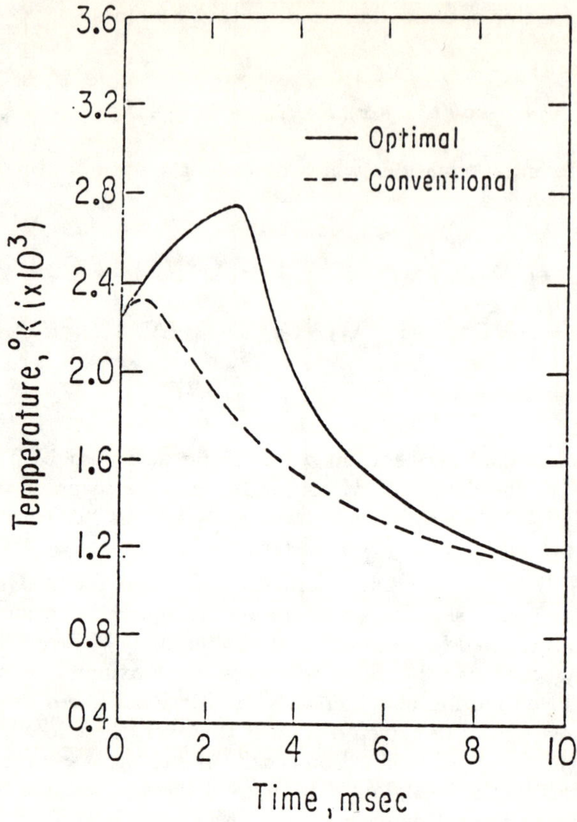

Figure 4. A comparison between the temperature profiles along the optimal and the conventional path for the power stroke of the Diesel engine. The higher temperature on the optimal path increases the availability of the energy delivered to the working fluid.

of becoming constant, the velocity decreases slowly along the power stroke as heat continues to leak to the environment. The relative improvements for the optimal engine cycle compared to the conventional one ranged between 6 and 19% for the net efficiency.

These investigations showed that the efficiency of internal combustion engines can be further improved by operating them along the optimal process paths which have been determined. The bounds provided by this analysis are also much more realistic than those

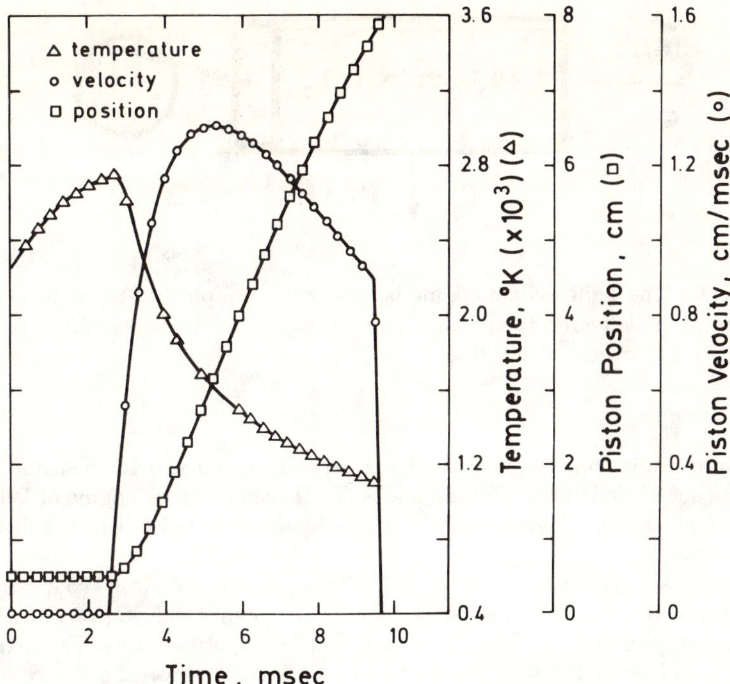

Figure 5. Working fluid temperature, piston position, and piston velocity along the optimal path for the power stroke of the Diesel engine. Note the initial part of the path during which the piston does not move.

determined by the use of reversible processes. While for instance the efficiency of an Otto cycle for typical temperatures is of the order of 70%, the efficiencies determined here are on the order of 50% and thus much closer to the observed efficiencies of real engines, which are usually below 40%.

3.2 Light driven engines

There is a class of heat engines called 'dissipative heat engines' for which the intrinsic irreversible losses are inextricably linked to the driving forces of their operation. For instance the engines discussed below are constantly and simultaneously coupled to the high and low temperature heat reservoir and thus no work producing reversible limit exists for them. It would be inappropriate to describe these systems that remain in states far from equilibrium by a Carnot–based model. Thus these engines are an instance where

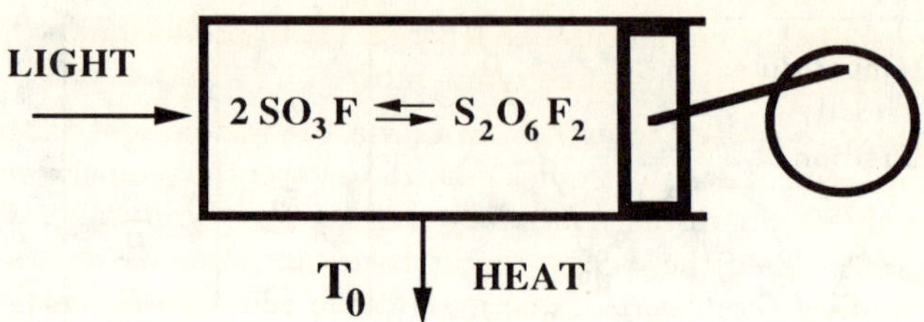

Figure 6. The light driven engine is constantly coupled to the cold heat bath. The energy is delivered to the engine as radiation, which is converted to heat by an exothermic chemical reaction.

only the use of tools from finite time thermodynamics can provide sensible performance bounds. One such 'dissipative heat engine' is the thermoacoustic engine of Wheatley *et al.* (Wheatley et al. 1983). Another one is the 'light driven' engine which will be presented in the following:

Ross and co–workers (Nitzan and Ross 1973, Zimmermann and Ross 1984, Zimmermann et al. 1984) proposed potentially oscillating chemical systems of N_2O_4 or $S_2O_6F_2$ and light, and later investigated the $S_2O_6F_2$ dimerization also experimentally. From these systems Mozurkewich and Berry (Mozurkewich and Berry 1983a) developed a theoretical model for a dissipative thermal heat engine, and showed that it could produce power. Based on this S. Watowich, R.S. Berry and I investigated the optimal paths for such engines first for a model with a very simple heating function (Watowich et al. 1985) and then for a system where the heating was based on the dimerization of $S_2O_6F_2$ (Watowich et al. 1989).

The engine is shown in figure 6. It consists of a working fluid in a cylinder of cross–sectional area A which is fitted at one end with a piston and at the opposite end with a transparent window. The cylinder is constantly coupled to the environment, which provides an infinite heat reservoir at temperature T_0. It operates endoreversibly, with the major irreversibilities being piston friction and heat conduction. The working fluid consists of the exothermic reacting system

$$2\,SO_3F \underset{\longrightarrow}{\overset{\longleftarrow}{}} S_2O_6F_2 \qquad\qquad (I)$$

in a non–reacting buffer gas. The excess buffer gas allows the working fluid to be treated as an ideal gas with constant heat capacity C_v and mole number N.

The fluorosulfate free radical selectively absorbs incident light, undergoes a rapid relaxation and releases the absorbed energy as heat:

$$SO_3F + \text{light} \longrightarrow SO_3F^* \longrightarrow SO_3F + \text{heat}. \qquad\qquad (II)$$

The temperature rise shifts the equilibrium of reaction (I) to produce more SO_3F, which in turn absorbs more light, which raises the temperature . . . , and so on.

The increased pressure is used to extract work via the piston, and at the end of the expansion stroke the movement has to be quick enough to lower the temperature sufficiently in order to undercut the feedback mechanism by shifting the equilibrium of reaction (I) back to the $S_2O_6F_2$ side. Then the compression stroke can proceed, but its velocity has to be slow enough to allow the absorbed energy and the compression energy to flow off into the environment, and only at the end of the stroke the temperature has to be raised enough to start the feedback mechanism again.

The dynamical equations describing the system are

$$\dot{x} = v \tag{3.2.1}$$

$$\dot{T} = \frac{1}{NC_v}[q + h - \frac{NRT\,v}{x}] \tag{3.2.2}$$

where x denotes the distance between the piston and the transparent window. The heat leak q is given by

$$q = [2A\kappa_1 + 2x\sqrt{\pi A}\kappa_2](T_0 - T) \tag{3.2.3}$$

where the heat conductancies κ_1 and κ_2 are constant. The heating function h describes the heat contribution from the radiation flow Φ:

$$h = \Phi\{1 - \exp[\frac{-\alpha_0}{4RTK_p}(-x + (x^2 + 16N_0RTx\frac{K_p}{A})^{\frac{1}{2}})]\}. \tag{3.2.4}$$

It is derived from Beer's Law, with α_0 being the absorption coefficient.

K_p is the pressure equilibrium constant of reaction I

$$K_p = K_0\, e^{-\frac{\Delta H - T\Delta S}{RT}} = \frac{p_{S_2O_6F_2}}{p_{SO_3F}^2} \tag{3.2.5}$$

with $K_0 = 1$ atm^{-1}. The change in enthalpy ΔH and the change in entropy ΔS are assumed to be constant over the operating temperatures and pressures of the engine. Finally N_0 is given by the number of moles of $S_2O_6F_2$ and of SO_3F:

$$N_0 = N_{S_2O_6F_2} + \frac{1}{2}N_{SO_3F}. \tag{3.2.6}$$

It is constant for reaction I.

We briefly describe how h is derived:
Beer's Law states that the absorbed fraction of the radiation flow Φ is given by

$$\frac{h}{\Phi} = 1 - e^{-\alpha_0 n_F} \tag{3.2.7}$$

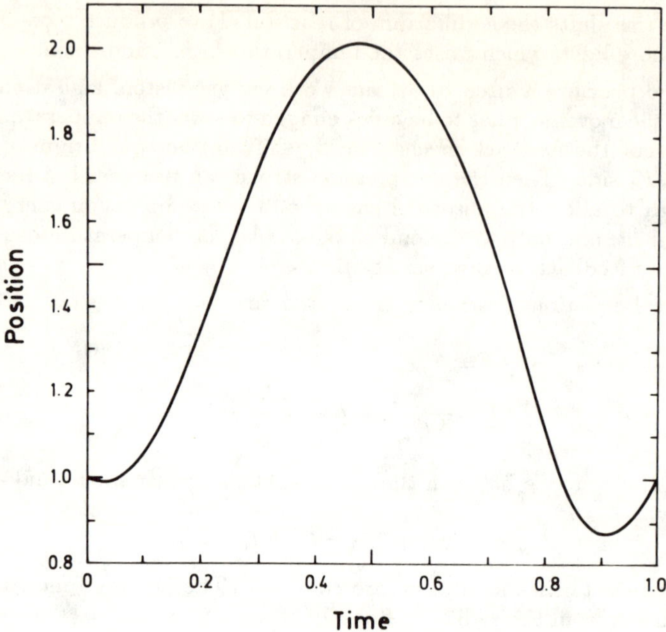

Figure 7. The position of piston for the maximum power path of the light driven engine is shown as a function of time. The trajectory is nearly sinusoidal, by eliminating the discontinuities between the branches it should be possible to guide the piston along a power optimal path with a properly adjusted conventional connecting nod–crankshaft linkage.

where n_F is the area density of the absorbant. Rewritten in the mole number of the absorbing SO_3F h becomes

$$h = \Phi(1 - \exp(-\alpha_0 \frac{N_{SO_3F}}{A})).\tag{3.2.8}$$

The mole numbers of SO_3F and $S_2O_6F_2$ are related by

$$\frac{N_{S_2O_6F_2}}{N_{SO_3F}^2} = \frac{RT}{xA} \cdot K_p \tag{3.2.9}$$

from eq. (3.2.5) and by eq. (3.2.6). Solving for N_{SO_3F} and inserting in eq. (3.2.8) gives h. Note the x and T dependence of h.

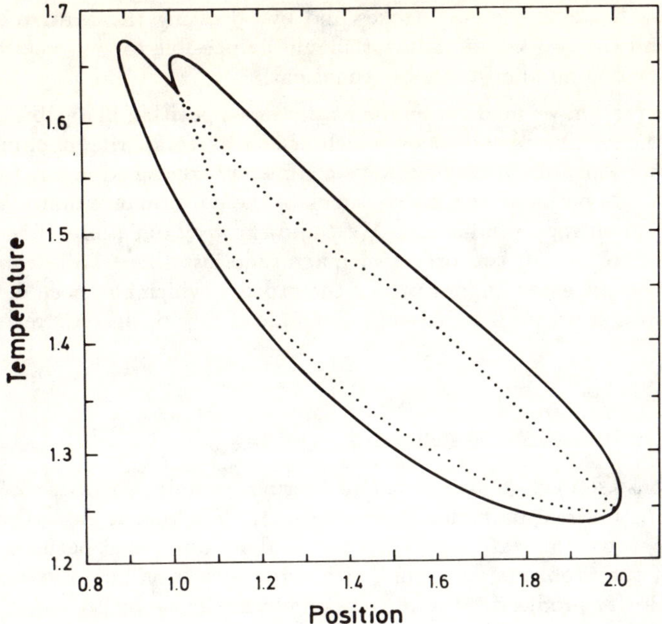

Figure 8. For the light driven engine the optimal cycles for maximum work (solid line) and for minimal entropy production (dotted line) are shown. Note that the former completely encloses the latter.

For fixed compression ratio and stroke times the optimal paths to maximize work

$$W = \int_0^\tau (\frac{NRT}{x}\frac{v}{} - \alpha v^2)dt \tag{3.2.10}$$

and to minimize the entropy production

$$\Delta S = \int_0^\tau [q \cdot (\frac{1}{T} - \frac{1}{T_0}) + \frac{\alpha v^2}{T_0}]dt \tag{3.2.11}$$

were determined. α is the coefficient of friction. We assume as stated that the entropy production results only from heat loss and friction. The energy dissipated by friction is assumed to be entirely transferred to the environment.

One interesting result for the optimized paths which yield maximum power is that they are nearly sinusoidal, see figure 7. By modifying the optimal path to eliminate the discontinuities between the two strokes and by adjusting the relative lengths of the connecting rod and the crankshaft radius it should be possible to construct a linkage that guides the piston along an almost work optimal path.

Another observation we made in temperature versus position plots, like the one shown in figure 8, was the nesting phenomenon which occurs by comparing maximum work and minimum entropy paths: the latter is always completely contained in the former. By determining the optimal paths we could also calculate the performance limits, i.e. maximum work and minimum entropy production. We could show by our finite–time analysis that the process limits are one to two orders of magnitude less than those from Carnot–like arguments from the extreme temperatures of the process, which had been determined previously. So this section provides a particular example of how misleading reversible bounds can be.

<u>3.3 The physical realization of optimal paths</u>

In the previous sections we have seen how thermodynamic processes can be optimized by determining the optimal paths for their operation. Whether it is worthwhile to build engines which operate – at least approximately – along optimized paths is still an open question. One of the problems connected with this question is how those paths and the corresponding velocity profiles for the piston can physically be realized.

There are in fact several ways of achieving those pathways of which we will discuss just two:
For a mechanical solution the problem consists in transforming the optimal paths into a rotation with constant angular speed, because this is the form of motion to which engineering is adapted. Usually such a motion is enforced by a flywheel. In the course of mechanical engineering there has been a wealth of devices to transform a constant angular motion into nearly any desired motion, but usually with the price of having many moving parts which would make such a solution costly and would increase frictional losses.

One mechanical solution which has not many moving parts is a properly designed camshaft as shown in figure 9. For such a camshaft the additional frictional losses have been estimated (Watowich 1986) to be only about half the gain from optimizing the motion of a typical diesel engine.

A completely different way to transform the optimized paths is the use of an electrical coupling. In this set–up the piston is connected to a generator as usual with connecting rod and crankshaft. Then the electrical excitation of the generator is electronically operated such that the driving force from the working fluid and the restraining forces from the generator balance in such a way that the piston moves along the optimized paths. In cases where the driving forces are too weak to overcome the inertia effects the generator has to operate as a motor for that part of the cycle.

The future (development of fuel prices) will tell whether using such couplings becomes economical.

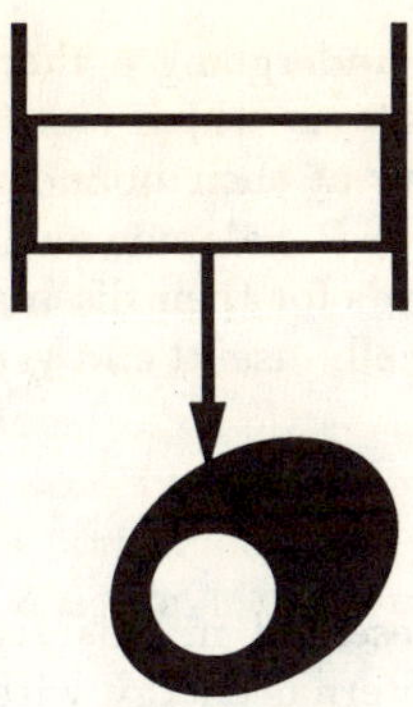

Figure 9. The optimized thermodynamic paths for heat engines can be obtained in real engines by using a cam shaft.

4. FURTHER BOUNDS FOR IRREVERSIBLE PROCESSES

In this chapter we leave the area of the applications and pursue further the problem of finding bounds and optima for irreversible thermodynamic processes from a more principal point of view.

The formulation of thermodynamics in geometrical terms has a long–standing tradition. Weinhold contributed to that, he introduced a metric in the space of thermodynamic equilibrium states using the second derivatives of the internal energy with respect to the extensive variables entropy, volume, mole number etc. (Andresen 1983, Ruppeiner 1979, Salamon et al. 1980a, Salamon et al. 1980b, Weinhold 1975a, Weinhold 1975b, Weinhold 1975c, Weinhold 1975d, Weinhold 1976, Weinhold 1978). This metric can be used to calculate a distance between two states or the length of a path in this space. Salamon and Berry (Salamon and Berry 1983) called this length 'thermodynamic length' and they showed that this length can be used to construct a bound for the loss of availability. This bound applies to a class of thermodynamic processes in which all the dissipated energy is taken out of the system. This approach for determining bounds is reviewed in section 4.1.

In 1988 B. Andresen, P. Salamon and I (Hoffmann et al. 1988) pursued this approach further: we studied thermodynamic processes in which part of the dissipated energy remains as heat in the system. As this is generally the case for real thermodynamic processes, this constitutes a considerable improvement over the previous studies. The results of our study are contained in section 4.2: it turns out that the bound provided by the thermodynamic length no longer bounds the loss of availability but a new quantity which we termed 'work deficiency'.

Most of the real physical systems undergoing a thermodynamic process can be described by a varying number of subsystems which can be considered to be in internal equilibrium with the dissipation occuring at their boundaries due to flows between them. For this general class of composite systems P. Salamon and I (Hoffmann and Salamon 1987) investigated the problem of finding bounds for their dissipation. In section 4.3 two bounds are presented which we think are physically useful and yet relatively easy to calculate.

4.1 Thermodynamic length

The bound for the dissipation presented in this section makes use of the second derivates $U_{ij} = \partial^2 U / \partial X_i \partial X_j$ of the internal energy with respect to the extensive variables X_i and X_j (entropy, volume, mole number, etc.). Usually the U_{ij} are interpreted as curvatures or second fundamental forms (Gibbs 1970, Lipschutz 1969, Tisca 1966). They can be used to calculate the change in the internal energy of a system if it is displaced for a small step in its extensive variables:

$$U - U_0 = \frac{1}{2} \sum_{ij} U_{ij}(X_i - X_i^0)(X_j - X_j^0). \tag{4.1.1}$$

This quantity can be identified with the availability of a system that can move from the state $\underline{X}$ to equilibrium with an environment at state $\underline{X}^0$.

In 1975 Weinhold (Weinhold 1975c, Weinhold 1978) used the U_{ij} in a completely different way: he showed that the first and second law of thermodynamics endow the U_{ij} with the positivity needed for a metric (or first fundamental form) on the surface of thermodynamic states. (The exception are points where U_{ij} is singular which happens for systems with coexistent phase equilibria. (Salamon and Berry 1983))

With the help of this metric one can define a length L for a path P between two thermodynamic states of a system:

$$L = \int_P (\sum_{ij} U_{ij} dX_i dX_j)^{\frac{1}{2}} \tag{4.1.2}$$

has been called 'thermodynamic length' and it is independent of the way in which the path is parametrized. The dimension of L is $\sqrt{\text{energy}}$.

In 1983 Salamon and Berry (Salamon and Berry 1983) established a connection between the thermodynamic length of a process and the availability dissipated in that process. To be precise they considered an endoreversible system which interacts with an environment by exchanging fluxes of extensive variables dX_k. Each of those fluxes occurs via a 'potential' difference, i.e. a difference between the respective conjugate variable Y_k of the system and Y_k^e of the environment. They assumed that all the potential work associated with those fluxes is completely dissipated and ends as heat in the environment which defines the availability. Salamon and Berry could show that for processes which proceed endoreversibly along equilibrium states the availability loss

$$-\Delta A^u = \int \sum_k (Y_k^e - Y_k) dX_k \qquad (4.1.3)$$

is bounded by

$$-\Delta A^u \geq L^2 \frac{\epsilon}{\tau} \qquad (4.1.4)$$

where L is the length of the path from the initial state i to the final state f, τ is the duration of the process and ϵ is a mean relaxation time for the system. For processes not proceeding endoreversibly but at least through states of local equilibrium the bound is given by the square of the distance, i.e. the length of the shortest connection between i and f times ϵ/τ. The geometry is illustrated in figure 10.

The main problem in applying this result to obtain bounds for thermodynamic processes lies with the determination of the mean relaxation time ϵ. For instance for a heat bath with a finite heat capacity c which is heated through a linear heat conductance κ the only possible relaxation time is obviously c/κ. But with increasing complexity of the system the relaxation times quickly become complicated and are hard to determine. Nonetheless the thermodynamic length provides an interesting and fruitful way to determine bounds for thermodynamic processes, and it has stimulated a number of further studies, e.g. (Feldmann et al. 1985, Nulton and Salamon 1985, Salamon et al. 1985, Schloegl 1985).

4.2 Work deficiency

In many of the finite time models for which process bounds or optima have been determined and which included friction as a source of dissipation all the work dissipated by the functional forces was put as heat into the environment and thus left the system. On the other hand for most real thermodynamic systems, this energy usually remains at least partly in the system. Including this feature in the modelling, however, would have increased calculational difficulties and thus was avoided, eventhough it was well–known.

After early attempts in 1984 B. Andresen, P. Salamon and I (Hoffmann et al. 1988) addressed this problem again in conjunction with determining availability loss and entropy production for such processes. As an example we considered the simple system shown in figure 11. Two subsystems 1 and 2 are connected via a cylinder with a piston, which in turn is connected to a work reservoir ∞. The subsystems are enclosed by an environment 0. The process we considered was a flow of volume dV from subsystem 1 to subsystem 2. If $p_1 > p_2$, then the work

$$dW^d = (p_1 - p_2)dV \qquad (4.2.1)$$

is delivered to the piston. Due to the friction part of the work is degraded to heat and goes as such to the different subsystems, while the other part is stored in the work reservoir

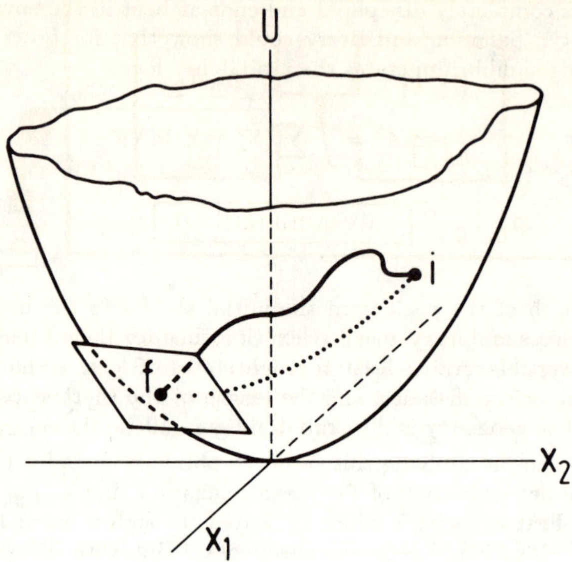

Figure 10. In this thermodynamic state space picture the internal energy U is plotted as a function of the extensive variables X_1 and X_2. The paraboloid represents equilibrium states of the system. A system initially in state i can move toward equilibrium with the environment at point f, represented by the tangent plane, along an infinite number of internal equilibrium paths. The length of the shortest of these (dotted) is determined by the Weinhold metric, but the actually traversed path (solid) may be longer. The availability dissipated in the process is related to the square of this distance divided by the duration of the process.

∞. The distribution of the energy between the different subsystems including the work reservoir will usually depend on the state of the system and can thus be time–dependent during a process.

Only if the entire work dW^d is degraded to heat and put into the environment 0, the loss of availability of the whole system equals dW^d, otherwise it is different. If for instance part of the energy goes as heat to subsystem 1 with $T_1 > T_0$, this heat has some residual availability, which can be gained by transporting it to the environment. dW^d is not even an upper bound for the loss of availability. As a counter example consider that part of the work goes as heat to subsystem 2 with a temperature T_2 lower than the environment temperature T. Then the loss of availability is larger than dW^d, as one has forgone a possible work gain from transporting that energy from T_0 to T_2. Note that here the work

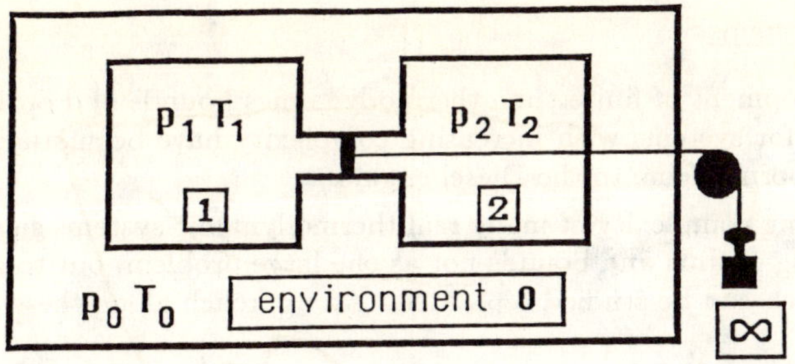

Figure 11. This example system consists of two subsystems 1 and 2, an environment 0 and a work reservoir ∞. For a process in which part of the work gained by a volume change of subsystem 1 and 2 is dissipated by friction the work defiency of that process is related to the bound derived from the thermodynamic length.

reservoir is included as part of the system, which differs from the usual set–up used in defining finite–time availability (Andresen et al. 1983).

In order to establish a connection to the thermodynamic length for the wider class of systems considered in this chapter we introduced the 'work deficiency' dW^d

$$dW^d = \frac{1}{2}\sum_{ij}(\underline{Y}_j - \underline{Y}_i)d\underline{X}_{ij}. \tag{4.2.2}$$

Here $d\underline{X}_{ij}$ is again the flux of extensities $\underline{X}$ from subsystem i to j and $\underline{Y}_i$ and $\underline{Y}_j$ are the vectors of the corresponding intensities of subsystems i and j, respectively. W^d is the work which in principle could have been extracted from the process.

When we investigated the relation between entropy production, loss of availability, work deficiency and the bound provided by the thermodynamic length we found that while the loss of availability is always connected to the entropy production

$$-\Delta A^u = T_0 \Delta S^u \tag{4.2.3}$$

the bound provided by the thermodynamic length is in general one for the work deficiency

$$W^d eq L^2 \frac{\epsilon}{\tau}. \tag{4.2.4}$$

This is a generalization of previous sections and it will hopefully help to deepen the understanding of the connection between irreversible processes and their geometrical formulation.

<u>4.3 Composite systems</u>

In the development of finite time thermodynamics bounds and optima of thermodynamic processes for systems with increasing complexity have been studied, ranging from the Curzon–Ahlborn–engine to the Diesel engine.

The increasing complexity of many real thermodynamic systems suggest to study the problem of finding optima and bounds not as one large problem, but to decompose it into subproblems which can be studied separately. An approach along these lines is presented in the following.

Most of the systems considered in finite time thermodynamics consisted of one endoreversible system which interacted irreversibly with its surroundings. The majority of real physical systems on the other hand is composed of <u>several</u> subsystems which interact irreversibly. The determination of bounds for these composite systems has long been on the agenda of finite time thermodynamics. One attempt into this direction is the previously mentioned research into the staging properties of endoreversible engines.

In 1984 P. Salamon and I (Hoffmann and Salamon 1987) started to look into this problem from a general point of view: what is the dissipation inherent in the time evolution of a composite system, i.e. what is the portion of the dissipation due to the constraint that the subsystems evolve in contact with each other. We choose our subsystems so as to insure that each one may be considered to be in internal thermal equilibrium during the entire process. This definition is equivalent to assuming separability of time scales inside the composite system, i.e., the relaxation processes inside each subsystem must be fast compared to the rates of relaxation between subsystems. For a given thermodynamic system, the choice of a subdivision depends on what relaxation processes are included in the dissipation, i.e., depending on what is considered a slow process. There is a whole range of divisions into ever more complicated structures.

The state of each subsystem is described by the state variables, some of which can be influenced from the outside. The thermodynamic variables which can be influenced are called *controllable*. A subsystem which has no controllable variable is called *uncontrollable*; a subsystem in which all variables are controllable is called an *environment*. We illustrate these definitions by considering a power station. In a first very simple subdivision the burner and the cooling tower are considered environments with controllable temperatures, while the working fluid is only partially controllable, e.g. by influencing its volume. If one desires a further subdivision, one might consider the material of the heat exchanger as a separate subsystem, which would be totally uncontrollable. As a further refinement, it would be possible to take into account the chemical reactions going on in the burner.

In general the subsystems undergo pairwise interaction in the form of fluxes $d\underline{X}_{ij}$ of extensive thermodynamic variables (particle flux, volume flux, energy flux) from subsystem i to j over 'potential' differences $\underline{Y}_i - \underline{Y}_j$, where the $\underline{Y}$ are the conjugate variables to the extensities $\underline{X}$.

The minimal dissipation ΔS_M^u of the process lasting time τ is given as

$$\Delta S_M^u = \text{Min } _{U \text{ subject } C} \frac{1}{2} \sum_{i,j} \int_0^\tau (\underline{Y}_i - \underline{Y}_j) d\underline{X}_{ij} \qquad (4.3.1)$$

where the minimum is to be taken subject to the constraints C which describe the time evolution of the system. U is the set of allowed controls. ΔS_M^u can in principle be determined by optimal control theory. However, in general this problem is very hard to solve – analytically as well as computationally. We thus addressed the far less ambitious task of finding bounds for ΔS_M^u. These bounds should be physically interesting, yet easy to calculate and ultimately useful. We proposed two bounds which we think fulfill these criteria. Both are defined as the solution to the optimization problem

$$B_{I,II}(\Delta S_M^n) = \frac{1}{2} \sum_i \text{Min } _{U_i^{I,II} \text{ subject } C_i^{I,II}} \sum_j \int_0^\tau (\underline{Y}_i - \underline{Y}_j) d\underline{X}_{ij}. \qquad (4.3.2)$$

Note that one has to solve a sum of independent optimization problems, one for each subsystem.

For both bounds the controls for each subsystem under study are taken to be the controllable variables of that subsystem and the state variables of all the other subsystems. The two bounds differ by the set of constraints under which the optimizations are performed:
For bound B_I the constraints are those which govern the subsystem under study and all the fluxes to and from it.
For bound B_{II} this set of constraint is enlarged by the requirement that the net fluxes during the process have the same values $\Delta \underline{X}_{ij}$ as in the original optimization problem,

$$\Delta \underline{X}_{ij} = \int_0^\tau d\underline{X}_{ij}. \qquad (4.3.3)$$

For these bounds it is easy to prove:

$$\Delta S_M^u \geq B_{II} \geq B_I. \qquad (4.3.4)$$

As an example we used the purely thermal system shown in fig. 4.3. For a certain process we calculated analytically the bounds B_I and B_{II} as well as the true minimum of the entropy production ΔS_M^u:
The system consists of three subsystems, each of them a heat bath. Two of them, A and B, have finite heat capacities k_A and k_B and are connected to each other with a finite heat conductivity γ. Both are connected to the third bath E with conductivities α and β, respectively. The state of the environment E is freely controllable. For the heatflows

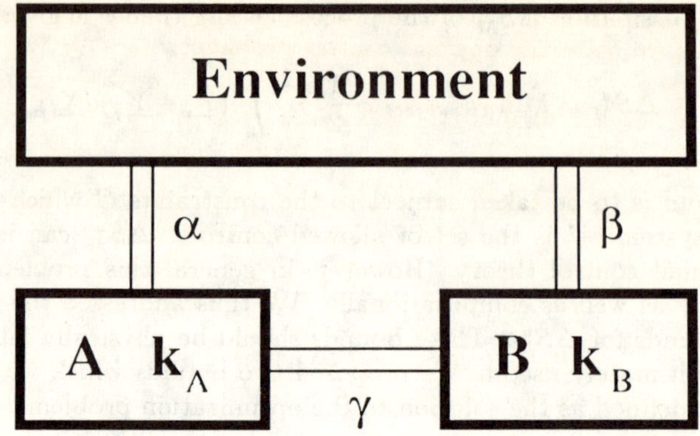

Figure 12. A simple, purely thermal composite system consisting of three heat baths. α, β, and γ are measures for the heat conductance between the baths, while k_A and k_B represent the heat capacities of baths A and B.

q through these thermal connections and the entropy production rate σ associated with them, we choose the usual expression given by linear nonequilibrium thermodynamics,

$$q_{EA} = \alpha\left[\frac{1}{T_A} - \frac{1}{T_E}\right], \tag{4.3.5}$$

$$q_{EB} = \beta\left[\frac{1}{T_B} - \frac{1}{T_E}\right], \tag{4.3.6}$$

$$q_{AB} = \gamma\left[\frac{1}{T_A} - \frac{1}{T_B}\right], \tag{4.3.7}$$

with T_j being the temperature of bath j, and

$$\sigma = \alpha\left[\frac{1}{T_A} - \frac{1}{T_E}\right]^2 + \beta\left[\frac{1}{T_B} - \frac{1}{T_E}\right]^2 + \gamma\left[\frac{1}{T_A} - \frac{1}{T_B}\right]^2. \tag{4.3.8}$$

The temperature dependence of the internal energies u_i of the baths is assumed to be

$$u_i = u_i^0 - k_i\frac{1}{T_i}. \tag{4.3.9}$$

The irreversible process considered is a temperature change from thermal equilibrium at temperature T_i to thermal equilibrium at temperature T_f in a given time τ. This change has to be achieved by suitably changing the temperature T_E of the environment in such a way that the entropy production during that process attains its minimum. In more mathematical terms we solve the optimal control problem:

$$\Delta S_M^u = min \ \Delta S^u = \int_0^\tau \sigma \ dt, \tag{4.3.10}$$

subject to the equations of motion (which are derived from the first law)

$$\frac{k_A}{T_A^2}\dot{T}_A = \alpha[\frac{1}{T_A} - \frac{1}{T_E}] + \gamma[\frac{1}{T_A} - \frac{1}{T_B}], \tag{4.3.11}$$

$$\frac{k_B}{T_B^2}\dot{T}_B = \beta[\frac{1}{T_B} - \frac{1}{T_E}] + \gamma[\frac{1}{T_B} - \frac{1}{T_A}], \tag{4.3.12}$$

and the boundary conditions

$$T_A(0) = T_B(0) = T_i, \tag{4.3.13}$$
$$T_A(\tau) = T_B(\tau) = T_f. \tag{4.3.14}$$

With the definitions

$$S = \gamma + \frac{\alpha\beta}{\alpha + \beta}, \tag{4.3.15}$$

$$L = [\frac{S}{\alpha + \beta}[\frac{\beta}{k_B} - \frac{\alpha}{k_A}]^2 + S^2[\frac{1}{k_A} + \frac{1}{k_B}]^2]^{1/2}, \tag{4.3.16}$$

$$x = L\tau, \tag{4.3.17}$$

$$D = x(e^x - e^{-x}) - 2(e^x + e^{-x} - 2), \tag{4.3.18}$$

we find for $\alpha k_B \neq \beta k_A$

$$\Delta S_M^u = (1/T_i - 1/T_f)^2 L^3 \frac{(e^x - e^{-x})}{D} \frac{1}{\alpha + \beta}[\frac{k_A k_B}{S}]^2. \tag{4.3.19}$$

For $\alpha k_B = \beta k_A$ we find

$$\Delta S_M^u = (1/T_i - 1/T_f)^2 \frac{(k_A + k_B)^2}{\tau(\alpha + \beta)}. \tag{4.3.20}$$

The latter case occurs when the time constants k_A/α and k_B/β for thermal relaxation of baths A and B connected separately to the environment are the same. In that case the temperatures in A and B remain equal for the whole process. Thus no heat flux occurs across γ and A and B can be treated as one bath with combined heat capacity.

We now show how bound B_I is determined:
Let us start with the *subsystem E*. All variables are freely controllable so the temperatures in all these baths are changed but held equal to one another. Therefor no heat flux occurs and thus no entropy production; $\Delta S_E = 0$.

For *subsystem A* we find that the optimal path has the same temperatures for E and B. Then they can be treated as one bath connected through conductance $\alpha + \gamma$, giving

$$\Delta S_A = \frac{1}{\alpha + \gamma}[\frac{1}{T_f} - \frac{1}{T_i}]^2 k_A^2 \frac{1}{\tau} \qquad (4.3.21)$$

for the entropy changes in systems A, B, and E. Due to the symmetry between A and B the result for *subsystem B* is obtained immediately,

$$\Delta S_B = \frac{1}{\beta + \gamma}[\frac{1}{T_f} - \frac{1}{T_i}]^2 k_B^2 \frac{1}{\tau}. \qquad (4.3.22)$$

Summing the three contributions up and dividing by two we find

$$B_I(\Delta S_M^u) = \frac{1}{2\tau}(1/T_i - 1/T_f)^2 \times [\frac{k_A^2}{(\alpha + \gamma)} + \frac{k_B^2}{(\beta + \gamma)}]. \qquad (4.3.23)$$

This example already shows the simplicity with which B_I can be determined.

A similar calculation gives for bound B_{II}

$$B_{II} = [\frac{1}{T_f} - \frac{1}{T_i}]^2 L^2 \frac{1}{\alpha + \beta}[\frac{k_A k_B}{S}]^2 \frac{1}{\tau}. \qquad (4.3.24)$$

Comparing B_{II} with the true ΔS_M^u we find

$$\frac{B_{II}}{\Delta S_M^u} = 1 - \frac{2}{x} \frac{e^x + e^{-x} - 2}{e^x - e^{-x}} \equiv F(x). \qquad (4.3.25)$$

This function is plotted in fig. 4.4. Note that for this example B_{II} provides the first term of a large time expansion of the minimum entropy production.

The first bound B_I is generally very easy to calculate especially since a great number of generic subsystems have already been analyzed as regards the minimum entropy production associated with transitions in time τ between given states.

The second bound B_{II} is more complicated to determine and in general the required values $\Delta \underline{X}_{ij}$ are not available without solving the original optimization problem. When using B_{II} for a process without any controllable variables, or if the structure of the composite system can give information about these net fluxes the determination of B_{II} is also relatively straightforward.

The decomposition of dissipation inherent in these bounds is very much in line with the program envisioned in earlier papers of finite time thermodynamics. A process is decomposed into three components: reversible, intrinsically irreversible, and waste (Andresen et al. 1983, Andresen et al. 1977b). Such decomposition of the dissipation is very useful for process design and an understanding of intrinsic irreversibility reveals interesting new physics (Wheatley and Cox 1985). The bound B_I measures the intrinsic irreversibility due to moving each subsystem. B_I is only achieved if each subsystem is in contact with

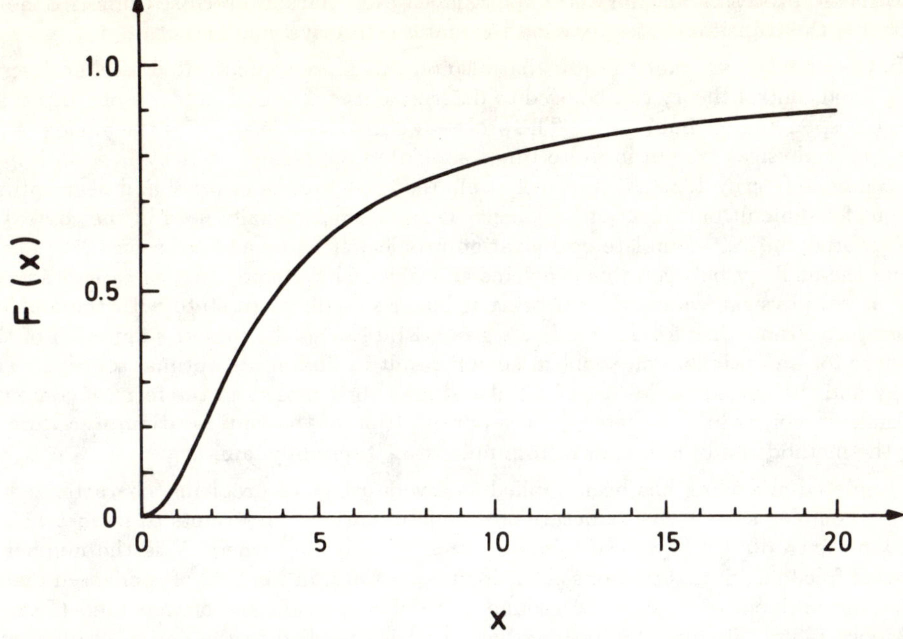

Figure 13. The bound $B_{II}(\Delta S_M^u)$ for the system displayed in figure 12 is shown in units of ΔS_M^u as a function of the rescaled process time x. Note that the long time behavior is proportional to $1/x$.

surroundings which are optimally suited to force the subsystem from the given initial state to the given final state in time τ. Accordingly, $\Delta S_M^u - B_I$ measures the mismatch in the time evolutions of the subsystems in the composite system in the sense that it measures how far other subsystems were from being each others' optimal environments. Similarly, $B_I - B_{II}$ measures the intrinsic irreversibility due to the fluxes which passed through all the subsystem boundaries. These bounds are thus useful in searching for potential savings of availability wasted in real processes.

5. SIMULATED ANNEALING

While in the previous chapters the thermodynamic processes have been descibed on a phenomenological level we now proceed to investigate bounds and optimal paths for irreversible processes descibed by statistical means. The models on this level are defined by stochastic processes and for the important case of Markov processes this is done by prescribing the transition rates between the states of the system.

In this chapter we want to show that also on this more sophisticated level of description optimal control theory can be used to determine useful bounds and – more important – the corresponding optimal paths. The process we use as an example is the analog of the cooling of a physical system in finite time: simulated annealing. It is by now an important technique (Cerny 1983, Kirkpatrick et al. 1983) for finding optimal and near optimal solutions for difficult optimization problems. It has been especially used in the context of combinatorial and NP–complete optimization problems (Garey and Johnson 1979) and it exploits the analogy between this problems and physical systems. Just as careful annealing of a real physical system should bring it into its equilibrium state with the ambient temperature T and thus for $T \to 0$ to its groundstate(s) so the proper simulation of this procedure for an optimization problem should result in finding its optimal solution. This analogy and the considerable cost of simulated annealing mostly in the form of computer time make it worthwhile to search for the optimization of this optimization procedure by using the methods and ideas known from finite time thermodynamics.

Simulated annealing has been applied to a wide range of problems, characterized by having a complex state space structure often due to various constraints to be met. These constraints introduce many local minima, often of order e^N where N is the number of degrees of freedom. Such situations occur in various areas, in the field of condensed matter physics the problem of finding the ground state of a spin glass is of that kind (Ettelaie and Moore 1985). Simulated annealing has also been applied in the design of integrated circuits (Kirkpatrick et al. 1983, Slarry and Dreyfus 1983, White 1984), for the partitioning problem as well as the wiring problem (Vecchi and Kirkpatrick 1982). It has been applied to the travelling salesman problem (Bonomi and Lutton 1984, Durbin and Willshaw 1987), as well as in graph partitioning (Salamon et al. 1988), and in the restoration of images (Geman and Geman 1984). And it has been used in parameter estimations, for instance in the seismic deconvolution problem (Jacobsen et al. 1988). This list is by no means exhaustive, but it should suffice to show that the problem attacked by simulated annealing are of great scientific and industrial importance.

5.1 Simulated annealing – a Markov process

Simulated annealing is based on the Monte Carlo simulation of physical systems. As such it uses the Metropolis algorithm (Metropolis et al. 1953) which involves a random walk through the state space of the system. Let $\Omega = \{\omega\}$ represent this state space, let $E : \Omega \to R$ be the cost function (energy) defined on the state space, and let T be the adjustable parameter in the algorithm representing the temperature of the heat bath in which the corresponding physical system is immersed. Moreover each state ω has a set of neighbors denoted by $N(\omega)$. This neighborhood relation is known as the move class and typically takes the form of an undirected graph structure on the state space. At each step of the algorithm a neighbor ω' of the current state ω_k is selected at random to become the candidate for the next state. It actually becomes the next state only with probability

$$P_{\text{acceptance}} = \begin{cases} 1 & \text{if } \Delta E \leq 0 \\ \exp(-\Delta E/T) & \text{if } \Delta E > 0, \end{cases} \qquad (5.1.1)$$

where $\Delta E = E(\omega') - E(\omega_k)$. If this candidate is accepted, then $\omega_{k+1} = \omega'$, else the next state is the same as the old state, $\omega_{k+1} = \omega_k$. Thus to complete the definition of the dynamics for the algorithm, we must specify the schedule of temperatures as a function of time T_k.

The matrix $\underline{\underline{\Pi}} = (\Pi_{\beta\alpha})$ of infinite–temperature transition probabilities from state α to β is defined by

$$\Pi_{\beta\alpha} = \begin{cases} 0 & \text{if } \beta \notin N(\alpha) \\ 1/\mid N(\alpha) \mid & \text{if } \beta \in N(\alpha), \end{cases} \qquad (5.1.2)$$

where $\mid N(\alpha) \mid$ is the number of neighbors of α. These are the transition probabilities if the algorithm automatically accepts each attempted move, i.e., if $T = \infty$. At finite temperature the acceptance decision is superimposed on $\underline{\underline{\Pi}}$ in eq. (5.1.2) to give $\underline{\underline{G}}(T)$ defined by

$$G_{\beta\alpha} = \begin{cases} \Pi_{\beta\alpha} \exp(-\Delta E/T) & \text{if } \Delta E > 0, \ \alpha \neq \beta \\ \Pi_{\beta\alpha} & \text{if } \Delta E \leq 0, \ \alpha \neq \beta \\ 1 - \sum_{\xi \neq \alpha} \Gamma_{\xi\alpha} & \text{if } \alpha = \beta, \end{cases} \qquad (5.1.3)$$

where now $\Delta E = E(\beta) - E(\alpha)$.

From a technical point of view the above introduced simulated annealing procedure is thus a discrete time Markov process with time dependent transition probabilities. For a given cooling schedule $T_k, \ k = 1, ...,$ its dynamics are given by

$$\underline{p}(k + 1) = \underline{\underline{G}}(T_{k+1})\underline{p}(k) \qquad (5.1.5)$$

and it describes the time evolution of the probability distribution $\underline{p}(k)$ of a random walker in the state space of the system.

In this picture the idea behind simulated annealing becomes clear:
As $\underline{p}(k)$ is constructed to become the Boltzmann distribution to the temperature T_k if that is kept fixed infinitely long, the probability of being in low energy states is increased for successively lower temperatures and long enough cooling time. The ultimate hope then is to have all the weight of the probability distribution in the gound state(s) for an appropriate cooling schedule in which T goes to zero (independent of the initial distribution).

For systems having local minima it is obvious that not every cooling schedule with $T \to 0$ can insure this, for instance a quench, i.e. immediately setting T to zero, will in general leave some probability in those local minima.

<u>5.2 Optimal annealing schedules</u>

Simulated annealing is applied as an optimization tool in many different areas. Used as such it incurs costs, mostly in the form of computer time. So it is very natural that from the very beginning people have tried to improve the simulated annealing procedure to lower those costs (Aarts and van Laarhoven 1985, Geman and Geman 1984, Huang et al. 1986, Lundy and Mees 1984, Morgenstern and Wuertz 1987, White 1984). There have been attempts to improve the results through changing the algorithm itself by avoiding steps in which new states are not accepted (Greene and Supowit 1984) or by choosing special move classes (Szu and Hartley 1987). The most important and – still – unanswered question is:

- **what is the optimal cooling schedule?**

However, before this question can be answered the yardstick must be defined with respect to which the optimality is determined. Indeed several criteria are possible:

- the probability to be in the ground state

- the mean final energy

- the mean B(est) S(o) F(ar) energy (Jacobsen et al. 1988)

Here we will use the mean final energy, because it is certainly the most natural choice when one analyses simulated annealing from the view point of its analogy to statistical mechanics. In addition it is simple from a computational point of view as it is only dependent on the distribution at the final time.

As already mentioned above simulated annealing is applied to systems which have a many valley state space. The very feature which makes simulated annealing a useful method is its hill climbing ability which allows it to leave local minima and overcome the bariers on the way to the global minimum. Peter Salamon and I studied thus studied small model systems (Hoffmann and Salamon 1988) with local minima and bariers. To describe in more detail what type of calculation we performed and of which kind the results are we discuss here our smallest model system:
It is a three state system which possesses one local and one global minimum.

Figure 14 shows this system. The state space consists of the states 1, 2, and 3 with energies:

$$\begin{aligned} E(1) &= 0 \\ E(2) &= 1 \\ E(3) &= D \geq 1 \end{aligned} \qquad (5.2.1)$$

Using the abbreviation $x = e^{\frac{1}{T}}$ the transition probability matrix $\underline{\underline{G}}$ implementing the metroplis algorithm is given by

$$\underline{\underline{G}}(x) = \begin{pmatrix} 1 - x^D & 0 & \frac{1}{2} \\ 0 & 1 - x^{D-1} & \frac{1}{2} \\ x^D & x^{D-1} & 0 \end{pmatrix} \qquad (5.2.2)$$

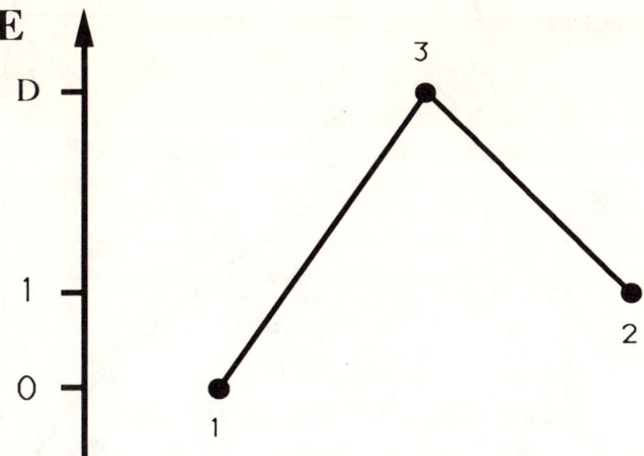

Figure 14. The simplest system with a move class represented by a regular graph and with one local and one global minimum is the four state system shown on the right. As the energies of states 3 and 3' are chosen to be equal one can lump the system to the three state system shown on the left.

The final probability distribution $\underline{p}(N)$ after N steps can be expressed as

$$\underline{p}(N) = \prod_{k=1}^{N} \underline{\underline{G}}(x_k)\, \underline{p}(0) \qquad (5.2.3)$$

with $\underline{p}(0)$ being the initial distribution. Using the energy vector $\underline{E} = (0, 1, D)$ the formal problem of finding an optimal schedule to minimize the mean energy in a finite number of Metropolis steps N can be written as

$$Min\ \bar{E} = Min_{\{x_k\}_{k=1,N}}\ \underline{E}\ \prod_{k=1}^{N} \underline{\underline{G}}(x_k)\, \underline{p}(0) \qquad (5.2.4)$$

This problem was first investigated numerically and then analytically. Optimal cooling schedules were determined for a varying number of steps N, for different activation energies D, and for different initial distributions.

Figure 15 shows the paths created by those optimal cooling schedules for different initial distributions. The most remarkable and unexpected feature is the 'turnpike' seen. By 'turnpike' we mean that the optimal path — after a short transition period lasting only a few steps — is independent of the initial distribution.

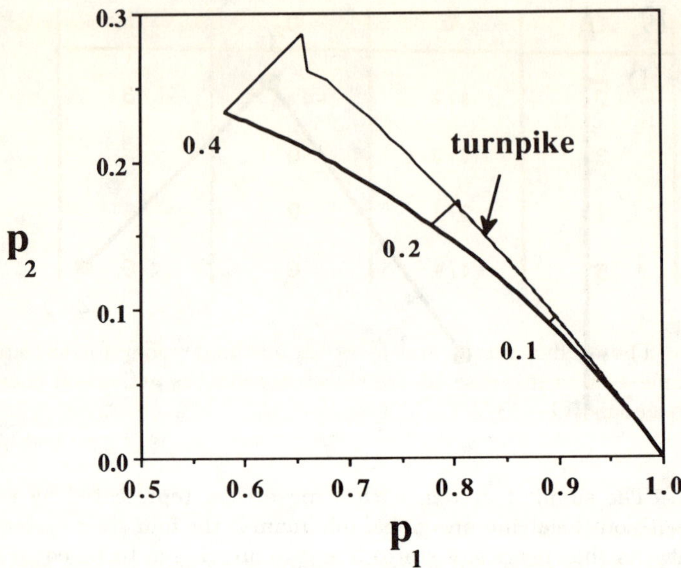

Figure 15. For the three state system of figure 14 the paths created by the optimal cooling schedules for different initial temperatures ($e^{-\frac{1}{T}} = 0.1$, 0.2, and 0.4) are shown. The heavy line marks the equilibrium states. Note the turnpike behaviour of the optimal paths: After the first few steps the optimal schedule leads the system always along the same curve.

We now discuss the optimal path and the corresponding optimal policy in more detail: The numerical results as well as an analytical calculation from

$$\bar{E} = \underline{E} \ \underline{\underline{G}}(x) \ \underline{p}(0) \tag{5.2.5}$$

show that the optimization policy for the last step should be $x = 0$, i.e. zero temperature. This can easily be understood: while all of the probability in state 3 is transported to states 1 and 2 independent of the temperature, any finite temperature will shift some probability from states 1 and 2 to state 3 and thus increases the final energy. From

$$Min_x \ \bar{E} = Min_x \ \underline{E} \ \underline{\underline{G}}(0) \ \underline{\underline{G}}(x) \ \underline{p}(0) \tag{5.2.6}$$

the policy for the last but one step can be determined:

$$x = \frac{D-1}{D} \frac{p_2(0)}{p_1(0)} \tag{5.2.7}$$

D	b_1	b_2	M
2	1/2	-1/4	3
3	2/3	0	4
4	3/4	0	5
5	4/5	0	6

Figure 16. The coefficients b_1 and b_2 of the optimal policy for the annealing of the example system in figure 14 are shown for various activation energies D. Note that b_1 equals $(D-1)/D$.

Note that x depends only on the ratio p_2/p_1.

The numerical results suggested to make the following assumptions for the policy for the remaining steps:

1) At any given point along the optimal path the optimal policy for the next step is only a function of the current state, i.e. it is a feed back control.

2) The probabilities enter the optimal policy only in the form p_2/p_1.

Thus we made the ansatz

$$x(i) = b_1 \frac{p_2(i-1)}{p_1(i-1)} + b_2 \left(\frac{p_2(i-1)}{p_1(i-1)}\right)^2. \qquad (5.2.8)$$

For computational simplicity we assumed that the initial distribution had already most of their weight in the low energy state

$$p_2 = \gamma p_1 \qquad (5.2.9)$$
$$p_3 = \gamma_3 \gamma^2 p_1 \qquad (5.2.10)$$

with $\gamma \ll 1$ and γ_3 of order one. The unknown constants b_1 and b_2 were determined by minimizing the average final energy $\bar{E}$ analytically for different step numbers N.

Expanding $\bar{E}$ in powers of γ and taking only terms up to order M gave the results displayed in figure 16 for the various activation energies. The b's do not depend on N.

Even from the small number of examples we give in figure 16 it can be seen that

$$b_1 = \frac{D-1}{D} \qquad (5.2.11)$$

which is the same prefactor of p_2/p_1 as in the policy for the last step. Another interesting feature which we found by persuing this approach further is the fact that the optimal policy is not a true feed back policy. It turns out that for higher orders in γ the b's do depend on the initial distribution.

Finally we want to show how the optimal policy can be obtained by optimal control theory. First we change the discrete time description of the dynamics eq.(5.1.5) into a continuous one (Hajek 1985):

$$\dot{\underline{p}} = \underline{\underline{A}}\,\underline{p} \tag{5.2.12}$$

with

$$\underline{\underline{A}} = \underline{\underline{G}} - \underline{\underline{1}}. \tag{5.2.13}$$

As the matrix $\underline{\underline{A}}$ has – apart from the stationary mode – a slow mode describing the flow of probability over the barrier and a quick mode which brings p_3 in equilibrium to given p_1 and p_2, we simplify the control problem further by eliminating p_3 adiabatically. Using the normalization of $\underline{p}$ we find for small temperatures $x \ll 1$:

$$\dot{p}_2 = -p_2 x^{D-1}\frac{1}{2} + \frac{1}{2}x^D(1 - p_2) \tag{5.2.14}$$

This is the usual dynamics for a two–level–system with Arhenius factors for jumping over the separating barrier of height D (Huse and Fisher 1986).

In this model the mean final energy is equal to the final p_2 and thus minimizing $\bar{E}$ is equivalent to

$$Min \int_0^\tau dt\, \dot{p}_2 \tag{5.2.15}$$

where τ is now the total time ($=$ total number of steps) for the process. The dynamics is given by eq. (5.2.14). Applying optimal control theory we find:

$$H = \dot{p}_2 + \lambda\dot{p}_2 = (1 + \lambda)\dot{p}_2 \tag{5.2.16}$$

The optimal policy is obtained by extremalizing H

$$\frac{\partial H}{\partial x} = \frac{\partial \dot{p}_2}{\partial x} = 0 \tag{5.2.17}$$

which for any dynamics $\dot{p}_2$ will lead to a feed back control, as the optimal x can only depend on p_2 (and not on λ). For our $\dot{p}_2$ we find

$$x = \frac{D-1}{D}\frac{p_2}{1 - p_2} = \frac{D-1}{D}\frac{p_2}{p_1} \tag{5.2.18}$$

which is the well known policy from our other analytical and numerical results.

We now explore the asymptotic form of our optimal schedule. Substituting the optimal x in eq.(5.2.18) into eq.(5.2.14) and integrating p_2 for $p_2 \ll 1$ we obtain

$$p_2(t) = [\frac{t}{2}\cdot(\frac{D-1}{D})^D + p_2(0)^{-(D-1)}]^{\frac{-1}{D-1}}. \tag{5.2.19}$$

For long times $t >> 2p_2(0)^{-(D-1)} \cdot (\frac{D}{D-1})^D$ this simplifies to

$$p_2(t) = (\frac{D}{D-1})^{\frac{D}{D-1}} \cdot (\frac{t}{2})^{-\frac{1}{D-1}}.$$

(5.2.20)

Within the approximations made above we find for the optimal schedule

$$x(t) = \frac{D-1}{D} \cdot p_2(t) = (\frac{D}{D-1})^{\frac{1}{D-1}} \cdot (\frac{t}{2})^{-\frac{1}{D-1}}$$

(5.2.21)

and thus for long times

$$T(t) \sim \frac{D-1}{\ln(t/2)}.$$

(5.2.22)

This schedule is also known (Geman and Geman 1984, Hajek 1985) to guarantee convergence to the groundstate with probability 1. The reason for this coincidence is that minimizing the mean final energy is equivalent to maximizing the likelihood $p_1(\tau)$ of occupying the groundstate at the final time.

We have shown here for a simple three state system how the optimal annealing schedule can be determined. We hope that further research in this direction will also lead to improvements for the commercially applied algorithms.

Eventhough it is certainly worthwhile to study simulated annealing as an stochastic optimization tool, in this paper the determination of the optimal schedule has just been an example that optimal control theory can successfully be used to find optimal paths and corresponding extrema for process values for systems descibed on a statistical level. Thus these methods are as well suited for instance to determine bounds on the entropy production for processes of small clusters or other mesoscopic systems where phenomenological laws no longer apply.

6. SUMMARY

In this paper we addressed the questions

— Under which conditions and how can realistic bounds for process variables of thermodynamic processes, which are performed in finite time, be determined and

— what are the optimal process paths to achieve the optimal process values.

In chapter 2 we first introduced the classical tools of finite time thermodynamics. After briefly reviewing finite time potentials and the tricycle formalism we discussed optimal control theory, which is *the* tool to determine the complete optimal pathways for processes.

Chapter 3 showed the results obtained by applying optimal control theory to heat engines. First we presented internal combustion engines which are probably still the most sophisticated models discussed so far by finite time thermodynamics. For those engines most of the results could only be determined numerically. Those results included not only the optimal process path but also much more realistic efficiencies which are much closer to the observed ones than the reversible ones. And finally we discussed the light engines as a particular example of how misleading reversible bounds can be for the so–called dissipative engines. For these intrinsically irreversible engines only optimal control theory provides the bounds which can serve as sensible estimates to real process values.

In chapter 4 further bounds for irreversible thermodynamic processes were introduced. By using the thermodynamic length a bound for the loss of availability was given. For processes in which not all dissipated heat is taken out of the system it becomes one for the work deficiency. For composite systems which are too complex for a complete process optimization two bounds were given which are relatively easy to determine but still physically interesting. They allow to quantify different parts of the dissipation in composite systems.

In chapter 5 we discussed how the methods and ideas from finite time thermodynamics can be used to determine optimal paths and process bounds for systems described on a statistical level. As an example we used the cooling of a system in a finite time. This process in analogy to the physical annealing of a system has commercially as well as scientifically highly interesting applications as an stochastic optimization method and is called simulated annealing. After a short discussion of possible optimization criteria we determined the optimal annealing schedule for a small model system by using optimal control theory.

In summary we have shown how bounds and optima for irreversible thermodynamic processes can be determined for a wide range of physical processes and how those methods can be put to use for an important artificial thermodynamic process: simulated annealing.

ACKNOWLEDGEMENTS

I would like to thank Bjarne Andresen, Steve Berry, Peter Salamon and Stan Watowich for the opportunity to work with them. Our collaborations have always been delightful and educational for me. In addition I would like to acknowledge the stimulating environment of the Telluride Summer Research Center where part of this work was done as well as partial travel support by the Deutsche Forschungsgemeinschaft.

REFERENCES

Aarts, E.H.L., and P.J.M. van Laarhoven. 1985. <u>Philips J. Res. 40</u>: 193.

Andresen, B. 1983. Finite-Time Thermodynamics. University of Copenhagen.

Andresen, B., R.S. Berry, A. Nitzan, and P. Salamon. 1977a. <u>Phys. Rev. A 15</u>: 2086.

Andresen, B., M.H. Rubin, and R.S. Berry. 1983. <u>J. Phys. Chem. 87</u>: 2704.

Andresen, B., P. Salamon, and R.S. Berry. 1977b. <u>J. Chem. Phys. 66</u>: 1571.

Band, Y.B., O. Kafri, and P. Salamon. 1981. J. Appl. Phys. 52: 3745.

Boltyanskii, V.G. 1971. Mathematical methods of optimal control. New York: Holt, Reinhart, and Winston.

Bonomi, E., and S.L. Lutton. 1984. SIAM Rev. 26: 551.

Bryson, J., A. E., and Y.-. Ho. 1975. Applied Optimal Control. New York: Wiley.

Callen, H.B. 1960. Thermodynamics. New York: Wiley.

Cerny, V. 1983. JOTA 45: 41.

Curzon, F.L., and B. Ahlborn. 1975. Am. J. Phys. 43: 22.

d'Isep, F., and L. Sertorio. 1982. Nuovo Cim. B 67: 41.

d'Isep, F., and L. Sertorio. 1983. Nuovo Cim. C 6: 305.

Durbin, R., and D. Willshaw. 1987. Nature 326: 689.

Ettelaie, R., and M.A. Moore. 1985. J. Physique Lett. 46: L-893.

Feldmann, T., B. Andresen, A. Qi, and P. Salamon. 1985. J. Chem. Phys. 83: 5849.

Garey, M.R., and D.S. Johnson. 1979. Computers and Intractability: A Guide to the Theory of NP-Completeness. New York: W. H. Freeman and Company.

Geman, S., and D. Geman. 1984. IEEE, PAMI 6: 721.

Gibbs, J.W. 1970. Collected Works. Cambridge, Mass.: MIT Press.

Greene, J.W., and K.J. Supowit. 1984. In ICCD 84 (IEEE), 658.

Gutkowics-Krusin, D., I. Procaccia, and J. Ross. 1978. J. Chem. Phys. 69: 3898.

Hajek, B. 1985. In Int. Conf. on Decision and Control (IEEE),

Hoffmann, K.H., B. Andresen, and P. Salamon. 1988. Phys. Rev. A 39: 3618.

Hoffmann, K.H., and P. Salamon. 1987. Phys. Rev. A 35: 369.

Hoffmann, K.H., and P. Salamon. 1988. Optimal cooling schedules for simulated annealing. Preprint.

Hoffmann, K.H., S.J. Watowich, and R.S. Berry. 1985. J. Appl. Phys. 58: 2125.

Hsu, J.C., and A.U. Meyer. 1968. Modern Control Principles and Applications. New York: McGraw-Hill.

Huang, M.D., F. Romeo, and A. Sangiovanni-Vincentelli. 1986. An Efficient General Cooling Schedule for Simulated Annealing. In ICCAD 86 (IEEE), 381.

Huse, D.A., and D.S. Fisher. 1986. Phys. Rev. Lett. 57: 2203.

Jacobsen, M.O., K. Mosegaard, and J.M. Pedersen. 1988. Global Model Optimization in
Reflection Seismology by Simulated Annealing. In Model Optimization
in Exploration Geophysics 2, ed. A. Vogel, 361. Braunschweig/Wiesbaden: Friedr.
Vieweg and Son.

Kirkpatrick, S., C.D. Gelatt Jr., and M.P. Vecchi. 1983. Science 220: 671.

Lipschutz, M.M. 1969. Differential Geometry. New York: McGraw-Hill.

Lundy, M., and A. Mees. 1984. Convergence of the annealing algorithm.
In Simulated Annealing Workshop, Yorktown Heights.

Metropolis, N., A. Rosenbluth, M. Rosenbluth, A. Teller, and E. Teller. 1953.
J. Chem. Phys 21: 1087.

Morgenstern, I., and D. Wuertz. 1987. Z. Phys. B 67: 397.

Mozurkewich, M., and R.S. Berry. 1981. Proc. Natl. Acad. Sci. U.S.A. 78: 1986.

Mozurkewich, M., and R.S. Berry. 1982. J. Appl. Phys. 53: 34.

Mozurkewich, M., and R.S. Berry. 1983a. J. Appl. Phys. 54: 3651.

Mozurkewich, M., and R.S. Berry. 1983b. J. Appl. Phys. 54: 3651.

Mullins, O.C., and R.S. Berry. 1984. J. Phys. Chem. 88: 723.

Nitzan, A., and J. Ross. 1973. J. Chem. Phys. 59: 241.

Nulton, J.D., and P. Salamon. 1985. Phys. Rev A 31: 2520.

Ondrechen, M.J., B. Andresen, and R.S. Berry. 1980a. J. Chem. Phys. 73: 5838.

Ondrechen, M.J., R.S. Berry, and B. Andresen. 1980b. J. Chem. Phys. 72: 5118.

Richter, P.H., P. Rehmus, and J. Ross. 1981. Prog. Theor. Phys. 66: 385.

Richter, P.H., and J. Ross. 1978. J. Chem. Phys. 69: 5521.

Ross, R.T., and J.M. Collins. 1980. J. Appl. Phys. 51: 4504.

Ross, R.T., and A.J. Nozik. 1982. J. Appl. Phys. 53: 3813.

Rubin, M. 1979a. Phys. Rev. A 19: 1277.

Rubin, M. 1979b. Phys. Rev. A 19: 1272.

Rubin, M. 1980. Phys. Rev. A 22: 1741.

Ruppeiner, G. 1979. Phys. Rev. A 20: 1608.

Salamon, P., B. Andresen, P.D. Gait, and R.S. Berry. 1980a. J. Chem. Phys. 73: 5407.

Salamon, P., B. Andresen, P.D. Gait, and R.S. Berry. 1980b. J. Chem. Phys. 73: 1001.

Salamon, P., Y.B. Band, and O. Kafri. 1982. J. Appl. Phys. 53: 197.

Salamon, P., and R.S. Berry. 1983. _Phys. Rev. Lett. 51_: 1127.

Salamon, P., and A. Nitzan. 1981. _J. Chem. Phys. 74_: 3546.

Salamon, P., J. Nulton, and R.S. Berry. 1985. _J. Chem. Phys. 82_: 2433.

Salamon, P., J. Nulton, J. Robinson, J. Pedersen, G. Ruppeiner, and L. Liao. 1988. _Comput. Phys. Commun. 49_: 423.

Schloegl, F. 1985. _Z. Phys. B 59_: 449.

Slarry, P., and G. Dreyfus. 1983. _J. Physique Lett. 45_: L-39.

Su, J.S., and A.J. Engel. 1980. _A. I. Ch. E. Symp. Ser. 76_: 6.

Szu, H., and R. Hartley. 1987. _Physics Letters A 122_: 157.

Taylor, C.F. 1966. _The Internal Combustion Engine in Theory and Practice_. Cambridge, Massachusetts: MIT Press.

Termonia, Y., and J. Ross. 1981. _J. Chem. Phys. 74_: 2339.

Tisca, L. 1966. _Generalized Thermodynamics_. Cambridge, Massachusetts: MIT Press.

Tolle, H. 1975. _Optimization Methods_. New York: Springer.

Vecchi, M.P., and S. Kirkpatrick. 1982. IBM Thomas J. Watson Research Center, Yorktown Heights, New York. RC9555.

Watowich, S.J. 1986. private communication.

Watowich, S.J., K.H. Hoffmann, and R.S. Berry. 1985. _J. Appl. Phys. 58_: 2893.

Watowich, S.J., K.H. Hoffmann, and R.S. Berry. 1989. _Nuovo Cim. B 104_: 131.

Weinhold, F. 1975a. _J. Chem. Phys. 63_: 2488.

Weinhold, F. 1975b. _J. Chem. Phys. 63_: 2496.

Weinhold, F. 1975c. _J. Chem. Phys. 63_: 2479.

Weinhold, F. 1975d. _J. Chem. Phys. 63_: 2484.

Weinhold, F. 1976. _J. Chem. Phys. 65_: 559.

Weinhold, F. 1978. In _Theoretical Chemistry, Advances and Perspectives_, ed. D. Henderson, and H. Eyring, New York: Academic Press.

Wheatley, J., and A. Cox. 1985. _Physics Today 38_:

Wheatley, J., T. Hofler, G. Swift, and A. Migliori. 1983. _Phys. Rev. Lett. 50_: 499.

White, S.R. 1984. Concepts of Scale in Simulated Annealing. In _ICCD '84 (IEEE)_, 646.

Zimmermann, E.C., and J. Ross. 1984. _J. Chem. Phys. 80_: 720.

Zimmermann, E.C., M. Schell, and J. Ross. 1984. _J. Chem. Phys. 81_: 1327.

Finite-Time Thermodynamics

Bjarne Andresen

Physics Laboratory, University of Copenhagen
DK-2100 Copenhagen Ø, Denmark

Abstract

Finite-time thermodynamics is the extension of traditional reversible thermodynamics to include the extra requirement that the process in question goes to completion in a specified finite length of time. As such it is by definition a branch of irreversible thermodynamics, but unlike most other versions of irreversible thermodynamics, finite-time thermodynamics does not require or assume any knowledge about the microscopics of the processes, since the irreversibilities are described by macroscopic constants such as friction coefficients, heat conductances, reaction rates and the like. Some concepts of reversible thermodynamics, such as potentials and availability, generalize nicely to finite time, others are completely new, e.g. endoreversibility and thermodynamic length. The basic ideas of finite-time thermodynamics are reviewed and several of its procedures presented, emphasizing the importance of power. Finally, its impact on the global optimization algorithm simulated annealing is outlined.

1. Introduction

1.1 Motivation

From its infancy over 150 years ago, thermodynamics has provided limits on work or heat exchanged during real processes. The first problem treated in a systematic way was how much work a steam engine can produce from the burning of one ton of coal. With true scientific generalization Sadit Carnot concluded that any engine taking in heat from a hot reservoir at temperature T_H has to deposit some of that heat in a cold reservoir (e.g. the surroundings), whose temperature we call T_L; the largest fraction of the heat which can be converted into work is

$$\eta_C = 1 - \frac{T_L}{T_H},\tag{1.1}$$

traditionally known as the Carnot efficiency. This expression contains the two basic ingredients of a thermodynamic limit: a) it applies to *any* process converting heat into work; and b) it is an *absolute* limit, i.e. no process, however ingenious, can do better.

As thermodynamic theory developed, emphasis changed from process variables like work and heat exchanged to state variables like entropy and chemical potential. A bridge between the two are the thermodynamic work potentials, such as Helmholtz free energy H for isothermal, isochoric processes or the Gibbs free energy G for isothermal, isobaric processes. They are defined such that their changes provide upper bounds on the work a process can supply or lower bounds on the work required to drive a process. Gibbs introduced the concept of 'available work' as the maximum work that can be extracted from a system allowed to go from a constrained, internally equilibrated state to a state in equilibrium with its surroundings. This quantity is used more and more frequently in engineering contexts (Keenan 1941; Gaggioli 1980) under the names 'availability' in the U.S. and 'exergy' in Europe. For a system relaxing to an ambient temperature T_0, pressure P_0, and chemical potentials μ_{0i} it is given by

$$A = U + P_0 V - T_0 S - \sum_i \mu_{0i} N_i \tag{1.2}$$

and is thus not a state function in the usual sense of depending only on variables of the system; the availability depends on the intensive variables of the environment as well.

Such criteria of merit have long been common currency for thermodynamic studies in physics, chemistry, and engineering. They all share one characteristic: The ideal to which any real process is compared is a *reversible process*. Stated in a different way, *traditional thermodynamics is a theory about equilibrium states and about limits on process variables for transformations from one equilibrium state to another.* Nowhere does time enter the formulation, so these limits must be the lossless, reversible processes which proceed infinitely slowly and thus take infinite length of time to complete. However, referring back to the original question addressed by Carnot, who is interested in an engine which operates infinitely slowly (and thus produces zero power) — or any other process with zero rate of operation, for that matter?

In order to obtain more realistic limits to the performance of real processes *finite-time thermodynamics is designed as the extension of traditional thermodynamics to deal with processes which have explicit time or rate dependences.* These constraints, of course, imply a certain amount of loss, or entropy production, which is at the heart of the question posed above.

1.2 Early developments

In the course of developing finite-time thermodynamics we discovered that a few isolated papers already had considered different aspects of processes operating at nonzero rates. The first of these was the important work of Tolman and Fine (1948) who put the Second Law of thermodynamics into equality form,

$$W = \Delta A - T_0 \int_{t_i}^{t_f} \dot{S}_{tot}\, dt \tag{1.3}$$

by subtracting the work equivalent of the entropy produced during the process from the reversible work, i.e. the decrease of system availability, as defined in eq. (1.2). The superscript dot indicates rate, and the integral limits are the initial and final times of the process. This is a quantification of the 'price of haste'.

Another model has evolved into almost a classic paradigm of systems operating in finite time. This is the model of Curzon and Ahlborn (1975), a Carnot engine with the simple constraint that it be linked to its surroundings through *finite* heat conductances. Figure 1 illustrates the slightly more general endoreversible system with the triangle signifying any reversible engine. (The term endoreversible means 'internally reversible', i.e. all irreversibilities reside in the coupling of flows to the surroundings. In this case that means resistance to heat transfer and possibly friction.) It turns out that the results derived by Curzon and Ahlborn explicitly for an interior Carnot engine are equally valid for a general endoreversible system. The maximum efficiency of their engine is of course $\eta_C = 1 - T_L/T_H$, obtained at zero rate so that losses across the resistors vanish, but these authors showed that, when the system operates to produce *maximum power*, the efficiency of the engine is only

$$\eta_w = 1 - \sqrt{\frac{T_L}{T_H}}. \tag{1.4}$$

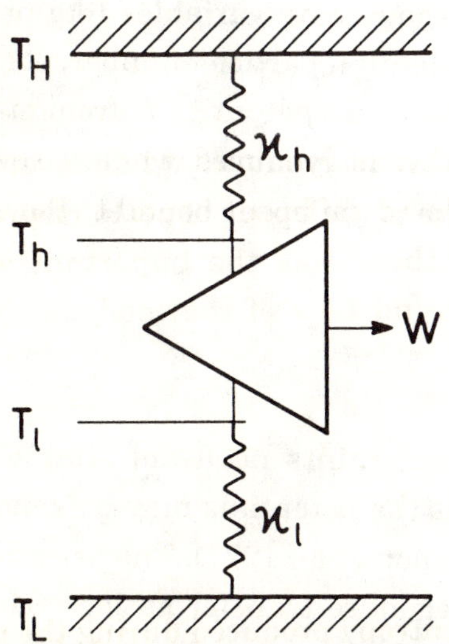

Figure 1. An endoreversible engine has all its losses associated with its coupling to the environment, there are no internal irreversibilities. This is illustrated here as resistances in the flows of heat to and from the working device indicated by a triangle. These unavoidable resistances cause the engine proper to work across a smaller temperature interval, $[T_h;T_l]$ than that between the reservoirs, $[T_H;T_L]$, one which depends on the rate of operation.

Besides the simplicity of the expression it is remarkable that it does not contain the value of the heat conductances. A closer analysis of this expression and its relationship to the effects of finite size heat reservoirs is presented by Jeff Gordon in this volume. The initial development of finite-time thermodynamics was primarily inspired by Curzon and Ahlborn's paper.

In the next section we will see how it is possible to derive bounds on finite-time processes without knowing their detailed time paths. Following that we derive the optimal paths for a few examples.

2. Performance bound without path

The smallest amount of information one can ask for concerning the performance of a system is a single number, e.g. the work or heat exchanged during the process, its efficiency, or any other figure of merit. In most cases this can be calculated without knowledge of the detailed path followed and is then computationally much simpler to obtain.

2.1 Generalized potentials

In traditional thermodynamics potentials are used to describe the ability of a system to perform some kind of work under given constraints. These constraints are usually the constancy of some state variables like pressure, volume, temperature, entropy, chemical potential, particle number, etc. Under such conditions the decrease in thermodynamic potential P from state i to state f is equal to the amount of work that is produced when a reversible process carries out the transition, and hence is the upper bound to the amount of work produced by any other process,

$$W \leq W_{rev} = P_i - P_f. \tag{2.1}$$

In this section we will show that the constraints need not simply be the constancy of some state variable, and that the potentials may be generalized to contain constraints involving time (Salamon et al. 1977). The procedure will be a straight forward extension of the Legendre transformations (Hermann 1973) used in traditional thermodynamics (Callen 1960; Tisza 1966), and we will start with such an example.

In a reversible process heat and work can be expressed as inexact differentials,

$$dQ = TdS, \qquad dW = PdV, \tag{2.2}$$

i.e. they cannot by themselves be integrated, further constraints defining the integration path are required. Such a constraint could be that the process is isobaric, $dP = 0$. One can then add a suitable integrating zero-term, xdP to make dW an exact differential. The obvious choice is $x = V$,

$$dW = PdV = PdV + VdP = d(PV), \tag{2.3}$$

such that the isobaric work potential becomes $P = PV$. Similarly the isobaric heat potential $U + PV$ is obtained from

$$dQ = TdS = dU + PdV = dU + PdV + VdP = d(U + PV), \tag{2.4}$$

where the First Law of thermodynamics

$$dU = TdS - PdV \tag{2.5}$$

has been used. Table 1 shows the results of this procedure for the classical examples.

Table 1. The classical thermodynamic potentials for the process variables of work $dW = PdV$ and heat $dQ = TdS$.

Process type	Zero along process	Integrating term	Work potential	Heat potential
Isobaric	dP	VdP	PV	U + PV
Isothermal	dT	SdT	TS – U	TS
Isochoric	dV	–PdV	0	U
Isentropic	dS	–TdS	–U	0

Now, the constraints need not be the constancy of one of the state variables. Consider a balloon with constant surface tension α. In equilibrium with an external pressure P_{ex} such a sphere of radius r has an internal pressure

$$P = P_{ex} + \frac{2\alpha}{r}, \tag{2.6}$$

which can be rearranged into

$$(P - P_{ex})V^{1/3} = 2\alpha \left(\frac{4\pi}{3}\right)^{1/3}. \tag{2.7}$$

Since the right hand side of this equation is a constant, this means that $(P - P_{ex})V^{1/3}$ is an integral of motion for the fluid inside the balloon. We can then add a suitable amount of $d[(P - P_{ex})V^{1/3}]$ $(=0)$ to dW to make it exact,

$$dW = PdV = PdV + \frac{3}{2}V^{2/3} \, d[(P - P_{ex})V^{1/3}]$$
$$= d[\tfrac{1}{2} V(3V - P_{ex})]. \tag{2.8}$$

Thus the work done by the coupled system surface + fluid is given by the decrease in the potential $P = \frac{1}{2} V(3V - P_{ex})$, regardless of path followed.

In its most general form the Legendre transformation can be used to calculate a potential P for the arbitrary process variable B, expressible as a path integral in terms of generalized forces f_i and displacements x_i,

$$B = \sum_i \int f_i \, dx_i = \int \mathbf{f} \cdot d\mathbf{x}; \tag{2.9}$$

B will usually be work, and vector notation is used for compactness. To find P, one adds to $\mathbf{f} \cdot d\mathbf{x}$ an integrating term $\mathbf{g} \cdot d\mathbf{y}$, where $d\mathbf{y}$ is necessarily zero as a result of the constraints defining the process. Note that $d\mathbf{y} = \mathbf{0}$ may involve time and could come from a condition in the form of a differential equation as well as from the more familiar thermodynamic condition of a constant variable, as used in the example above. Hence the differential form $d\mathbf{y} = \mathbf{0}$ is used rather than the integrated form $\mathbf{y} = \mathbf{constant}$, since $\mathbf{y}$ itself may not exist. The mathematical problem of finding P has two steps, finding a function $\mathbf{g}$ which makes $d\omega = \mathbf{f} \cdot d\mathbf{x} + \mathbf{g} \cdot d\mathbf{y}$ an exact differential dP, and then integrating to get P itself. The first step involves the Cauchy-Riemann condition that $d\omega$ has equal cross derivatives with respect to the free state variables, e.g. a and b:

$$\frac{\partial}{\partial b}\left[\mathbf{f}\cdot\left(\frac{\partial \mathbf{x}}{\partial a}\right)_b + \mathbf{g}\cdot\left(\frac{\partial \mathbf{y}}{\partial a}\right)_b\right] = \frac{\partial}{\partial a}\left[\mathbf{f}\cdot\left(\frac{\partial \mathbf{x}}{\partial b}\right)_a + \mathbf{g}\cdot\left(\frac{\partial \mathbf{y}}{\partial b}\right)_a\right] \tag{2.10}$$

or

$$\left(\frac{\partial \mathbf{g}}{\partial a}\right)_b\cdot\left(\frac{\partial \mathbf{y}}{\partial b}\right)_a - \left(\frac{\partial \mathbf{g}}{\partial b}\right)_a\cdot\left(\frac{\partial \mathbf{y}}{\partial a}\right)_b = \left(\frac{\partial \mathbf{f}}{\partial b}\right)_a\cdot\left(\frac{\partial \mathbf{x}}{\partial a}\right)_b - \left(\frac{\partial \mathbf{f}}{\partial a}\right)_b\left(\frac{\partial \mathbf{x}}{\partial b}\right)_a. \tag{2.11}$$

With $\mathbf{f}$, $d\mathbf{x}$, and $d\mathbf{y}$ known, this is the equation from which $\mathbf{g}$ may be obtained. In the usual case of $\mathbf{f} = P$, $d\mathbf{x} = dV$, $a = V$, and $b = P$, the right hand side of eq. (2.11) simplifies to 1. The second step in finding P, the integration of dP, is, of course, only unique within a constant of the motion; i.e. two methods of integration may yield two different potentials P and P', but their variations will always be the same, $\Delta P = \Delta P'$.

Whereas for reversible processes there is no question that thermodynamic potentials exist, because the processes can always be reversed or go by way of an arbitrary third state, this is not obvious for generalized potentials with built-in time dependence and possible loss terms. In generalizing the Legendre transformation above, we have implicitly assumed the existence of a potential or, equivalently, a solution to eq. (2.11). The conditions for existence are (Salamon et al . 1977):

a) the process is quasistatic, and

b) the process variable can be expressed as a path integral [eq. (2.9)].

The quasistaticity is equivalent to saying that the relaxation times of the system are negligible compared to the time scale of the process. In other words, the system is not required to be in equilibrium with its surroundings at all times, but all state variables must be defined (make sense) at a countable pointset along the path. The second condition is similar in content, because this seemingly trivial condition is violated when one tries to describe the extraction of work from a system at a rate faster than that system can equilibrate internally (think of a combustion process). In such a case thermodynamic variables lose their meaning, and one must go to a definition of work in terms of energy transfer at the microscopic level, which sometimes can be too complex to be useful. Nevertheless, work and availability can be defined for some systems, such as simple lasers whose operation depends on changes in populations of specific quantum states (Geusic et al. 1967). Another approach relies on the information-theoretic 'maximum entropy' formalism (see e.g. Levine and Tribus 1979).

2.2 Finite-time availability

One of the more powerful results in finite-time thermodynamics is the definition of a finite-time availability (Andresen et al. 1983). As mentioned in Sec. 1.1, the traditional availability A of a system in contact with given surroundings is a state function with the quality that the decrease in its value in going from state i to state f is the maximum (and hence reversible) work that can be extracted during that process. The finite-time availability A retains this property and simply adds that the process is restricted to operate (go to completion) during time $\tau = t_f - t_i$. Then

$$A = W_{max}(\tau) = \max\left[A(t_i) - A(t_f) - T_0 \int_{t_i}^{t_f} \dot{S}_{tot}\, dt \right], \qquad (2.12)$$

where the last equality uses the Tolman-Fine (1948) form of the Second Law of thermodynamics, eq. (1.3).

The extension may seem trivial, but the principal content lies in the way the maximization is carried out, or rather restricted. It must be carried out within the constraints imposed on the process, temporal or otherwise. These constraints in effect define a generic model which constitutes the confines within which we expect to be able to modify our real system in order to improve its performance. The constraints in such a generic model should not be excessively detailed, but only contain the essential loss terms and limiting factors in the process, otherwise the calculations will become unwieldy. Of course, if one is going to use finite-time availability to compare the performance of two different processes, they must be represented by the same generic model. Otherwise a path allowed in one may not be available to the other process, and such a restriction always costs performance. Losses are not always detrimental to the performance of a system if they open up new pathways — actually some processes depend on irreversibilities for their very existence (e.g. Wheatley et al. 1983).

In addition to the above considerations, the maximum search in eq. (2.12) can either be constrained to exactly reach a given final state at time t_f (the initial state is always considered known), in which case ΔA is fixed, and the optimization becomes one of minimizing the entropy production, or also the final state may be included in the optimization, in which case A must be evaluated by optimal control. If the final state is specified, a solution may not exist if τ is too short, since only a certain set of states can be reached from a given initial state in time τ. In addition, the finite-time availability does not necessarily have ΔA as its limit for very long times, because the system may contain internal relaxation processes which remain irreversible even for very slow operation. If there is a direct heat leak from the system reservoir to the surroundings, then a long process time may even reduce A to zero.

As defined, the finite-time availability is as general as the traditional availability, i.e. it can be applied to *any* thermodynamic process. Andresen et al. (1983) report an optimal control calculation of the finite-time availability for a work-producing system with competing internal relaxation which can be interpreted as heat engines, internal molecular degrees of freedom, a hydraulic system, or a chemical reaction, simply by changing nomenclature.

2.3 *The tricycle formalism*

An entirely different way to assess the cost of finite-time operation is through the 'tricycle' formalism (Andresen et al. 1977), a construction based on conservation equations for the process in question. A heat exchange system is represented pictorially in Fig. 2 by a triangle with heat flow rates q_1, q_2, q_3 into reservoirs with temperatures T_1, T_2, T_3. A conventional heat engine or refrigerator is a special case with one of the temperatures, e.g. T_1, infinite, such that no entropy flow is associated with this energy flow, and q_1 is identifiable as power. Any such process can then be divided into its reversible part with zero rate of entropy production, s = 0, and a totally irreversible component (see Fig. 2). Nothing new, of course, is learned from such a decomposition per se, but by putting in specific loss mechanisms like heat resistance, friction, and heat leak, the rate dependencies of such irreversibilities can be deduced. It may seem that these three loss mechanisms should be treated individually, but they are in fact interdependent and can be solved simultaneously.

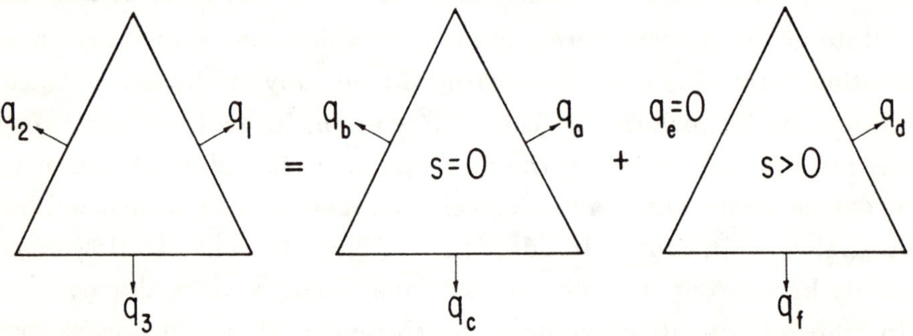

Figure 2. Tricycle decomposition of a thermal process into a reversible component and a totally irreversible component. The quantities q_1, q_2, q_3 are the heat flows into the reservoirs with temperatures T_1, T_2, T_3.

A convenient way to represent the results is a contour diagram or a three dimensional plot like Fig. 3, where the height represents power output for a heat engine/heat pump. Reduced variables are used such that the ordinate is a measure of rate, and the abscissa has friction, relative to thermal resistance, increasing to the right. First of all one observes that the rate coordinate is divided into three distinct regions. The system operates as a heat pump for all negative values and as a heat engine for positive rates, but only up to a certain limit. The region above that limit (9 with the parameters used in Fig. 3) is inaccessible because such large rates would require negative temperatures to overcome the resistances. The loci of maximum power are indicated by heavy line, and for negligible friction (to the far left) there is indeed one such maximum, as observed by Curzon and Ahlborn (1975). This remains unchanged as friction increases until a point where a classical bifurcation splits the locus of maximum power into two with a minimum in between. Thus in this region there are two quite different heat rates which yield (actually identical) maxima in the power production, obviously at vastly different efficiencies. The friction parameter divides the space into distinct heat resistance- and friction-dominated regions. No such structure exists in the heat pump region, not even a single maximum, since there is no limit to how much power can be put in to speed up the pump.

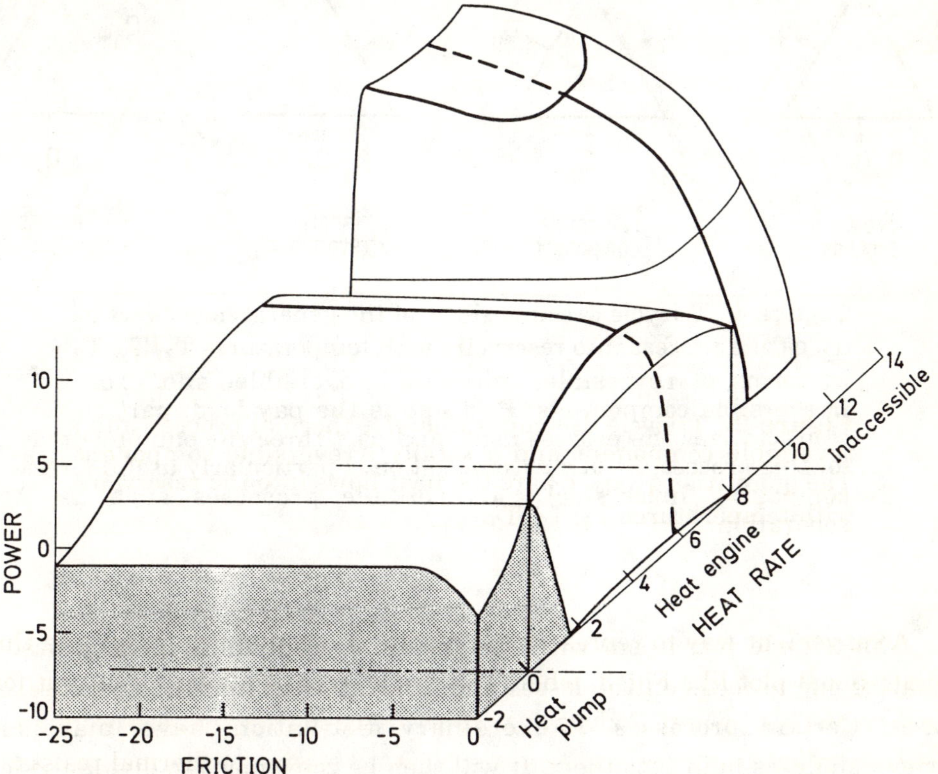

Figure 3. The power output for an endoreversible engine/heat pump with heat resistances to its reservoirs and friction as function of its rate of operation and frictional coefficient (increasing to the right); everything is scaled with the heat resistances. The heat rate divides the mode of operation into three regions: acting as a heat pump, acting as a heat engine, and an inaccessible region where heat transfers exceed the physically possible. Extrema are indicated by heavy line. Thus the mode of operation of the heat engine is also divided into a heat resistance dominated region with one maximum (to the left) and a friction dominated region with two maxima (to the right).

The advantage of such diagrams over a single optimum-performance number is that one can also read the penalty for off-optimum operation — in the low-friction region of Fig. 3 it is, for example, very slight. Similar diagrams can be drawn for the efficiency (Andresen et al. 1977) or any other objective function.

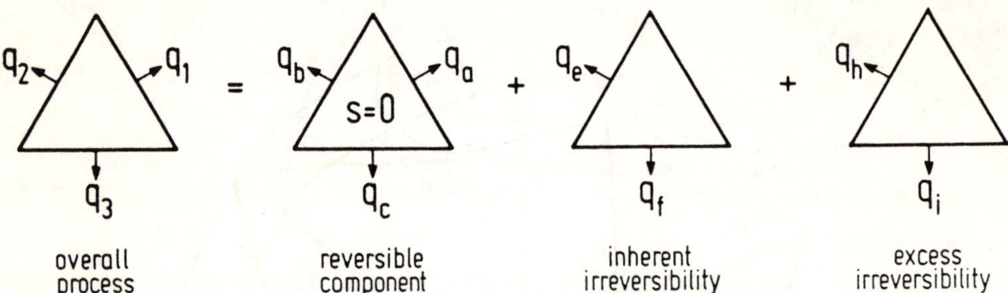

Figure 4. Tricycle decomposition of the general heat flows q_1, q_2, q_3 of a process into reservoirs with temperatures T_1, T_2, T_3 in terms of reversible, inherent irreversible, and excess irreversible components. Part one is the pay load, part two characterizes the process used, and part three the equipment for this process. Such a decomposition is particularly useful for comparing intrinsically irreversible processes such as separation.

The tricycle formalism with the reversible contributions separated off enables one to focus on the losses and calculate which are the most serious ones. Certain processes, like ordinary distillation, have unavoidable irreversibilities built into them. It will then be convenient to divide the loss-tricycle further into one for the unavoidable losses and one for the excess losses, as shown in Fig. 4. Comparing different (e.g. separation) processes, the latter tells how much room there is for improvement of this particular process, and the former how much energy can be saved by developing an entirely new, more nearly reversible process.

In an effort to develop a more direct and transparent way of calculating all the usual partial derivatives in traditional thermodynamics Weinhold (1975abcd, 1978) proposed using vector products between vectors in the abstract space of equilibrium states of a system, represented by all its extensive variables X_i. The products were defined relative to

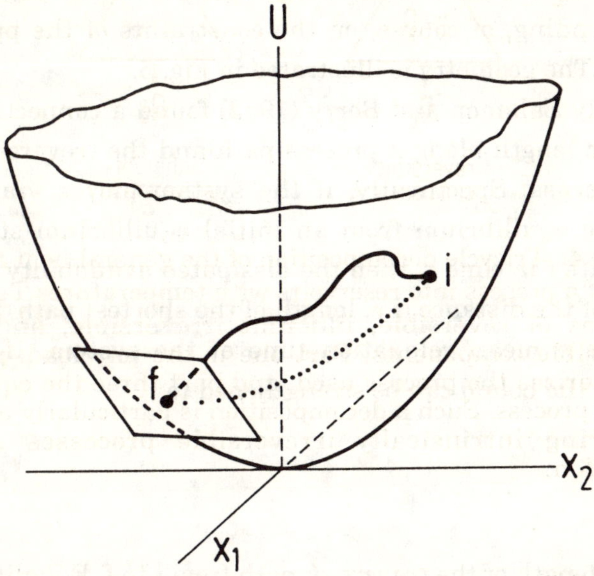

Figure 5. The state space of a thermodynamic system represented by the internal energy U and the other extensive variables X_1 and X_2. The paraboloid illustrates the equilibrium states of the system, while a tangent plane represents an environment with its constant intensive variables. A system initially in state i can move toward equilibrium with the environment at the tangent point f along an infinite number of internal equilibrium paths. The length of the shortest of these (dotted) is determined by the Weinhold metric, but the actually traversed path (solid) may be longer. The availability dissipated in the process is bounded by the square of this distance times an internal relaxation time and divided by the duration of the process.

$$\mathbf{M}_U = \left\{ \frac{\partial^2 U}{\partial X_i \partial X_j} \right\} \tag{2.13}$$

as the metric, where U is the internal energy. However, second derivatives are usually identified as curvatures and, as such, should be interpreted as availabilities in Gibbs space (U as a function of all the other extensive variables). This lead us to seek another interpretation of pathlengths calculated with Weinhold's metric, now called thermodynamic lengths, and we (Salamon et al. 1980a) found that they always are changes in some molecular velocities, depending, of course, on the constraints of the process (isobaric, isochoric, etc.). The geometry is illustrated in Fig. 5.

Subsequently Salamon and Berry (1983) found a connection between the thermodynamic length along a process path and the (reversible) availability lost in the process. Specifically, if the system moves via states of local thermodynamic equilibrium from an initial equilibrium state i to a final equilibrium state f in time τ, then the dissipated availability $-\Delta A$ is bounded by the square of the distance (i.e. length of the shortest path) from i to f times ε/τ, where ε is a mean relaxation time of the system. If the process is endoreversible, the bound can be strengthened to

$$-A \le \frac{L^2 \varepsilon}{\tau}, \tag{2.14}$$

where L is the length of the *traversed* path from i to f. Equality is achieved at constant thermodynamic speed $v = dL/dt$. For comparison, the bound from traditional thermodynamics is only

$$-A \le 0. \tag{2.15}$$

An analogous expression exists for the total entropy production during the process:

$$\Delta S^u \le \frac{L^2 \varepsilon}{\tau}. \tag{2.16}$$

The length L is then calculated relative to the entropy metric

$$\mathbf{M_S} = -\left\{\frac{\partial^2 S}{\partial X_i \partial X_j}\right\} \qquad (2.17)$$

which (when expressed in identical coordinates!) is related to $\mathbf{M_U}$ by (Salamon et al. 1984)

$$\mathbf{M_U} = -T_0 \mathbf{M_S}, \qquad (2.18)$$

where T_0 as usual is the environment temperature. In statistical mechanics, where entropy takes the form

$$S(\{p_i\}) = -\sum_i p_i \ln p_i, \qquad (2.19)$$

the metric $\mathbf{M_S}$ is particularly simple, being the diagonal matrix (Feldmann et al. 1985)

$$\mathbf{M_S} = -\left\{\frac{1}{p_i}\right\}. \qquad (2.20)$$

The same procedure of calculating metric bounds for dynamic systems has been applied to coding of messages (Flick et al. 1987) and to economics (Salamon et al. 1987).

3. Optimal path

3.1 Optimal path calculations

A knowledge of the maximum work that can be extracted during a given process, e.g. calculated by one of the procedures described in the previous section, may not by itself be sufficient. One may also want to know *how* this maximum work can be achieved, i.e. the time path of the thermodynamic variables of the system. The primary tool for obtaining this path is optimal control theory.

This is not the place to repeat the mechanics of optimal control calculations (see e.g. Boltyanskii 1971; Tolle 1975; Leondes 1964); a more thorough presentation has been given by Karl Heinz Hoffmann in this volume. Let it suffice to point out that in order to set up the optimal control problem one must specify

- the controls, i.e. the variables that can be manipulated by the operator (they may be a volume, rate, voltage, heat conductance, etc.);
- limits on the controls and on the state variables, if any (in order to avoid unphysical situations such as negative temperatures and infinite speeds);
- the equations that govern the time evolution of the system (they will usually be differential equations describing heat transfer rates, chemical reaction rates, friction, and other loss mechanisms);
- the constraints that are imposed on the system (e.g. conserved quantities, the quantities held constant, or any requirements on reversibility. The constraints may either be differential, instantaneous i.e. algebraic, or integral i.e. not obeyed at each point but over the entire interval);
- the desired quantity to be maximized, called the objective function (usually expressed as an integral); and finally
- whether the duration of the process is fixed or part of the optimization.

Typical manipulation usually leads to a set of coupled, non-linear differential equations for which a qualitative analysis and a numerical solution are the only hope. Thus answering the more demanding question about the optimal time path rather than the standard question about maximum performance requires a considerably larger computational effort. On the other hand, once the time path is calculated, all other thermodynamic quantities may be calculated from it, much like the wave function is the basis of all information in quantum mechanics.

3.2 Criteria of performance

Efficiency, the earliest criterion of performance for engines, measured how much water could be pumped out of a mine by burning a ton of coal. Other familiar criteria include effectiveness, change of thermodynamic potential, and loss of availability, all of which are measures of work. Potentials for heat can also be defined (see Sect. 2.1) but are less common, and we have used the minimization of entropy production in a separate study (Salamon et al. 1980b).

The Curzon-Ahlborn analysis and most of our own analyses use a quite different criterion, that of *power*. This quantity is of course zero for any reversible system, and maximizing power forces us to deal with systems operating at finite rates. Other criteria of performance are the rate of entropy production and the rate of loss of availability. Entropy production was a function introduced in the earliest thinking about irreversible thermodynamics (Onsager 1931; Prigogine 1962, 1967), but more from the differential, local, instantaneous viewpoint than from the global, integral view of entire optimized processes. Under some circumstances, optimizing one of these quantities is equivalent to optimizing another. For example, minimizing the entropy production is equivalent to minimizing the loss of availability, at least in those cases in which the irreversibilities can be represented as spontaneous heat flows (Salamon et al. 1980b).

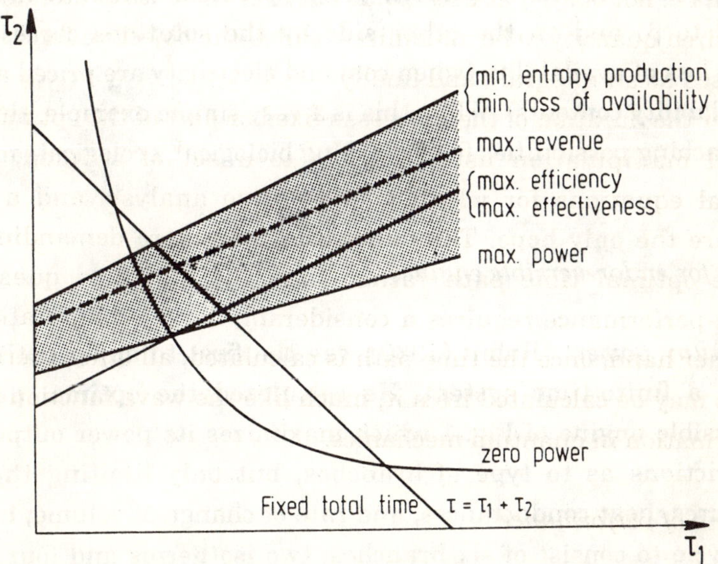

Figure 6. If an endoreversible engine (Fig. 1) spends time τ_1 in contact with the hot reservoir and τ_2 in contact with the cold reservoir, the optimal proportioning between τ_1 and τ_2 depends on what one chooses to optimize, as indicated. The locus of maximum revenue for a power producing system falls in the shaded area for any choice of prices, as described in the text. Only contact times above the hyperbola marked 'zero power' actually correspond to positive power production.

Salamon and Nitzan (1981) have optimized the Curzon-Ahlborn engine for a number of these objective functions. Assuming the working fluid to be in contact with the hot reservoir for the period τ_1 and the cold reservoir for the period τ_2, the optimal time distributions are shown in Fig. 6. The diagonal $\tau = \tau_1 + \tau_2$ indicates fixed total cycle time, and only processes above the curve labeled 'zero power' produce positive power. It is quite obvious that different criteria of merit dictate different operating conditions for the process. Even when not knowing the precise objective function but only that it belongs to a specified class, one can sometimes say a good deal about the possible optimal behaviour of the system. If one considers the Curzon-Ahlborn engine to be a model of a power plant which buys heat (coal) at the unit price α and sells work (electricity) at the unit price β, its net revenue is $\Pi = \beta w - \alpha q_1$. All solutions to the problem of maximizing this revenue are bounded on one side by the solutions to the maximum power problem (when α is significant compared to β) and on the other side by the solutions corresponding to minimum loss of availability (when coal and electricity are priced according to their availability contents). While this is a very simple example, this approach has far reaching possibilities for describing biological, ecological, and economic systems.

3.3 Paths for endoreversible engines

Maximum power. Rubin (1980) was the first to apply optimal control theory to a finite-time system. He calculated the optimal path for the endoreversible engine of Fig. 1 which maximizes its power output. Without any restrictions as to type of branches, but only limiting the reservoir temperatures, heat conductances, and rate of change of volume, he found the optimal cycle to consist of six branches: two isotherms and four maximum-power branches, but no adiabats. The optimal controls (reservoir temperature, heat conductance, and relative rate of volume change) are shown in Fig. 7, and the optimal state variables (temperature and pressure) in Fig. 8. Note that only the extreme available temperatures, T_H and T_L, are required, as is the maximum conductance to the reservoirs. The adiabats of a Carnot cycle are replaced by the maximum-power branches along which the volume changes at its maximum permissible rate while the working fluid is still in contact with one or the other heat reservoir. Thus the optimal cycle is never isolated from both its reservoirs. If the limitation on the rate of volume change is lifted, these new branches proceed instantaneously and in effect become adiabats. This optimal cycle is the finite-time equivalent of the Carnot Cycle in reversible thermodynamics and serves the same purpose, viz. as an idealized reference processes for evaluation of real cycles. Curzon and Ahlborn's (1975) assumption of an interior Carnot cycle in their system (see Sec. 1.2) is now justified by this optimization.

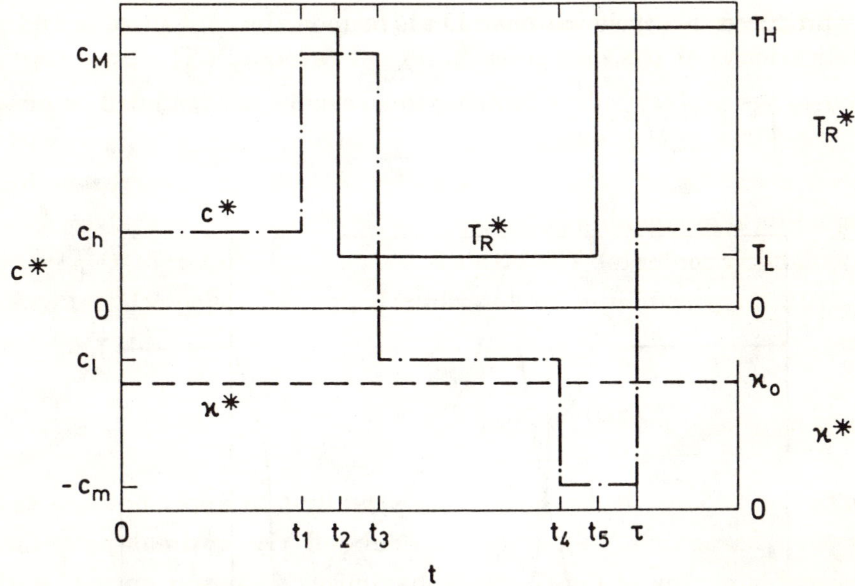

Figure 7. The optimal controls over a period τ for an unrestricted optimization of the endoreversible engine (Fig. 1).

$c^*(t) = \dot{V}/V$ is the relative rate of change of volume, and $\varkappa^*(t)$ is the heat conductance to the reservoir with temperature $T_R^*(t)$. Note that, except for the rate of change of volume on the isothermal branches $[0;t_1]$ and $[t_3;t_4]$, only the extreme values of the controls are required.

Minimum entropy production. About the same time we (Salamon et al. 1980b) analyzed the performance of endoreversible engines in terms of their entropy production. Minimizing that is equivalent to minimizing the loss of availability, but not to maximizing efficiency or power, as most other studies have done. The system exchanges generalized fluxes with the surroundings at a rate which depends on the generalized forces in a completely arbitrary way, but not on time or any time derivative of the forces. Then, regardless of whether the internal or the external temperature can be controlled by the operator, minimum total entropy production always implies constant rate of entropy production on each branch of the cycle. If the fluxes are linear in the forces, this constant entropy production rate must be the same on all branches of the cycle. If the class of systems is further restricted to admit only Fourier heat flow through a conductance κ to the reservoir, one finds that the total entropy produced in a cycle with period τ is bounded by

$$\Delta S \le \left(\sum_i |\sigma_i| \right)^2 \big/ \kappa\tau, \tag{3.1}$$

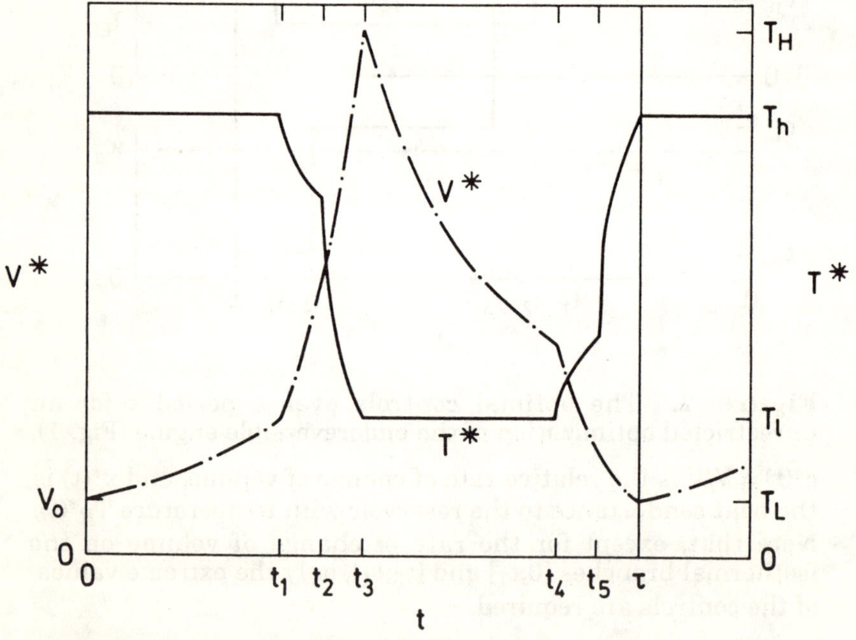

Figure 8. The optimal trajectories associated with the controls in Fig. 7. V*(t) and T*(t) are the volume and temperature of the working fluid, respectively.

where σ_i is the entropy change of the working fluid on branch i. Combine this expression with the equality form of the Second Law of thermodynamics, eq. (1.3), the work W produced by such an engine is bounded by

$$W \le W_{rev} - T_0 \left(\sum_i |\sigma_i| \right)^2 / \kappa\tau, \tag{3.2}$$

where T_0 is the temperature of the environment with respect to which availability is defined. This expression is a particularly clear exposition of one kind of cost of finite-time operation.

The main result, that the 'best process' is associated with a constant rate of entropy production, is reminiscent of Prigogine's (1967) theorem of irreversible thermodynamics which states that the entropy production rate is minimum at near-equilibrium steady states. Both results are derived from variational principles. However, the approach of Prigogine ('irreversible thermodynamics') involves instantaneous quantities while our formalism ('finite-time thermodynamics') investigates the extrema of integrals over time.

Staging. Endoreversible engines have the same staging property as Carnot engines: If one puts two or more engines of the same kind in sequence, then the whole system behaves as a single engine of the same kind (Rubin and Andresen 1982). In our unconstrained optimization the interface between the two stages, which for the present model is the intermediate temperature and the relative timing of the two engines, is arbitrary and can be used to satisfy other, non-thermodynamic constraints. A possible constraint which could be added to make the problem more realistic concerns the volume swept by the two engines, e.g. the total volume or the sum of the compression ratios. However, if we want to go into such detail, we must also specify the amount of working fluid in each engine and its equation of state, and those new parameters precisely balance the added constraints, so the interface is still arbitrary. This conclusion implies that one of the stages could shrink to zero without it being detected from the outside. There is thus no *thermodynamic* reason for the usual practice of dividing turbines, compressors, refrigerators, and the like into stages.

This staging property of endoreversible engines makes it a unique building block for constructing and analyzing more complex finite-time thermodynamic systems — much like the Carnot engine is in reversible thermodynamics.

3.4 Simulated annealing.

The algorithm for the recent global optimization procedure simulated annealing (Kirkpatrick et al. 1983) along with an interpretation in terms of Markov chains are described in the chapter by Karl Heinz Hoffmann in this volume and shall not be repeated here. Instead I would like to make the analogy between the abstract simulated annealing process and a real thermodynamic process. Specifically, if the correspondence with statistical mechanics implied in the simulated annealing procedure involving phase space and state energies is valid, then further results from thermodynamics will probably also be possible to carry over to simulated annealing.

So far all suggested simulated annealing temperature paths (annealing schedules) have been of the a priori type and thus have not adjusted to the actual behaviour of the 'system' as the annealing progresses. Examples of such paths are

$$T(t) = a \exp(-t/b) \tag{3.3}$$

$$T(t) = a/(b+t) \tag{3.4}$$

$$T(t) = a/\ln(b+t). \tag{3.5}$$

The real annealing of physical systems often has rough parts where the surrounding temperature must be decreased slowly due to phase transitions or regions of large heat capacity or slow internal relaxation. The same behaviour is seen in the abstract systems, so annealing schedules which take such variation into account are preferable in order to keep computation time at a minimum for a given accuracy of the final result (Ruppeiner 1988). Since asking a question (= one evaluation of the energy function) in information theoretic terms is equivalent to producing one bit of entropy, the temperature path $T(t)$ which produces minimum entropy, as calculated with thermodynamic length [c.f. eq. (2.16)], at once suggests itself as the optimal simulated annealing schedule:

$$\frac{dT}{dt} = -\frac{vT}{\varepsilon\sqrt{C}} \tag{3.6}$$

or equivalently

$$\frac{\langle E \rangle - E_{eq}(T)}{\sigma} = v. \tag{3.7}$$

In these expressions v is the (constant) thermodynamic speed, C and ε are the heat capacity and internal relaxation time of the system, respectively, $\langle E \rangle$ and σ the corresponding mean energy and standard deviation of its natural fluctuations, and finally $E_{eq}(T)$ is the internal energy the system would have if it were in equilibrium with its surroundings at temperature T. The physical interpretation of eq. (3.7) is that the environment should at all times be kept v standard deviations ahead of the system. Similarly eq. (3.6) indicates that the annealing should slow down where internal relaxation is slow and where large amounts of 'heat' has to be transferred out of the system. In case C and ε do not vary with temperature, eq. (3.6) integrates to the standard schedule eq. (3.3).

The extra temperature dependent variables of the constant thermodynamic speed schedule of course require additional computational effort. Since systems often change considerably in a few steps, ergodicity is not fulfilled, so the use of time averages to obtain $\langle E \rangle$, σ, C, and ε is usually not satisfactory. Instead we (Andresen et al. 1988) suggest to run an ensemble of systems in parallel, i.e. with the same annealing schedule, in the true spirit of the analogy to statistical mechanics. Then these variables can be obtained anytime as true ensemble averages based on the system degeneracies $p_i = p(E_i)$:

$$Z(T) = \sum_i p_i \, \exp(-E_i/T) \tag{3.8}$$

$$E(T) = T^2 \frac{d \ln Z}{dt} \tag{3.9}$$

$$C(T) = \frac{dE}{dT} = \frac{\langle (\Delta E)^2 \rangle}{T^2} \tag{3.10}$$

$$\varepsilon(T) = \frac{-1}{\ln\lambda_2} \approx \frac{T^2\, C(T)}{\sum\limits_i p_i \sum\limits_{j>i} (E_j - E_i)^2\, P_{ji}\, \exp(-E_i/T)}, \qquad (3.11)$$

where λ_2 is the second largest eigenvalue of the thermalized version of the transition probability matrix $\mathbf{P}$ among all the energy levels ($\lambda_1 = 1$ corresponds to equilibrium).

But where does one get the degeneracies p_i from? Actually (Andresen et al. 1988), information to calculate the temperature-independent (or infinite-temperature, if you prefer) transition probability matrix $\mathbf{P}$ can be accumulated during the annealing run by simply adding up in a matrix $\mathbf{Q}$ the number of attempted moves (not just the accepted ones) from level i to j as the calculation progresses. Normalization of $\mathbf{Q}$ yields a good estimate of $\mathbf{P}$,

$$P_{ji} = Q_{ji} \Big/ \sum_k Q_{ki}. \qquad (3.12)$$

The degeneracies $\mathbf{g}$ are then the eigenvector of $\mathbf{P}$ corresponding to the eigenvalue 1.

This use of ensemble annealing is particularly well suited for implementation on present day parallel computers. A further analysis of its performance has been carried out by Ruppeiner, Pedersen, and Salamon (1990), and the trade-off between ensemble size and duration of annealing for fixed total computation cost has been addressed by Pedersen et al. (1990).

4. Summary

Finite-time thermodynamics was 'invented' in 1975 by R. S. Berry, P. Salamon, and myself as a consequence of the first world oil crisis. It simply dawned on us that all the existing criteria of merit were based on reversible processes and therefore were totally unrealistic for most real processes. That made an evaluation of the potential for improvement of a given process quite difficult.

Finite-time thermodynamics is developed from a macroscopic point of view with heat conductances, friction coefficients, overall reaction rates, etc. rather than based on a microscopic knowledge of the processes involved. Consequently most of the ideas of traditional thermodynamics have been assimilated, e.g. the notions of thermodynamic potential (Sect. 2.1) and availability (Sect. 2.2). At the same time we have seen new concepts emerge, e.g. the non-equivalence of well-honored criteria of merit (Sect. 3.2), the importance of power as the objective (Sect. 1.2 and 3.2), the generality of the endoreversible engine (Sect. 3.3), and in particular thermodynamic length (Sect. 2.4). Several of these abstract concepts have been successfully applied to practical optimizations, some of which are reviewed by Karl Heinz Hoffmann in this volume (Hoffmann et al. 1985; Mozurkewich and Berry 1981, 1982). In line with the global philosophy of finite-time thermodynamics Jeff Gordon in his chapter is describing calculations of the global wind pattern (Gordon and Zarmi 1989).

Lately the notions and results of finite-time thermodynamics, at times in connection with statistical mechanics and information theory, have been used to perfect the global optimization method simulated annealing (Sect. 3.4). However, the surface in only scratched, there is still plenty of room for inspiration from such well-known concepts as state entropy, free energy, and transition state theory.

5. Acknowledgments

Much of the work reported in this review has been done in collaboration with colleagues near and far. In particular I want to express my gratitude to profs. R. Stephen Berry, Peter Salamon, Jim Nulton, and Karl Heinz Hoffmann.

References

Andresen, B.; Hoffmann, K. H.; Mosegaard, K.; Nulton, J.; Pedersen, J. M.; and Salamon, P. 1988. On lumped models for thermodynamic properties of simulated annealing problems, *J. Phys. (France)* **49**, 1485.

Andresen, B.; Rubin, M. H.; and Berry, R. S. 1983. Availability for finite-time processes: General theory and a model, *J. Chem. Phys.* **87**, 2704.

Andresen, B.; Salamon, P.; and Berry, R. S. 1977. Thermodynamics in finite time: Extremals for imperfect heat engines, *J. Chem. Phys.* **66**, 1571.

Boltyanskii, V. G. 1971. *Mathematical methods of optimal control* (Holt, Reinhart, and Winston, New York).

Callen, H. B. 1960. *Thermodynamics* (Wiley, New York).

Curzon, F. L.; and Ahlborn, B. 1975. Efficiency of a Carnot engine at maximum power output, *Am. J. Phys.* **43**, 22.

Feldmann, T.; Andresen, B.; Qi, A.; and Salamon, P. 1985. Thermodynamic lengths and intrinsic time scales in molecular relaxation. *J. Chem. Phys.* **83**, 5849

Flick, J. D.; Salamon, P.; and Andresen, B. 1987. Metric bounds on losses in adaptive coding. *Info. Sci.* **42**, 239.

Gaggioli, R. A. (ed.) 1980. *Thermodynamics: Second law analysis*, ACS Symposium series **122** (American Chemical Society, Washington, D. C.).

Geusic, G.; Schulz-Dubois, E. O.; and Scovil, H. E. D. 1967. Quantum equivalent of the Carnot cycle, *Phys. Rev.* **156**, 343.

Gordon, J. M.; and Zarmi, Y. 1989. Wind energy as a solar-driven heat engine: A thermodynamic approach, *Am. J. Phys.* **57**, 995.

Hermann, R. 1973. *Geometry, physics, and systems* (Marcel Dekker, New York).

Hoffmann, K. H.; Watowitch, S.; and Berry, R. S. 1985. Optimal paths for thermodynamic systems: The ideal diesel cycle, *J. Appl. Phys* **58**, 2125.

Keenan, J. H. 1941. *Thermodynamics* (MIT Press, Cambridge, Massachusetts).

Kirkpatrick, S.; Gelatt, C. D. Jr.; and Vecchi, M. P. 1983 Optimization by simulated annealing, *Science* **220**, 671.

Leondes, C. T. (ed.). 1964 and onward. *Advances in control systems* (Academic Press, New York) [especially articles in the first volumes].

Levine, R. D.; and Tribus, M. 1979. *The maximum entropy formalism* (MIT Press, Cambridge, Massachusetts).

Mozurkewich, M.; and Berry, R. S. 1981. Finite-time thermodynamics: Engine performance improved by optimized piston motion, *Proc. Natl. Acad. Sci. USA* **78**, 1986.

Mozurkewich, M.; and Berry, R. S. 1982. Optimal paths for thermodynamic systems: The ideal Otto cycle, *J. Appl. Phys* **53**, 34.

Onsager, L. 1931a. Reciprocal relations in irreversible processes. I, *Phys. Rev.* **37**, 405.

Onsager, L. 1931b. Reciprocal relations in irreversible processes. II, *Phys. Rev.* **38**, 2265.

Pedersen, J. M.; Mosegaard, K.; Jacobsen, M. O.; and Salamon, P. 1990. Optimal degree of parallel implementation in optimization. *J. Phys. A.*

Prigogine, I. 1962. *Non-equilibrium statistical mechanics* (Wiley, New York).

Prigogine, I. 1967. *Introduction to thermodynamics of irreversible processes* (Wiley, New York), pp. 76 ff.

Rubin, M. H. 1980. Optimal configuration of an irreversible heat engine with fixed compression ratio, *Phys. Rev. A* **22**, 1741.

Rubin, M. H.; and Andresen, B. 1982. Optimal staging of endoreversible heat engines, *J. Appl. Phys.* **53**, 1.

Ruppeiner, G. 1988. *Nucl. Phys. B (Proc. Suppl.)* **5A**, 116.

Ruppeiner, G.; Pedersen, J. M.; and Salamon, P. 1990. Ensemble approach to simulated annealing. *J. Phys. (France)*.

Salamon, P.; and Berry, R. S. 1983. Thermodynamic length and dissipated availability, *Phys. Rev. Lett.* **51**, 1127.

Salamon, P.; and Nitzan, A. 1981. Finite time optimizations of a Newton's law Carnot cycle, *J. Chem. Phys.* **74**, 3546.

Salamon, P.; Andresen, B.; and Berry, R. S. 1977. Thermodynamics in finite time. II. Potentials for finite-time processes, *Phys. Rev. A* **15**, 2094.

Salamon, P.; Andresen, B.; Gait, P. D.; and Berry, R. S. 1980a. The significance of Weinhold's length, *J. Chem. Phys..* **73** 1001; erratum *ibid.* **73**, 5407E.

Salamon, P.; Komlos, J.; Andresen, B.; and Nulton J. 1987. A geometric view of welfare gains with non-instantaneous adjustment. *Math. Soc. Sci.* **13**, 153.

Salamon, P.; Nitzan, A.; Andresen, B.; and Berry, R. S. 1980b. Minimum entropy production and the optimization of heat engines, *Phys. Rev. A* **21**, 2115.

Salamon, P.; Nulton, J.; and Ihrig, E. 1984. On the relation between entropy and energy versions of thermodynamic length. *J. Chem. Phys.* **80**, 436.

Tisza, L. 1966. *Generalized thermodynamics* (MIT Press, Cambridge, Massachusetts).

Tolle, H. 1975. *Optimization methods* (Springer, New York).

Tolman, R. C.; and Fine, P. C. 1948. On the irreversible production of entropy, *Rev. Mod. Phys.* **20**, 51.

Weinhold, F. 1975a. Metric geometry of equilibrium thermodynamics, *J. Chem. Phys.* **63**, 2479.

Weinhold, F. 1975b. Metric geometry of equilibrium thermodynamics. II. Scaling, homogeneity, and generalized Gibbs-Duhem relations, *J. Chem. Phys.* **63**, 2484.

Weinhold, F. 1975c. Metric geometry of equilibrium thermodynamics. III. Elementary formal structure of a vector-algebraic representation of equilibrium thermodynamics, *J. Chem. Phys.* **63**, 2488.

Weinhold, F. 1975d. Metric geometry of equilibrium thermodynamics. IV. Vector-algebraic evaluation of thermodynamic derivatives, *J. Chem. Phys.* **63**, 2496.

Weinhold, F. 1978. Geometrical aspects of equilibrium thermodynamics, in *Theoretical chemistry, advances and perspectives*, Vol. 3, edited by D. Henderson and H. Eyring (Academic Press, New York).

Wheatley, J.; Hofler, T.; Swift, G. W.; and Migliori, A. 1983. Experiments with an intrinsically irreversible acoustic heat engine, *Phys. Rev. Lett.* **50**, 499.

Nonequilibrium Thermodynamics for
Solar Energy Applications

J.M. Gordon

Applied Solar Calculations Unit, Jacob Institute for Desert Research
Ben-Gurion University of the Negev
Sede Boqer Campus, Israel
and
The Pearlstone Center for Aeronautical Engineering Studies,
Department of Mechanical Engineering
Ben-Gurion University of the Negev
Beersheva, Israel

ABSTRACT

Three problems in solar energy and solar power -producing systems are reviewed, in which applying the formalism of nonequilibrium thermodynamics has resulted in practical design progress and improved physical insight: (1) Modeling natural convection as an irreversible thermodynamic heat engine for the specific example of the Earth's winds; (2) Optimizing solar-driven heat engines for conventional power generation; and (3) Examining the impact of the specifically radiative functional form of heat transfer inherent to solar energy on the performance of heat engines operating at maximum power.

1. INTRODUCTION

Applying the techniques of nonequilibrium thermodynamics - in particular the recently-developed procedures referred to as finite-time thermodynamics - to problems in solar energy and solar power-producing systems has yielded conceptual and practical design progress in several areas. The different topics covered in this chapter are tied together by three common threads. First is the application of the concepts and formalism of finite-time thermodynamics toward understanding fundamental characteristics of solar-driven heat engines. Section 2 offers a brief review of the relevant conceptual elements of finite-time thermodynamics. Second is the impact of the specifically radiative nature of solar energy on the performance of solar-driven heat engines. Third is an examination of the performance of these heat engines at maximum power point - as distinct from reversible - operation.

One area is modeling natural convection as an irreversible thermodynamic heat engine, in particular the solar-driven convection process known as the Earth's winds (Section 3). Via the formalism of finite-time thermodynamics, an upper bound can be calculated for the annual average energy in the Earth's winds.

The Earth's atmosphere is viewed as the working fluid of a heat engine where the heat input is solar radiation, the heat rejection is to the surrounding universe, and the work output is the energy in the Earth's winds. The upper bound of 17 W/m^2 can be contrasted with the actual estimated annual average wind power of 7 W/m^2. Model predictions of the average extreme temperatures of the Earth's atmosphere are also in good agreement with observation.

A second category is the optimization of solar-driven heat engines for conventional power generation. Applications include solar thermal electricity generation, mechanical power for pumps, etc. Heat engines will usually be designed somewhere between the two limits of: (1) maximum efficiency (reversible) operation, albeit at zero power; and (2) maximum power point. Each of these limits implies a specific dependence of heat engine efficiency on the temperatures of the hot and cold reservoirs between which the heat engine operates.

We demonstrate how the principles of nonequilibrium thermodynamics are applied to determining the net overall maximum efficiency operating point. Such an optimization arises because solar collector efficiency decreases, whereas heat engine efficiency increases, with increasing solar collector temperature.

It turns out that the energetically optimal operating temperature for solar-driven heat engines is relatively insensitive to the engine design point. This is shown to pertain to solar collectors whose heat loss can range from predominantly linear (conductive/convective) to primarily radiative.

A third area is the impact of the functional temperature dependence of heat transfer on the performance of heat engines operating at maximum power. The radiative heat transfer inherent to solar energy is but one special case in this analysis. Prior finite-time thermodynamic analyses of cyclic heat engines have been restricted to Newtonian (linear) heat transfer laws.

It is demonstrated that the functional form of heat transfer has a profound effect on the performance of heat engines operated at maximum power, with the most pronounced differences occurring for radiative heat transfer. A broader spectrum of solutions emerges for efficiency at maximum power, as a function of the principal system parameters, than was originally realized. These findings may have implications for the design of certain types of man-made and naturally-occurring solar-driven heat engines.

2. REVIEW OF FINITE-TIME THERMODYNAMICS

2A. Fundamental Concepts

Traditional thermodynamics typically considers equilibrium states and limits on processes between equilibrium states. Time does not enter the formulation and reversible limits are treated, for which rates are infinitely slow and losses vanish.

In irreversible thermodynamics, one usually tries to model the effect of sources of irreversibility on the instantaneous behavior of systems. In many applications, it is of interest to consider how system performance depends on the constraint that the process must be completed in a fixed finite time (e.g., heat engine performance over one cycle). This has been called "finite-time thermodynamics".

Finite-time thermodynamics was originally developed toward placing realistic bounds on processes that are irreversible and proceed in finite (rather than infinite) time. The approach incorporates the constraints of finite-time operation; constraints on system variables (such as maximum attainable velocity or acceleration in reciprocating engines); and generic models for sources of irreversibility and hence entropy production (such as finite rate heat transfer, friction, heat leaks, etc.). One then calculates an extremum or optimum of a thermodynamically meaningful variable, such as minimization of entropy production, maximization of energy availability, maximization of power, maximization of efficiency, etc. The paper that initiated this discipline, for example, addressed maximization of power in simplified irreversible heat engines [Curzon and Ahlborn 1975].

Successful applications of finite-time thermodynamics include the analysis of heat engines of assorted cycle types [Andresen et al 1977a, Andresen 1977b, Hoffman et al 1985, Mozurkewich and Berry 1982, Ondrechen et al 1981, Ondrechen et al 1983, Rubin and Andresen 1982, Salamon et al 1980, Salamon and Nitzan 1981, Salamon et al 1982]; chemical reaction systems [Ondrechen et al 1980]; and separation processes [Brown et al 1986]. The validity and value of the formalism are as good or as realistic as the generic models for irreversibility mechanisms and as the limitations on system variables that serve as input to the optimization procedure.

In finite-time thermodynamics, one (a) expresses the laws of thermodynamics as they pertain to the system under analysis; (b) introduces models for irreversibilities; (c) states the constraints on the process variables; (d) states the variable(s) to be optimized (e.g., power, entropy production, etc.); and (e) applies the calculus of variations to arrive at the desired solution.

2B. Heat Engines Operating At Maximum Power

Heat engines are usually designed to function between the two limits of: (1) reversible, maximum efficiency operation, and (2) irreversible, maximum power operation. The actual design point is usually dictated by tradeoffs between the costs of fuel (efficiency) and the costs of installed capacity (power) (e.g., the costs of heat exchangers relative to the costs of turbines in conventional power plants).

The upper bound on the efficiency of a heat engine is the so-called Carnot efficiency η_C of a reversible engine:

$$\eta_C = 1 - (T_C/T_H) \tag{2.1}$$

where T_C and T_H are the temperatures of the cold and hot reservoirs, respectively, between which the heat engine operates. An upper bound can also be placed on the efficiency of a heat engine operating at its maximum power point, with the upper bound depending on the nature of the irreversibilities.

One of these upper bounds is the so-called Curzon-Ahlborn (CA) efficiency η_{CA}:

$$\eta_{CA} = 1 - \sqrt{T_C/T_H} \tag{2.2}$$

where the sole source of irreversibility in the CA analysis is linear (Newtonian) finite-rate heat transfer (as opposed to reversible heat transfer) between the engine and its two heat reservoirs [Curzon and Ahlborn 1975]. The engine cycle is Carnot-like in that it consists of two finite-time isotherms and two essentially instantaneous adiabats (see Figure 1).

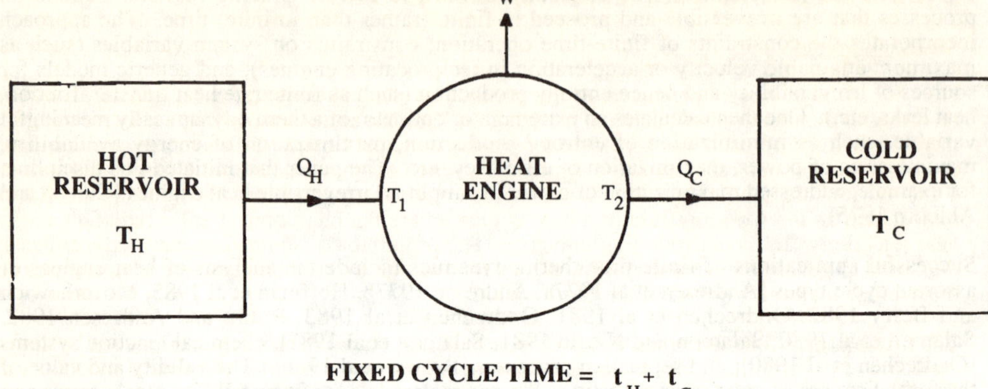

Figure 1. Schematic of Curzon-Ahlborn cyclic heat engine.

For the finite-time heat engine depicted schematically in Figure 1, the power (work per cycle time) P can be expressed as [Curzon and Ahlborn 1975, Gordon 1990]

$$P = \frac{\eta \{ T_H(1 - \eta) - T_C \}}{t_0 (1 - \eta) \{ 1/(K_H t_H) + 1/(K_C t_C) \}} \tag{2.3}$$

where η = heat engine efficiency = work output/heat input; K_H and K_C = heat transfer coefficients between reservoir and engine working fluid on the heat input and heat rejection branches, respectively; t_H and t_C = time spent on hot and cold isotherms, respectively; and t_0 = cycle time = $t_H + t_C$. Maximizing P as a function of the two variables η and t_H, for example, yields the maximum power P(max)

$$P(max) = \left(\sqrt{T_H} - \sqrt{T_C} \right)^2 / \left\{ 1/\sqrt{K_H} + 1/\sqrt{K_C} \right\}^2 \tag{2.4}$$

for which the times spent on the isothermal branches are related by

$$t_C/t_H = \sqrt{K_H/K_C} \tag{2.5}$$

and the efficiency at maximum power is given by Equation (2.2). The above expressions will serve as a basis for comparison in Section 5 when the mode of heat transfer is treated as a variable and different formulae are derived.

3. WIND ENERGY AS A SOLAR-DRIVEN HEAT ENGINE

3A. Background

Can the average power produced in the process of natural convection be predicted without resorting to detailed hydrodynamical models? In [Gordon and Zarmi 1989], a relatively simple, thermodynamic approach was presented for placing an upper bound on the average work produced in one of the few natural convective systems for which estimates of the work associated with the fluid motion are available: the Earth's winds. This approach entails treating the Earth's atmosphere as the working fluid of a heat engine where the heat input is solar radiation and the heat rejection is to the surrounding universe. The principles of finite-time thermodynamics are then used to generate quantitative expressions for average atmospheric extreme temperatures and the average maximum power of the Earth's winds.

The driving force for the Earth's winds is solar radiation or, more properly, the uneven heating of the Earth at different latitudes by solar radiation. Via assorted detailed dynamical models of the Earth's atmosphere [Brunt 1939, Ellsaesser 1969, Gustavson 1979, Lorenz 1967], the annual average wind energy has been estimated as 3×10^{20} J/day or, equivalently, 7 W/m^2 of Earth surface area. By "average" wind energy, we refer to the average over both time and space, i.e., over the entire year, the entire planet, and all altitudes at which there is an atmosphere.

The yearly average solar radiation flux per unit area of the Earth's surface, $\bar{q}_S$, can be expressed as

$$\bar{q}_S = I_{SC} (1 - \rho)/4 = 223 \text{ W/m}^2$$

where I_{SC} is the yearly average solar constant (1373 W/m^2) and ρ is the effective average albedo (reflectance) of the Earth's atmosphere (0.35). The factor of 1/4 arises from a factor of 1/2 to account for the day/night difference, and a geometric factor of 1/2 to account for the Earth's cross section, which is intercepted by solar radiation, as opposed to the corresponding hemispherical surface area of the Earth.

The Sun-Earth-Wind system is viewed as a heat engine. The heat input is solar radiation. The working fluid is the Earth's atmosphere, and the energy in its motion - winds - is the work produced. The cold reservoir to which the engine rejects heat is the 3-K surrounding universe.

The treatment is essentially one of viewing free convection flow as a heat engine - an idea that has been proposed qualitatively [Tritton 1977]. The four branches of the oversimplified heat engine correspond very approximately to the global-scale motion of winds in convective cells (Hadley, Ferrel and polar cells) namely: (a) two isothermal branches, one in which the atmosphere absorbs solar radiation at low altitudes, and one in which the atmosphere rejects heat at high altitudes to the universe; and (b) two intermediate instantaneous adiabats with rising and falling air currents (see Figure 2).

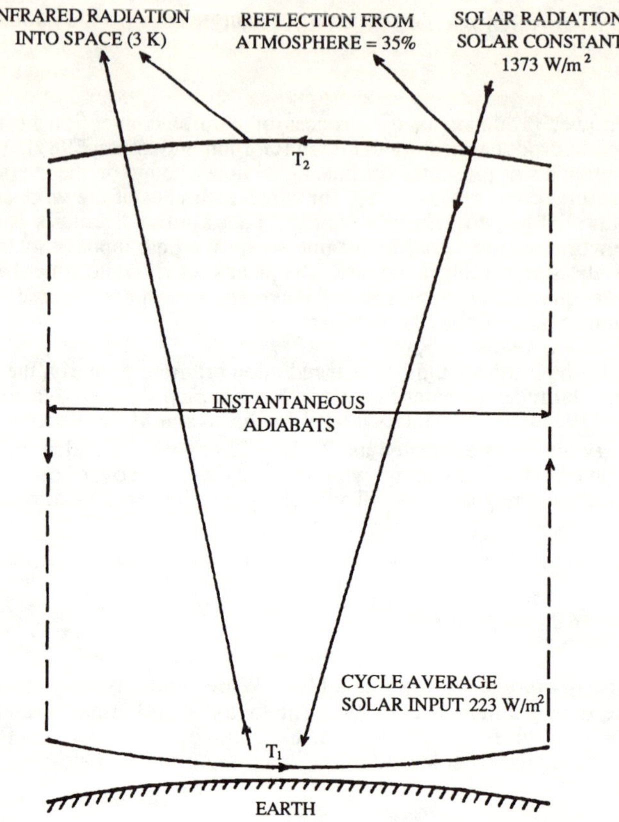

Figure 2. Schematic of simplified Sun-Earth-Wind system as a heat engine cycle.

The work output of this oversimplified heat engine is the maximum wind energy that <u>could</u> be harnessed if perfectly converting wind machines could be placed throughout the Earth's atmosphere. In reality, however, essentially none of the wind energy is harnessed or extracted. Rather, wind energy is dissipated as heat via mechanisms such as internal friction and frictional interactions with the Earth's surface. These frictional losses are <u>not</u> accounted for in the model because they represent the avenue via which the wind-work output of the atmospheric heat engine is dissipated. This is one reason (of several) that the calculation corresponds to an upper bound on yearly average wind energy. Convection is not modeled explicitly since the model, by its very nature, views convection as the work output of a heat engine.

The rotation of the Earth, with its consequent day-night difference, is a secondary, relatively small factor in the determination of the yearly energy in winds. Its primary effect is in influencing wind direction. Specifically, a detailed calculation of the heat fluxes, communicated via the Earth's atmosphere, required to prevent net heating or cooling at the different latitudes of the planet, indicate that the isolated effect of uneven solar illumination of the Earth accounts for approximately 70% of the actual estimated average wind energy. Furthermore, the order of magnitude of the isolated contribution of the Earth's rotation to average wind energy is 10%. As such, its effect is not explicitly incorporated into the heat engine picture.

The wind heat engine operates cyclically. Although locally winds may be turbulent and noncyclic, on a global scale they exhibit behavior on a time scale that is very short compared to 1 year. This observation simplifies the analysis of the wind heat engine in that the changes in thermodynamic state functions, such as internal energy and entropy, are zero in cyclic processes. In addition, it is important to note that other sources of heat generation, such as internal heat production from the Earth, energy in rivers and streams, etc., are orders of magnitude smaller than the average solar heat input.

Since the heat engine considered is highly irreversible, and since the objective is to establish an upper bound for the heat engine's average power, the approach of finite-time thermodynamics is employed [Gordon & Zarmi 1989]. In this procedure, the maximum average power (i.e., maximum work per cycle time) is derived for the wind heat engine subject to thermodynamic constraints.

3B. <u>Model Formulation and Maximization of Power</u>

The wind heat engine model, illustrated schematically in Figure 2, is oversimplified in many regards in order to facilitate presentation of the basic concepts, and to arrive at simple quantitative results. The stress is on a new approach rather than on very accurate predictions.

The spatial dependence of the working fluid (the wind) is effectively ignored, i.e., averaged over. The working fluid is viewed as absorbing solar radiation for half the cycle, during which the working fluid temperature is T_1. During the second half of the cycle, heat is rejected via blackbody radiation only from the working fluid at temperature T_2 to the cold reservoir at temperature T_{ex}. The two adiabatic branches in the engine's cycle are treated as instantaneous and therefore do not contribute to the energy and entropy equations below.

The objective is to maximize the work per cycle (average power) subject to thermodynamic constraints. Application of the first law of thermodynamics yields:

$$\Delta E = -w + \int_0^{t_0} \{q_S(t) - \sigma[T^4(t) - T_{ex}^4(t)]\, dt\} = 0 \tag{3.1}$$

where ΔE = change in the internal energy of the working fluid in one cycle, which is zero because E is a state function; t = time; t_0 = time of one cycle; w = work performed by the working fluid in one cycle; σ = Stefan-Boltzmann constant = 5.67×10^{-8} W/(m^2 -K^4); and T = temperature of the working fluid. The variables E, w, and q_S are expressed per unit area of Earth's surface. In principle, q_S, T, and T_{ex} are functions of time. In the simplified approach, however, with two isothermal heat exchange branches, the time dependences of these variables are taken as

$$T(t) = \begin{cases} T_1 & 0 \le t \le t_0/2 \\ T_2 & t_0/2 \le t \le t_0 \end{cases} \tag{3.2a}$$

$$q_S(t) = \begin{cases} I_{SC}(1-\rho)/2 & 0 \le t \le t_0/2 \\ 0 & t_0/2 \le t \le t_0 \end{cases} \tag{3.2b}$$

$$T_{ex}(t) = 3\,K \qquad 0 \le t \le t_0 \tag{3.2c}$$

Average values, denoted by an overhead bar, are then

$$\bar{T} = (T_1 + T_2)/2$$

$$\overline{T^n} = (T_1^n + T_2^n)/2 \tag{3.3}$$

$$\overline{q_S} = I_{SC} \, (1 - \rho)/4 \, .$$

From Equations (3.1) and (3.2), the average power per unit area P can be expressed as

$$P = w/t_0 = \overline{q_S} + \overline{\sigma \, T_{ex}^4} - \overline{\sigma T^4} \approx \overline{q_S} - \overline{\sigma \, T^4}. \tag{3.4}$$

the approximation following from the observation that

$$\overline{q_S} >> \sigma \, T_{ex}^4 \qquad (223 \text{ W/m}^2 >> 4.59 \times 10^{-6} \text{ W/m}^2).$$

The change in entropy per unit area ΔS can be expressed as

$$\Delta S = \int_0^{t_0} \left(\frac{q_S(t) - \sigma\{T^4(t) - T_{ex}^4(t)\}}{T(t)} \right) dt = 0 \tag{3.5}$$

For a cyclic process ΔS for an entire cycle is zero because entropy is a state function. With Equation (3.2), Equation (3.5) reduces to

$$\overline{q_S/T} = \overline{\sigma T^3} \, , \quad \text{or}$$

$$I_{SC}(1 - \rho)/(4 \, T_1) = \sigma \, (T_1^3 + T_2^3)/2. \tag{3.6}$$

Toward maximizing P (Equation (3.4)) subject to the constraint of Equation (3.6), Lagrange multiplier λ is introduced, and the modified Lagrangian L is defined

$$L = T^4(t) + \lambda\{q_S(t)/T(t) - \sigma \, T^3(t)\} \tag{3.7}$$

Finding the extremum then corresponds to solving the equation

$$\partial L/\partial T(t) = 0 \tag{3.8}$$

subject to Equation (3.6). This results in

$$0 = T^5(t) - 3 \, \sigma \, \lambda T^4(t)/4 - \lambda \, q_S(t)/4 \tag{3.9}$$

or, for the two-branch model of the heat engine

$$0 = T_1^5 - 3 \, \sigma \, \lambda \, T_1^4 /4 - \lambda \, I_{SC}(1 - \rho)/8 \tag{3.10}$$

$$0 = T_2^5 - 3 \, \sigma \, \lambda \, T_2^4 /4. \tag{3.11}$$

(The calculation of one additional derivative confirms that the extremum is a maximum in P, rather than a minimum.)

Equations (3.9) and (3.10), subject to Equation (3.6), are solved numerically. With solutions for T_1 and T_2, Equation (3.4) yields

$$P_{max} = w_{max}/t_0 = I_{SC}(1 - \rho)/4 - \sigma(T_1^4 + T_2^4)/2. \qquad (3.12)$$

The solution is

$$\lambda = 4.514 \times 10^9 \ m^2 \ K^5/W; \quad T_1 = 277 \ K; \quad T_2 = 192 \ K; \quad P_{max} = 17.1 \ W/m^2.$$

3C. <u>Analysis of Results</u>

The model prediction of $P_{max} = 17 \ W/m^2$ is of the same magnitude as the actual annual wind energy of 7 W/m^2. The simplifying assumptions introduced ensure that the calculation represents an upper limit in predicting annual average wind energy for several reasons.

First, approximating two branches of the heat engine's cycle as instantaneous adiabats yields a P_{max} that is higher than P_{max} for the more realistic case of nonzero time adiabats, particularly in the case of atmospheric motions where the approximately adiabatic branches are a nonnegligible fraction of the total cycle time.

Second, heat input to the wind heat engine is modeled as blackbody radiation only. However, in the actual wind heat engine, a significant amount of the heat transfer occurs by absorption of solar radiation in the ground, followed by convective heat transfer to the atmosphere. The latter implies a less efficient heat engine than the idealized model.

A knowledge of the cycle time t_0 is not required - only the knowledge that t_0 is short compared to 1 year since annual average wind energy only is being calculated.

It is intriguing to note that the average maximum and minimum temperatures of the Earth's atmosphere (as a function of altitude) are around 290 K (at ground level) and 195 K (at an altitude of around 75-90 km), respectively - which are remarkably close to the respective wind engine predictions of $T_1 = 277 \ K$ and $T_2 = 192 \ K$.

Next, consider the sensitivity of the results to the assumption that the heat input and heat rejection branches each occur over 50% of the total cycle time t_0. For example, when the time division for the two isothermal branches is altered to 75%-25% or 25%-75%, the results are $T_1 = 261 \ K$, $T_2 = 190K$, $P_{max} = 7 \ W/m^2$, for heat input during 75% of t_0 and heat rejection during 25% of t_0; and $T_1 = 307 \ K$, $T_2 = 194 \ K$, $P_{max} = 36 \ W/m^2$, for heat input during 25% of t_0 and heat rejection during 75% of t_0. T_1 and T_2 are relatively insensitive to this large but not unreasonable time division. The order of magnitude of P_{max} remains unchanged.

3D. <u>Summary</u>

It appears that a reasonable upper bound on the annual energy in the Earth's winds can be derived from fundamental, dynamics-independent, thermodynamic considerations. The Earth's winds are viewed as the working fluid of a heat engine whose heat input is solar radiation and whose heat rejection reservoir is the surrounding universe.

The model invoked is a modest step toward the general modeling of the process of natural convection as an irreversible thermodynamic heat engine. The specific case of the Earth's winds - a solar-driven heat engine - happens to be one of the few instances in which an accurate estimate of the average power of a convective engine is available.

Via the formalism of finite-time thermodynamics, an upper bound of 17 W/m^2 can be placed on the annual average wind power, which is to be compared with the actual estimated value of 7 W/m^2. The model's prediction of average extreme atmospheric temperatures turns out to be in agreement with observation. The analysis is restricted to a somewhat idealized and oversimplified irreversible wind heat engine in order to illustrate clearly the concepts involved in the approach.

The global thermodynamic analysis employed estimates wind _energy_, as opposed to wind _dynamics_. The basic point is that, as in many other problems in the physical sciences, a thermodynamic approach can often yield a reasonable estimate of the energy change or work output of a process without consideration of the detailed internal dynamics.

There are a limited number of destinations for the solar radiation that enters the Earth's atmosphere. One of the nonnegligible destinations is the production of wind energy (even if essentially all of this wind energy is eventually dissipated as heat). Thermodynamics place an upper bound on the achievable work output from the conversion of incident solar radiation into wind energy - a detailed calculation of which has been the objective here.

The vehicle toward this end is viewing convection as the atmospheric work output of a solar-driven heat engine. The liberal simplifying assumptions in the analysis ensure that the calculations represent an upper bound. The fact that quantitative predictions for average wind energy and extremal atmospheric temperatures are reasonably close to actual estimated and observed values simply indicates that the loss mechanisms we have ignored are secondary factors. The more general issue of whether a better understanding or modeling capability of free convection can be gained via the kind of thermodynamic analysis presented above remains an open question.

4. OPTIMIZED CONVENTIONAL SOLAR-DRIVEN HEAT ENGINES

4A. _Background_

Solar power generation can be direct or indirect. Direct power production refers to an immediate, intrinsic conversion of sunlight into power, such as in photovoltaic cells. Indirect conversion refers to using solar thermal collectors to convert sunlight into heat, and then using the heat to drive conventional heat engines. The power block is essentially the same as in conventional power plants; only the source of the heat is different.

Of all the commercial solar power produced today, the overwhelming majority is indirect, primarily via concentrating solar thermal collectors providing the heat input to conventional heat engines. An essential element in the optimal design of solar-driven heat engines is a knowledge of the dependence of the heat engine's efficiency on the temperatures of its cold reservoir T_C (typically ambient conditions) and hot reservoir T_H (essentially solar collector outlet temperature). Knowing the heat engine's efficiency as a function of T_C and T_H enables determination of the optimal operating temperature. Solar collector efficiency decreases with operating temperature T_H whereas heat engine efficiency increases with T_H, which implies an energetically optimal value of T_H.

The Carnot efficiency (Equation (2.1)) is far larger than the efficiency of real heat engines. Nonetheless, it is often used in determining optimum operating temperature [Kreith and Kreider 1978]. The Curzon-Ahlborn efficiency (Equation (2.2)) appears to correspond more realistically to both the magnitude and functional dependence of the types of efficient, conventional heat engines used in existing solar power producing systems [Barber and Prigmore 1981]. The question is therefore raised as to how sensitive the _location_ of the energetic optimum is to the functional dependence of heat engine efficiency on reservoir temperatures.

The analysis here considers the two limiting cases of maximum efficiency (reversible limit) and maximum power toward studying the sensitivity of the optimal operating temperature for solar-driven heat engines to heat engine design. Assuming that heat engine design will fall between the two limits of reversible vs. maximum power operation, the optimal operating temperature for solar-driven heat engines should also lie between the optimal temperatures as determined in these two limiting cases.

For example, data presented in [Caputo 1981] indicate that the dependence on collector operating temperature (= effective hot reservoir temperature for the heat engine) lies between these two limits. It is illustrated below that in the calculation of the energetically optimal operating temperature, there are negligible differences in these two limiting cases, and hence that either can serve as a basis for estimating optimal operating conditions.

4B. <u>Energetic Optimization</u>

The objective is to assess the accuracy of the determination of optimal T_H values, T_H (opt), based on the Carnot and Curzon-Ahlborn heat engine efficiencies, for solar thermal collectors [Gordon 1988]. The limiting cases of solar collector heat loss being predominantly radiative or conductive/convective will be considered. Overall power system efficiency η (system) is

$$\eta \text{ (system)} = \eta \text{ (collector)} \; \eta \text{ (engine).} \tag{4.1}$$

T_H (opt) is obtained by solving

$$d\,\eta \text{ (system)}/dT_H = 0. \tag{4.2}$$

Two types of solar collectors are considered:

(1)　Collector heat loss is proportional to $T_H - T_a$ (where T_a = effective ambient temperature for the solar collectors), characteristic of collectors whose heat loss is dominated by conduction and/or convection.

(2)　Collector heat loss is proportional to $T_H^4 - T_a^4$, characteristic of collectors dominated by radiative heat loss (e.g., evacuated receivers and very high temperature systems).

Real solar systems will fall between these two limiting cases. In the case of linear collector heat loss, solar collector efficiency is expressed as

$$\eta \text{ (collector)} = (U/I)(T_S - T_H) \tag{4.3}$$

where U = linearized collector heat loss coefficient (W/K-m^2); I = incident solar radiation (W/m^2); and T_S = solar collector stagnation temperature at solar irradiance I (i.e., T_S is the collector temperature under no-flow conditions). For the maximum efficiency (Carnot) case,

$$\eta \text{ (system)} = (U/I)(T_S - T_H)(1 - \{T_C/T_H\}). \tag{4.4}$$

Applying Equation (4.2), one obtains

$$T_H \text{ (opt)} = \sqrt{T_C\,T_S} \tag{4.5}$$

In the maximum power (Curzon-Ahlborn) limit,

$$\eta \text{ (system)} = (U/I)(T_S - T_H)(1 - \sqrt{T_C\,T_H}) \tag{4.6}$$

which, via Equation (4.2) yields T_H (opt) as the solution to the equation

$$0 = 2\,T_H^{3/2}\,(\text{opt}) - T_H\,(\text{opt})\,\sqrt{T_C} - T_S\,\sqrt{T_C} \tag{4.7}$$

The results for this linear heat loss case are illustrated in Figure 3 for a wide range of T_S values.

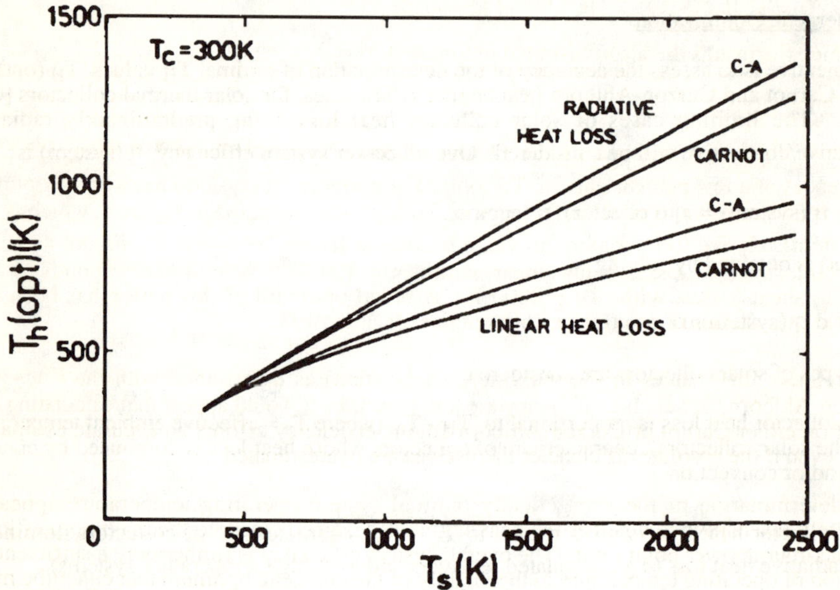

Figure 3. Optimal operating temperature T_H (opt) vs. solar collector stagnation temperature T_S for solar-driven heat engines. "Carnot" denotes calculation with Carnot efficiency, and "C-A" denotes calculation with Curzon-Ahlborn efficiency. Lower and upper pairs of curves correspond to solar collector heat losses proportional to $T_H - T_a$ and $T_H^4 - T_a^4$, respectively. $T_a = 300$ K in all cases.

In the case of radiative collector heat loss, solar collector efficiency is expressed as

$$\eta \text{ (collector)} = (\epsilon\,\sigma/I)\,(T_S^4 - T_H^4) \tag{4.8}$$

where ϵ = collector effective emissivity; and σ = Stefan-Boltzmann constant. In the maximum efficiency (Carnot) limit,

$$\eta \text{ (system)} = (\epsilon\,\sigma/I)(T_S^4 - T_H^4)\,(1 - \{T_C/T_H\}) \tag{4.9}$$

and T_H (opt) is the solution to the equation

$$0 = 4 T_H^5 (opt) - 3 T_C T_H^4 (opt) - T_C T_S^4. \qquad (4.10)$$

In the maximum power (Curzon-Ahlborn) limit,

$$\eta \, (system) = (\in \sigma / I)(T_S^4 - T_H^4)(1 - \sqrt{T_C/T_H}) \qquad (4.11)$$

and T_H (opt) is the solution to the equation

$$0 = 8 T_H^{9/2} (opt) - 7 T_H^4 (opt) \sqrt{T_C} - T_S^4 \sqrt{T_C} \qquad (4.12)$$

for which the results are also illustrated in Figure 3.

4C. <u>Conclusions</u>

Figure 4 illustrates that use of the Carnot vs. the Curzon-Ahlborn efficiency gives rise to differences of a few percent only in T_H (opt). Furthermore, η (system) has a broad optimum and hence is not sensitive to small differences in T_H (opt), as illustrated in Figure 4, which is a plot of η (system) relative to its maximum value vs. T_H, at fixed $T_C = 300$ K. Figure 4a pertains to solar collector heat losses being linear, with $T_S = 400$ K. Figure 4b corresponds to radiative collector heat losses, with $T_S = 2000$ K. A broad optimum of this nature has been noted in specific case studies of solar-driven heat engines [Caputo 1981].

The relative differences in the maximum η (system) as determined with the Carnot vs. the Curzon-Ahlborn heat engine efficiency are less than 1%. It would appear that calculating T_H (opt) based on either Carnot or Curzon-Ahlborn engine efficiency affords an accurate estimate of T_H (opt), even if the absolute efficiencies themselves are overestimated.

The determination of the energetically optimal system operating temperature appears to be relatively insensitive to whether the heat engine is designed closer to the maximum efficiency or the maximum power point limit. The overall system efficiency is furthermore a sufficiently broad function of operating temperature in the vicinity of the energetic optimum that either the maximum efficiency or maximum power point limit can be used for accurate estimation of optimal operating conditions.

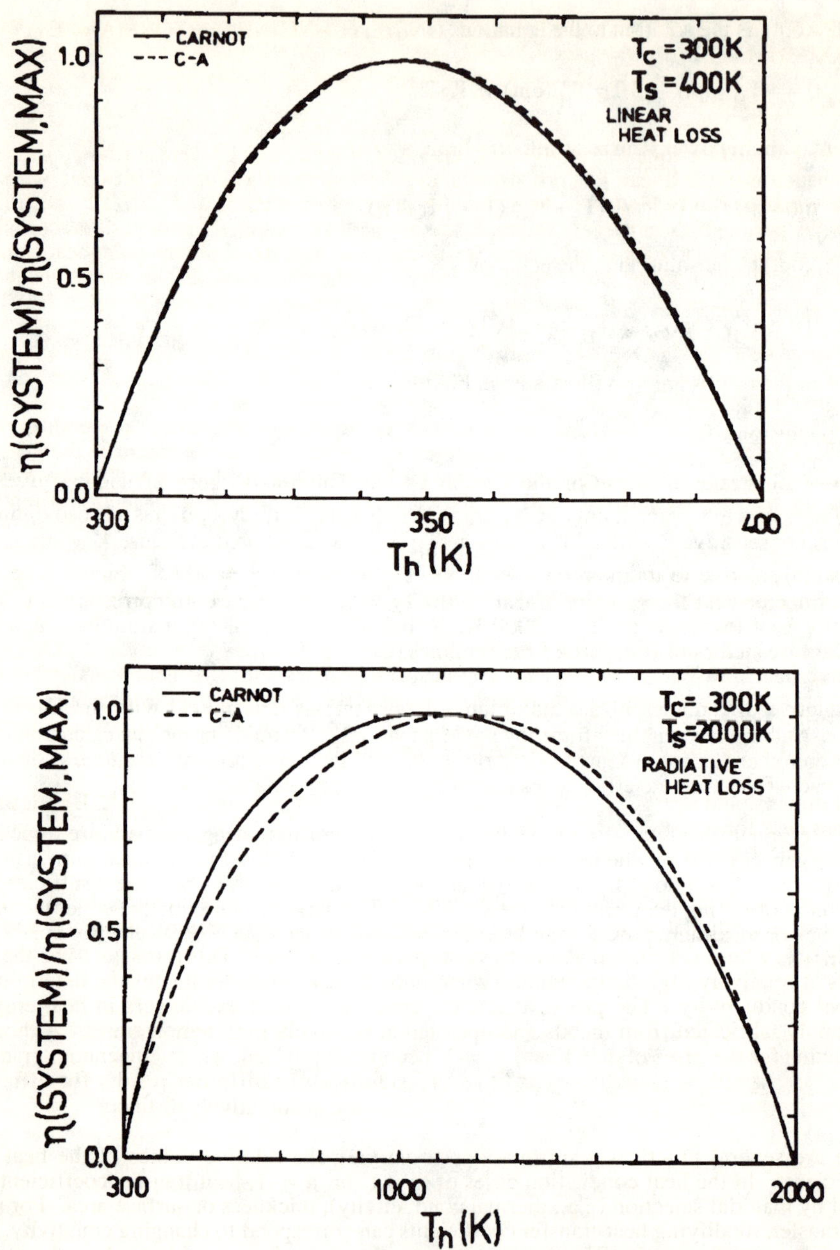

Figure 4. System efficiency relative to its maximum value, $\eta(\text{system})/\eta(\text{system,max})$, vs. T_H, calculated with the Carnot and Curzon-Ahlborn heat engine efficiencies. $T_C = 300$ K.
(a) Linear solar collector heat losses, with $T_S = 400$ K.
(b) Radiative solar collector heat losses, with $T_S = 2000$ K.

5. IMPACT OF HEAT TRANSFER MODE ON HEAT ENGINE PERFORMANCE AT MAXIMUM POWER

5A. <u>Background</u>

Until recently, finite-time thermodynamic analyses of heat engines were restricted to systems in which heat transfer is linear, i.e., proportional to the temperature difference between reservoir and engine working fluid. However, heat transfer in many man-made and naturally-occurring heat engines is radiative-dominated. Sections 3 and 4 provide selected examples. Accordingly, the question is raised as to how sensitive the performance of maximum-power-point heat engines is to the functional form of the heat transfer. It will be shown below that in many instances, this sensitivity is substantial, being most pronounced for radiative heat transfer [Gordon 1990].

The type of heat engine problem originally posed and solved by Curzon and Ahlborn [Curzon and Ahlborn 1975], as well as in subsequent finite-time thermodynamic analyses, can be cast with the functional temperature dependence of heat transfer considered as a variable [De Vos 1985]. As shown below, a richer spectrum of solutions emerges for efficiency at maximum power. Specifically, the efficiency at maximum power is in general <u>not</u> independent of the heat transfer coefficients, as it is in the linear heat transfer problem solved by Curzon and Ahlborn (Equation (2.2)). A type of "symmetry" will be shown to exist relative to the functional temperature dependence of linear (Newtonian) heat conduction. Furthermore, the Curzon-Ahlborn efficiency turns out <u>not</u> to be a fundamental upper limit on heat engine efficiency at maximum power point. Rather, this upper limit depends both on the functional temperature dependence of the heat transfer, and on the relative value of the hot and cold side heat transfer coefficients.

Two special cases are illustrated here that may be of special practical interest. One is that of radiative heat transfer. In many heat engines, the primary mode of heat transfer between the engine and either or both of its heat reservoirs is radiative. Also, certain systems in nature that can be analyzed effectively as heat engines - such as the generation of the Earth's winds [Gordon and Zarmi 1989] - are also dominated by radiative heat transfer.

In a d-dimensional world, blackbody radiative power is proportional to T^{d+1}. Besides the most relevant case for real heat engines of $d = 3$ $(n = 4)$, the case of $d = 1$ $(n = 2)$ is applicable to radiative heat transfer along thin transmission lines [De Vos 1985].

The other case, from linearized nonequilibrium thermodynamics, is heat conduction for which heat transfer is proportional to the gradient of the <u>reciprocal</u> of the temperature, $n = -1$. It also pertains to materials whose thermal conductivity is proportional to $1/T^2$. This turns out to be the case for metals at relatively high temperatures, when phonon scattering dominates the determination of thermal conductivity. The practical interest arises since heat exchangers in heat engines are commonly fabricated from metals and operated at relatively high temperatures. Although heat conduction for the cases of $n = 1$ and $n = -1$ becomes equivalent when temperature gradients are sufficiently small, it turns out that there are significantly different results for efficiency at maximum power when the $n = 1$ and $n = -1$ modes are quantitatively distinct.

There are several practical interpretations to varying the relative values of the heat transfer coefficients. In the heat conduction cases of $n = 1$ or $n = -1$, heat transfer coefficients can be varied by material selection (e.g., thermal conductivity), thickness or surface area. For radiative heat transfer, modifying heat-transfer coefficients can correspond to changing emissivity, index of refraction, or surface area. These parameters can be viewed as variables in heat engine design, and the question can be raised as to how heat engine efficiency varies with them.
Since material selection and material dimensions are independent of the choice of heat reservoir temperatures, separate sensitivity studies are performed for heat engine efficiency, at maximum power point, to these variables. These are the results that will be illustrated in Figures 6 to 11.

In addition, the Curzon-Ahlborn efficiency (Equation (2.2)) is actually not a fundamental upper limit on the efficiency of cyclic heat engines operating at maximum power point. Rather, this upper limit depends both on the functional temperature dependence of the heat transfer and on K_H/K_C. Unlike the Curzon-Ahlborn efficiency, even at $T_C/T_H = 0$, efficiency at maximum power point can be (considerably) less than unity, depending on system parameter values.

As in the original Curzon and Ahlborn problem, we consider a heat engine in which the only irreversibilities are in finite-rate heat transfer between the heat reservoirs and the engine working fluid (see Figure 5). Heat transfer is taken to be proportional to T^n where, for example $n = 1$ yields the original Curzon-Ahlborn problem, and $n = 4$ corresponds to (3-dimensional) radiative (blackbody) heat transfer.

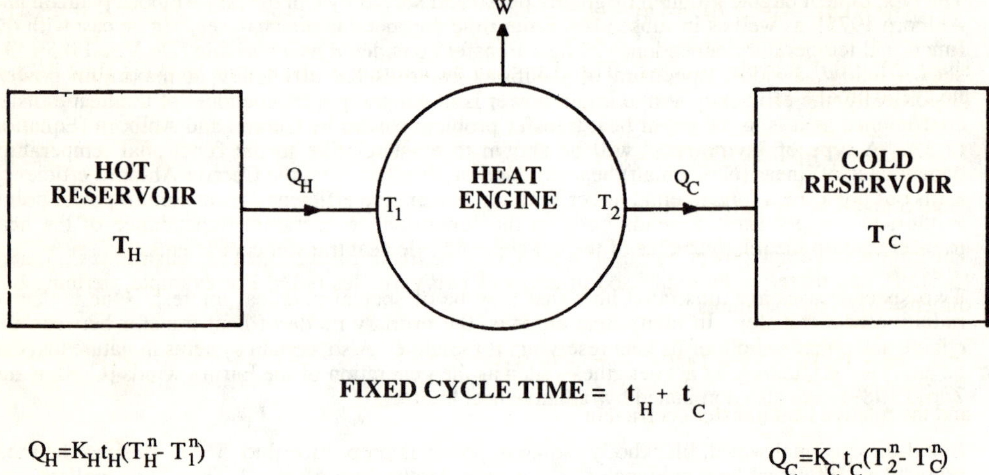

Figure 5. Schematic diagram of generalized cyclic finite-time heat engine.

The findings summarized below indicate that the functional temperature dependence of heat transfer plays a role in "preferring" different heat transfer modes for maximizing efficiency at maximum power point, depending on system parameter values. Although the heat engine designer may not have much leeway in selecting the mode of heat transfer, it may be interesting to understand quantitatively nature's built-in tendencies.

5B. <u>Statement and Solution of the Problem</u>

Consider the cyclic finite-time heat engine of Figure 5. The only irreversibilities are the finite rate of heat transfer between the reservoirs and the engine working fluid. It is straightforward to first prove that the engine cycle that maximizes power is again Carnot-like, comprised of two finite-time isotherms and two instantaneous adiabats [Chen and Tan 1989]. Heat transfer between the engine working fluid and the reservoirs Q is expressed as

$$Q_H = K_H t_H (T_H^n - T_1^n) \tag{5.1}$$

$$Q_C = K_C t_C (T_2^n - T_C^n) \tag{5.2}$$

where K is a heat transfer coefficient in units of W/K^n (note that $K < 0$ for $n < 0$). The total engine cycle time is fixed at t_0

$$t_H + t_C = t_0. \tag{5.3}$$

{A more general problem is to have different exponents for the temperature dependence of Q on the hot and cold isothermal branches. However, in the spirit of illustrating a new set of results, and in order to express the results in terms of the parameter K_H/K_C, attention is restricted here to the same functional temperature dependence for Q_H and Q_C.}

The first law of thermodynamics for the change in the internal energy of the engine working fluid ΔE requires that the work output per cycle w is

$$w = Q_H - Q_C - \Delta E = Q_H - Q_C \tag{5.4}$$

since $\Delta E = 0$ in a cyclic process. The change in the entropy of the engine working fluid ΔS for the cyclic process is

$$\Delta S = (Q_H/T_1) - (Q_C/T_2) = 0 \tag{5.5}$$

(both S and E being state functions).

The engine's average power output, $P = w/t_0$, is to be maximized subject to constraints (5.3) and (5.5). P can therefore be expressed in terms of two variables only. For example, defining the dimensionless temperatures

$$x \equiv T_1/T_H \qquad y \equiv T_2/T_H \qquad \tau \equiv T_C/T_H$$

and the relative heat transfer coefficient

$$k \equiv K_H/K_C$$

one can express P as

$$P = \frac{K_H T_H^n (1 - x^n)(y^n - \tau^n)(x - y)}{k\, y\, (1 - x^n) + x\, (y^n - \tau^n)} \tag{5.6}$$

The values of x and y that maximize P, x_{mp} and y_{mp}, for given values of τ and k, are then calculated (in general this must be done numerically). Efficiency at maximum power η_{mp} is then

$$\eta_{mp} = 1 - (y_{mp}/x_{mp}). \tag{5.7}$$

The reversible Carnot limit corresponds to $x = 1$ and $y = \tau$, with $P = 0$ (albeit at the maximum permissible efficiency of $1 - \tau$). The thermal short-circuit limit corresponds to $x = y$, with $P = 0$ and an efficiency of zero. The maximum power point is intermediate between these two limiting cases.

Alternatively, P can be expressed in terms of the heat engine efficiency η and the time spent on isothermal branches t_H and t_C

$$P = \frac{K_H T_H^n \, \eta \, \{ (1 - \eta)^n - \tau^n \}}{t_o (1 - \eta) \, \{ K_H/(K_C t_C) + (1 - \eta)^{n-1}/t_H \}} \tag{5.8}$$

(t_H and t_C are not independent variables due to Equation (5.3)). Maximizing P with respect to η and t_H yields the solution (in general a numerical solution) [Chen and Tan 1989]

$$P = \frac{K_H T_H^n \, \{ (1 - \eta')^n - \tau^n \}}{(1 - \eta')^n \, \{ 1 + \sqrt{k} \, (1 - \eta')^{n+1}/2 \}^2} \tag{5.9}$$

where η' is the solution to the equation

$$0 = (1 - \eta')^{3n+1}/2 + n \sqrt{k} \, (1 - \eta')^{n+1} - (1 - n)\tau^n (1 - \eta')^{n+1}/2$$

$$- (1 - \eta')^{n-1}/2 \, \{ n \, \tau^n - (1 - n) \sqrt{k} \} - \tau^n \sqrt{k} \tag{5.10}$$

and the times spent on the isothermal branches are related by

$$t_C/t_H = \sqrt{K_H/K_C} \, (1 - \eta')^{1-n}/2 \tag{5.11}$$

For $n = 1$ the Curzon-Ahlborn solution is recovered analytically. The above formulae can be contrasted with the corresponding results for $n = 1$, which are noted in Equations (2.3)-(2.5). A special property of the $n = 1$ case is that η_{mp} is independent of K_H and K_C and depends on τ only (Equation (2.2)). For n different from unity, η_{mp} depends on the relative values of the heat transfer coefficients, $k = K_H/K_C$.

5C. <u>Observations</u>

There is a type of "symmetry" for η_{mp} about $n = 1$. For $n > 1$, η_{mp} is a decreasing function of k, while for $n < 1$, η_{mp} is an increasing function of k. This point is illustrated in Figures 6 to 8, which are plots of η_{mp} vs. $K_H/(K_H + K_C)$ (or, equivalently, vs. $k/(1 + k)$) at 3 fixed values of τ. (The abscissa form in Figures 6 to 8 is selected to permit covering the range $0 \le k \le \infty$ on an abscissa scale of zero to unity. Also note the expanded ordinates in Figures 7 and 8.) In the calculations, the specific n values of -1 and 4 are selected for practicality and succinctness.

The values $n = 2$ and $n = 3$ correspond to radiative heat transfer in one- and two-dimensional systems, respectively. Since qualitatively their behavior is similar to the $n = 4$ case, and quantitatively their behavior is not as pronounced as the $n = 4$ case, they are omitted here in the spirit of conciseness. Also note that in the case $n = -1$, K_H and K_C are negative.

It is equally important to compare maximum power, and not just efficiency at maximum power, for these different cases. However, since the heat transfer coefficients are fundamentally and quantitatively different for different values of n, efficiency at maximum power affords a non-arbitrary basis for comparison.

Another format for comparing these results is a plot of η_{mp} vs. τ at a fixed value of k, which is shown in Figures 9 to 11. Of all n values of practical interest, radiative heat transfer (n = 4) represents the most pronounced differences, relative to linear heat transfer, for the maximum-power-point performance of heat engines. In the limit of $\tau \to 0$, whether the maximum power point approaches the reversible limit (and hence maximum power approaches zero) and the associated efficiency of unity, or whether the maximum power point remains at a distinguishable irreversible operating point at non-zero power and an efficiency less than unity, depends on both n and k. Observations on assorted limiting behaviors of η_{mp} have been noted in [Gordon 1990, Chen and Tan 1989].

For $\tau \to 1$, it is easy to show that $\eta_{mp} \to (1 - \tau)/2$ (half the Carnot efficiency) for all n. This limiting behavior can be seen in Figures 9 to 11, and the limit can be approached either from above or from below. The reason is that when T_C and T_H are sufficiently close, then all governing equations can, to an excellent approximation, be linearized, and hence reduce to the Curzon-Ahlborn (linear heat transfer) problem.

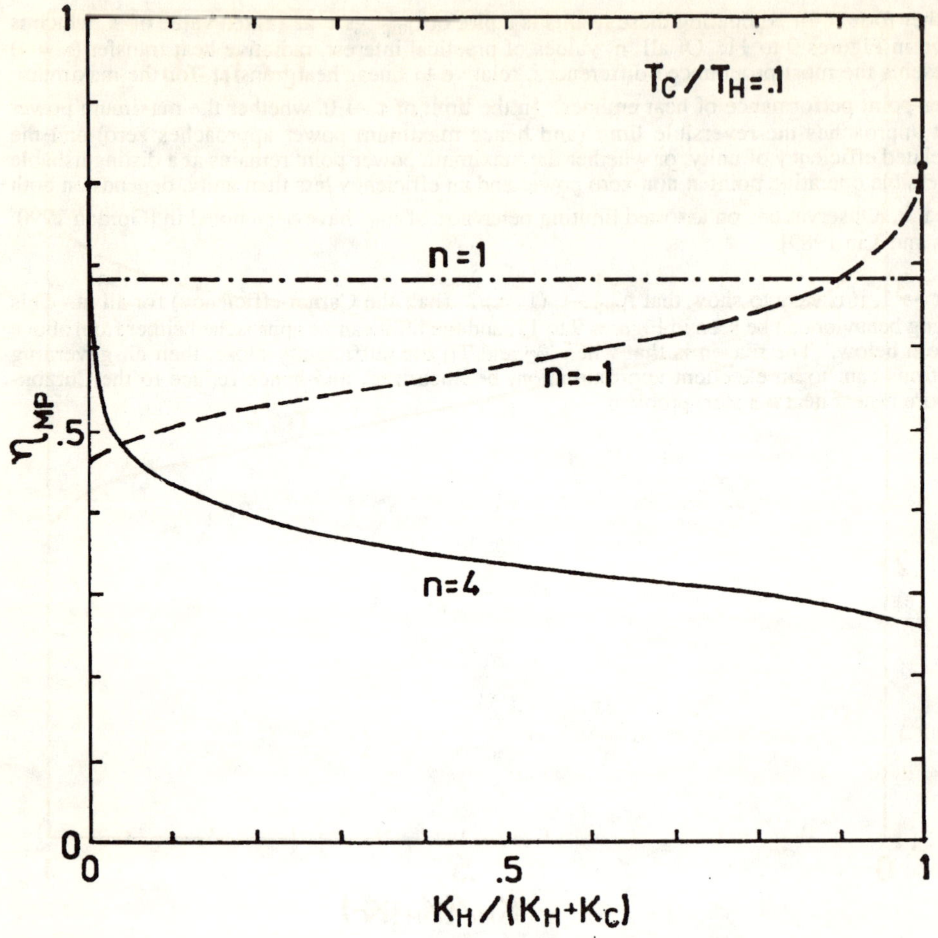

Figure 6.   Efficiency at maximum power η_{mp} vs. $K_H/(K_H + K_C)$ at fixed $\tau = T_C/T_H = 0.1$.
The abscissa form of $k/(1 + k)$ is selected in order to permit including the range
$0 \leq k = K_H/K_C \leq \infty$ on an abscissa scale of zero to unity (for Figures 7 and 8 as well).

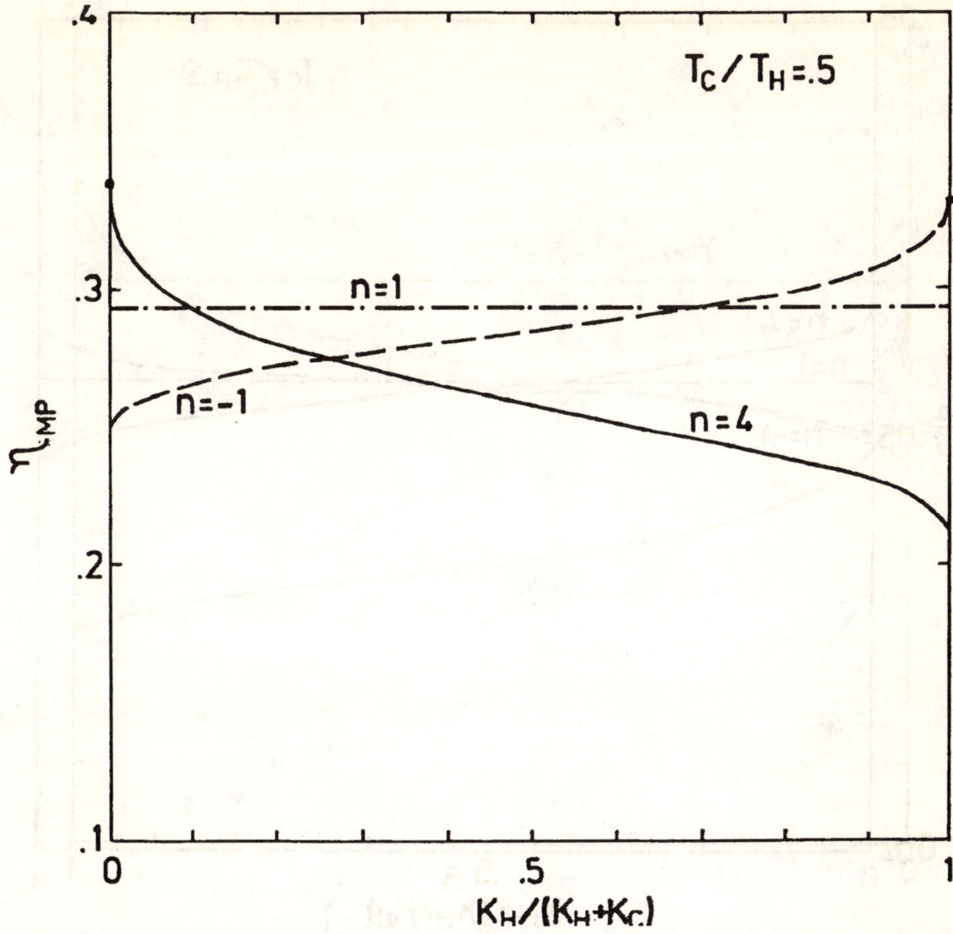

Figure 7. Efficiency at maximum power point η_{mp} vs. $K_H/(K_H + K_C)$ at fixed $\tau = T_C/T_H = 0.5$. Note expanded ordinate.

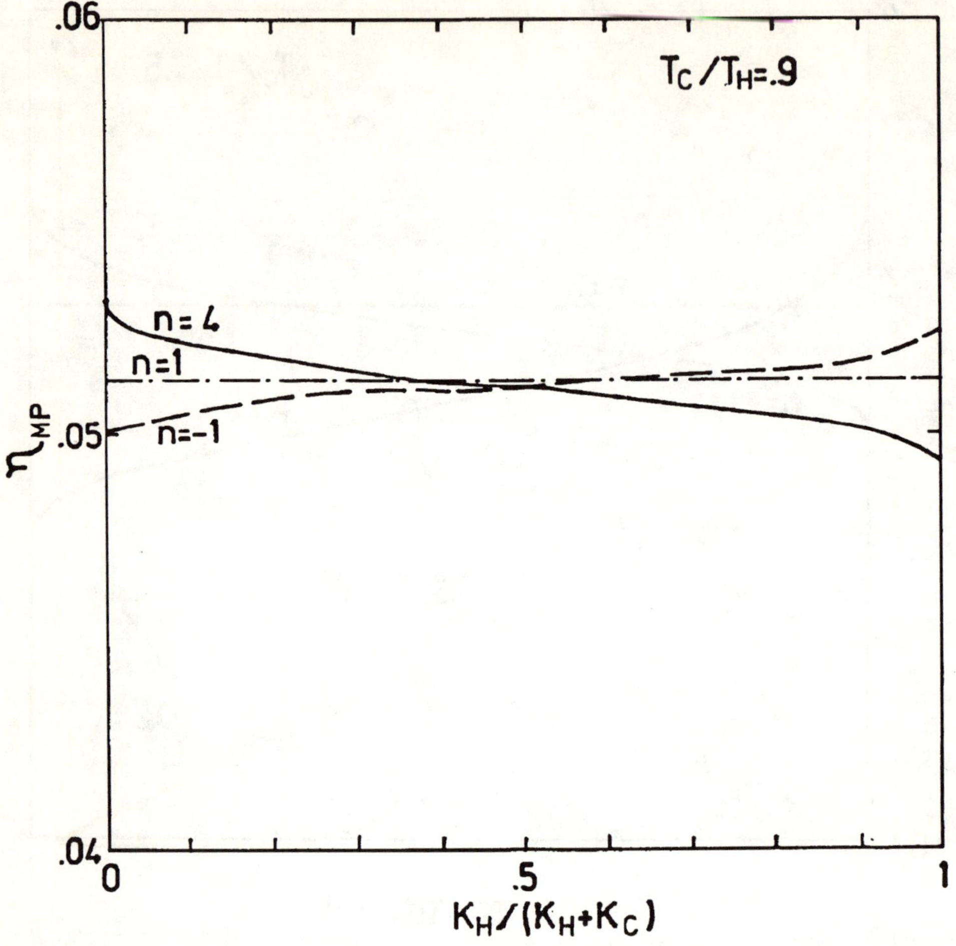

Figure 8. Efficiency at maximum power point η_{mp} vs. $K_H/(K_H + K_C)$ at fixed $\tau = T_C/T_H = 0.9$. Note expanded ordinate.

116

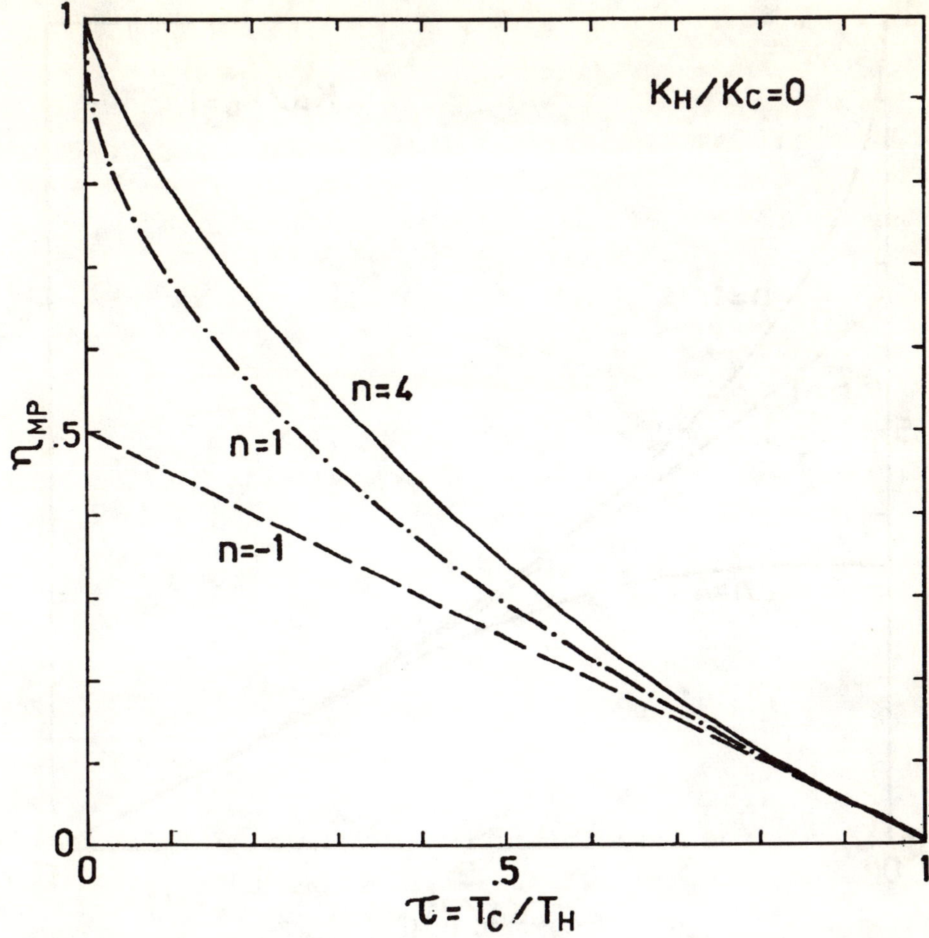

Figure 9. Efficiency at maximum power point η_{mp} vs. $\tau = T_C/T_H$ at fixed $k = K_H/K_C = 0$.

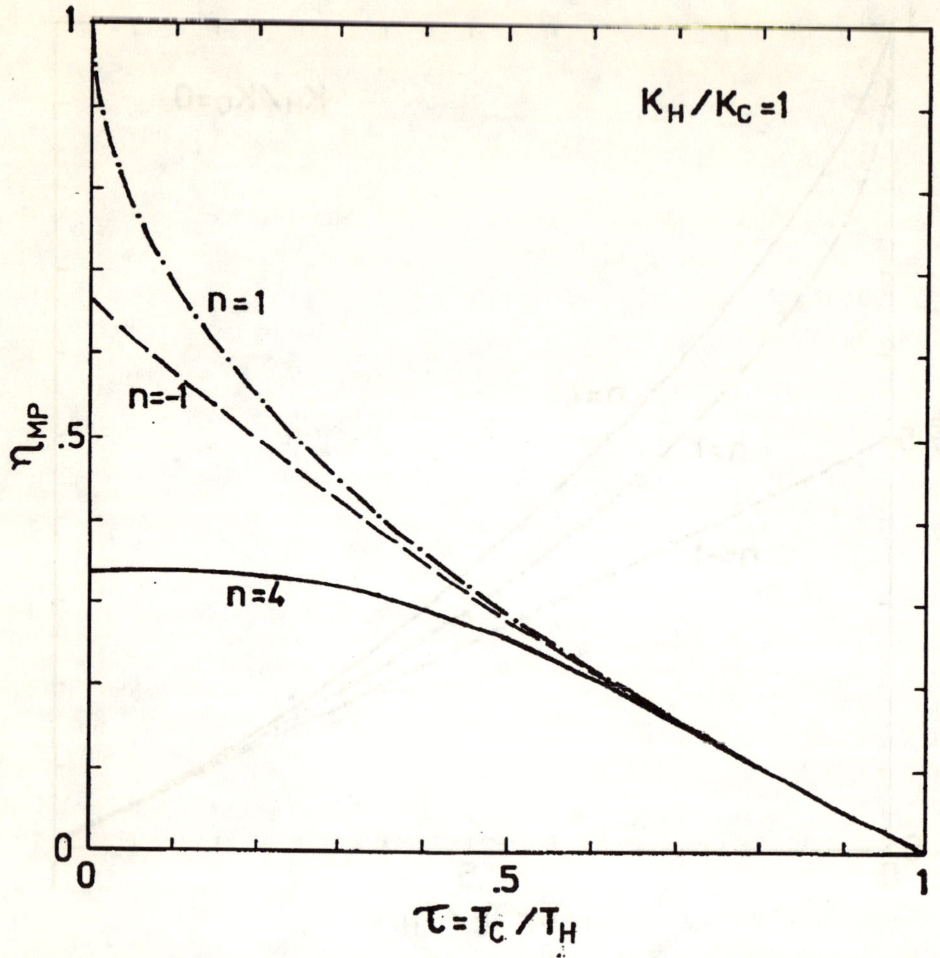

Figure 10. Efficiency at maximum power point η_{mp} vs. $\tau = T_C/T_H$ at fixed $k = K_H/K_C = 1$.

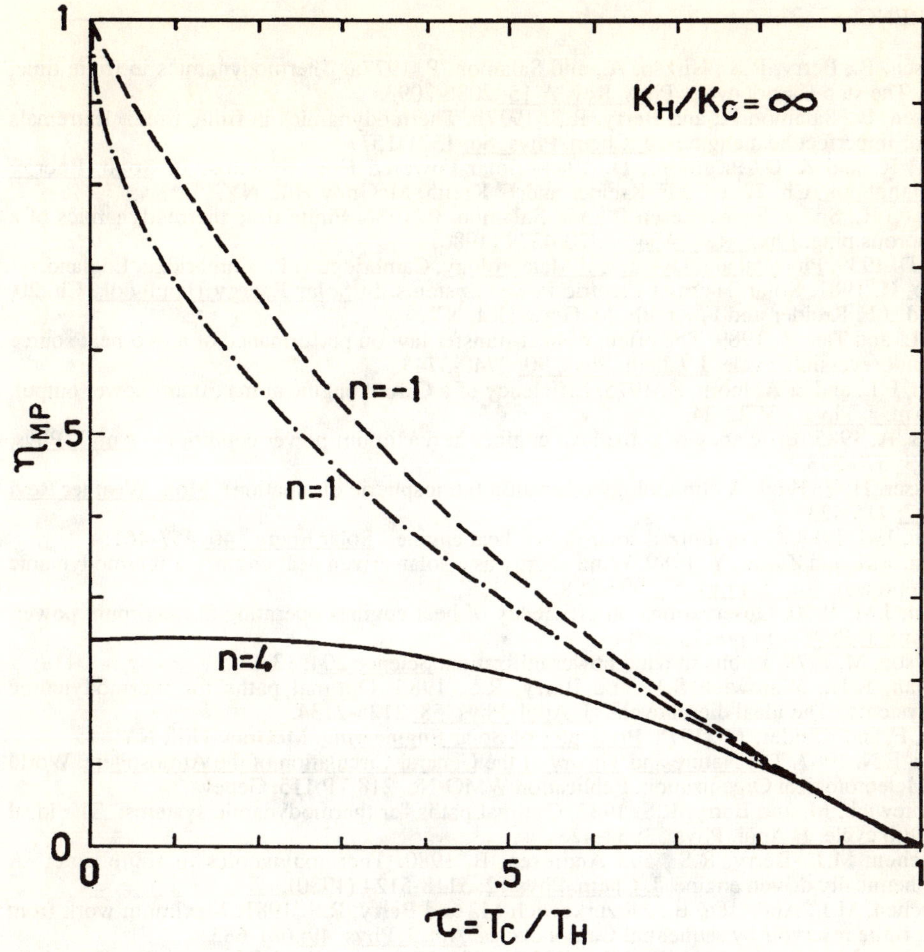

Figure 11. Efficiency at maximum power point η_{mp} vs. $\tau = T_C/T_H$ at fixed $k = K_H/K_C = \infty$.

REFERENCES

Andresen, B., Berry, R.S., Nitzan, A., and Salamon, P. 1977a. Thermodynamics in finite time. I. The step-Carnot cycle. Phys. Rev. A 15: 2086-2093.

Andresen, B., Salamon P., and Berry, R.S. 1977b. Thermodynamics in finite time: Extremals for imperfect heat engines. J. Chem. Phys. 66: 1571-1577.

Barber, R. and & D. Prigmore, D. 1981. Solar-Powered Heat Systems. In Solar Energy Handbook, Ch. 22, ed. J.F. Kreider and F. Kreith, McGraw-Hill, NY.

Brown, G.R., Snow, S., Andresen B., and Salamon, P. 1986. Finite-time thermodynamics of a porous plug. Phys. Rev. A34: 4370-4379 (1986).

Brunt, D. 1939. Physical and Dynamical Meteorology, Cambridge U.P., Cambridge, England.

Caputo, R. 1981. Solar-Thermal Electric Power Systems. In Solar Energy Handbook, Ch. 20, ed. J.F. Kreider and F. Kreith, McGraw-Hill, NY.

Chen, L. and Tan, Z. 1989. The effect of heat-transfer law on performance of a two-heat source endoreversible cycle. J. Chem. Phys. 90: 3740-3743.

Curzon, F.L. and & Ahlborn, B. 1975. Efficiency of a Carnot engine at maximum power output. Am. J. Phys. 43: 22-24.

De Vos, A. 1985. Efficiency of some heat engines at maximum-power conditions. Am. J. Phys. 53: 570-573.

Ellsaesser, H.W. 1969. A climatology of epsilon (atmospheric dissipation). Mon. Weather Rev. 97: 415-423.

Gordon, J.M. 1988. On optimized solar-driven heat engines. Solar Energy 40: 457-461.

Gordon, J.M. and Zarmi, Y. 1989. Wind energy as a solar-driven heat engine: a thermodynamic approach. Am. J. Phys. 57: 995-998.

Gordon, J.M. 1990. Observations on efficiency of heat engines operating at maximum power. Am. J. Phys. - in press.

Gustavson, M. 1979. Limits to wind power utilization. Science 204: 13-17.

Hoffman, K.H., Watowich S.J., and Berry, R.S. 1985. Optimal paths for thermodynamic systems: The ideal diesel cycle. J. Appl. Phys. 58: 2125-2134.

Kreith, F. and Kreider, J.F. 1978. Principles of Solar Engineering, McGraw-Hill, NY.

Lorenz, E.N. 1967. The Nature and Theory of the General Circulation of the Atmosphere, World Meteorological Organization, Publication WMO-No. 218 TP.115, Geneva.

Mozurkewich, M. and Berry, R.S. 1982. Optimal paths for thermodynamic systems: The ideal Otto cycle. J. Appl. Phys.53: 34-42.

Ondrechen, M.J., Berry, R.S., and Andresen, B. 1980. Thermodynamics in finite time: A chemically driven engine. J. Chem. Phys.72: 5118-5124 (1980).

Ondrechen, M.J., Andresen, B., Mozurkewich M., and Berry, R.S. 1981. Maximum work from a finite reservoir by sequential Carnot cycles. Am. J. Phys. 49: 681-685.

Ondrechen, M.J., Rubin, M.H., and Band, Y.B. 1983. The generalized Carnot cycle: A working fluid operating in finite time between finite heat sources and sinks. J. Chem. Phys. 78: 4721-4727.

Rubin, M.H. and Andresen, B. 1982. Optimal staging of endoreversible heat engines. J. Appl. Phys. 53: 1-7.

Salamon, P., Nitzan, A., Andresen, B., and Berry, R.S. 1980. Minimum entropy production and the optimization of heat engines. Phys. Rev. A 21: 2115-2129.

Salamon, P. and Nitzan, A. 1981. Finite time optimization of a Newton's law Carnot cycle. J. Chem. Phys. 74: 3546-3560.

Salamon, P., Band, Y.B., and Kafri, O. 1982. Maximum power from a cycling working fluid. J. Appl. Phys. 53: 197-202.

Tritton, D.J. 1977. Physical Fluid Dynamics, Van Nostrand, Cambridge.

Applications of Finite-Time Thermodynamics to Solar Power Conversion

Kurt O'Ferrell Lund
Department of Mechanical Engineering
San Diego State University
San Diego, CA

ABSTRACT
Thermal conversion of solar power depends on a heat engine
operating between the receiver and sink temperatures. The
principle of operation of a solar-thermal power plant is
presented in terms of finite heat transfer rates and an
internally reversible heat engine. The theory of finite-time
thermodynamics and maximum power is utilized to determine the
upper limit to power output for terrestrial and space-based
solar-thermal engines. Parametric equations are derived for
determining optimum performance variables. In terrestrial systems
an optimum receiver temperature is determined for maximum power
output: whereas for space-based systems, maximum power is
achievable at any receiver temperature.

1. INTRODUCTION
Classical thermodynamics shows how the transfer of thermal energy
may be converted into mechanical work by a heat engine operating
between two thermal reservoirs of temperatures T_{ih} and T_{ic}, with

$T_{ih} > T_{ic}$, and how heat transfer at infinitesimal temperature
differences between the reservoirs and the heat engine lead to a
reversible process and the well-known Carnot efficiency as the
upper limit to conversion of heat to mechanical work [e.g.,
Reynolds and Perkins 1977]:

$$\eta_c \equiv \frac{W}{Q_h} = 1 - \frac{T_{ic}}{T_{ih}} \tag{1.1}$$

The requirement of heat transfer at vanishing temperature
differences implies an infinitely slow <u>rate</u> of heat transfer, or
an infinite heat transfer area, both of which are unacceptable in
practical engines. In essence, we seek not from Nature to
produce work, but to produce power, or work in a finite period of
time. Thus, all thermal power plants utilize thermodynamics in
converting rates of heat transfer at finite temperature

differences, $\dot{Q}_h$, to mechanical power, $\dot{W}$, but without ever closely
approximating the theoretical upper limit, usually written as the
rate ratio

$$\eta_c = \dot{W}/\dot{Q}_h$$

(In fact, most power plants are designed for maximum power
output, which tends to minimize the all-important capital costs
of the plant, rather than for maximum thermal efficiency, which
would "only" save on fuel costs; this becomes especially
important for solar power plants because in this application the
capital costs far outweigh operation and maintenance costs, the
"fuel" being "free").

This discrepancy led Curzon and Ahlborn [1975] to consider a heat
engine whose only irreversibility was finite-time heat transfers
across finite temperature differences between the reservoirs and
the heat engine:

$$\dot{Q}_h = K_h(T_h - T_{ih}) \ , \quad \dot{Q}_c = K_c(T_{ic} - T_c) \tag{1.2,3}$$

where now T_{ih} and T_{ic} are the hot and cold temperatures internal

to the engine (i.e., the working fluid temperatures), and T_h and
T_c are the reservoir temperatures "external" to the engine. The
remarkable result was that with T_{ih} and T_{ic} optimized for
maximum power output, and with the internally reversible
conversion cycle (1.1), the overall conversion efficiency was
independent of K_h and K_c, and dependent only on the reservoir
temperature ratio, as previously:

$$\eta_e \equiv \frac{\dot{W}_{max}}{\dot{Q}_h} = 1 - \sqrt{T_c/T_h} \tag{1.4}$$

This cycle is here referred to as an _endoreversible_ cycle
[Callen 1985] and is utilized subsequently to predict maximum
possible solar power conversion. It was indicated that practical
power plants can have efficiencies in close agreement with (1.4),
but not with (1.1) [Curzon and Ahlborn 1975; Callen 1985]. The
derivation of (1.4) required the processes, (1.2) and (1.3) to be
isothermal, even though finite. Subsequent analyses have shown
these processes to indeed yield maximum power [Salamon & Nitzan,
1981]; have been applied to the sequences of combustion engines
[Mozurkewich and Berry 1982; Band, Kafri, and Salamon 1982ab;
Andresen, Salamon, and Berry 1984]; to fluctuating external
temperatures [Salamon, Band, and Kafri 1982]; and recently to
refrigeration cycles [Chen and Yan 1988; Yan and Chen 1989]. The
result (1.4) was also derived for a continuous, steady-flow cycle
which is more typical of stationary power plants [Bejan 1988];
in that analysis, which also allowed for a leakage heat flow, a
further maximization of the power output yielded equal values for
the optimum hot-side and cold-side conductances: $K_h = K_c = K$.

In conventional power plants, T_h in (1.4) is usually a
metallurgical upper limit for the material transferring heat to
the working fluid. As such, it is a constant that is maintained
by combustion (or other processes) regardless of the magnitude of
$\dot{Q}_h$. In solar plants, by contrast, T_h represents the temperature
of the receiver, which varies according to the magnitude of $\dot{Q}_h$.
Therefore, the highest permissible T_h does not lead to the

maximum conversion rate of available energy to mechanical power
as in conventional plants, as has been shown for the Carnot cycle
[Reynolds and Perkins 1977, p. 236; Howell and Bannerot 1977;
Bejan 1987a], and for the endoreversible cycle [Gordon 1988].
In following sections, these methods are utilized in deriving
formulae for the optimum temperature ratio T_h/T_C for greatest
power conversion. Conversion cycles with diathermal hot-side and
cold-side heat exchange processes have also been considered
[Bejan, Kearney, and Kreith 1981; Wu 1988]. Here the objective
is to provide upper bounds for power output so that (1.4)
applies; this equation is approachable in real systems where
there is phase change or only small temperature variations in the
heat exchangers.

The low-temperature reservoir usually has T_C as the environment
temperature to which any heat rate $\dot{Q}_C$ can be rejected. This is
so for both conventional thermal plants and terrestrial solar
plants. However, for space-based solar-thermal power plants
under current development the low temperature "reservoir" at T_C
is the space radiator where $\dot{Q}_C$ depends on T_C, thus requiring
special treatment.

In what follows, a typical solar-thermal power conversion system
is described and characterized, and investigated for maximum
possible endoreversible power conversion, for both terrestrial
and space applications. This leads to the best receiver and
radiator temperatures, and system efficiencies, for various types
of designs.

2. DESCRIPTION OF SOLAR HEAT ENGINE
The principle of operation of a solar-thermal power plant is
shown schematically in Fig. 1, where the solar irradiance, $\dot{q}_S$
(W/m^2), is incident on projected or aperature area, A_p, of the
reflector, or concentrator. In general, $\dot{q}_S$ varies with time
(diurnally and annually as well as due to weather patterns) so
that various mechanisms are required for the concentrator to
track $\dot{q}_S$ [e.g., Duffie and Beckman 1980; Kreith and Kreider

1978]. This time variation is rather slow compared to engine
processes; thus, $\dot{q}_s$ may be taken as a constant for present
purposes.

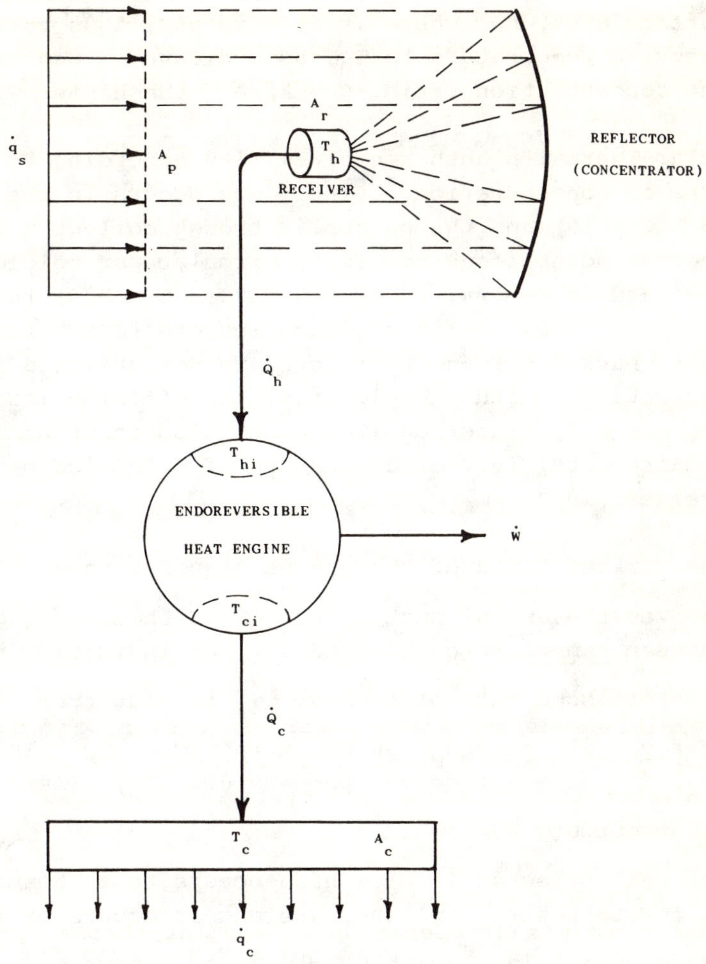

Figure 1. Conceptual Solar-Thermal Power Plant.

Although the source of $\dot{q}_s$ is the sun, and thermodynamic analyses
have been applied with T-sun as the source temperature [e.g.,
Bejan 1987b], and conversion of radiant energy has been
considered [De Vos 1988], here we take the more pragmatic, or

engineering approach which assumes that $\dot{q}_s$ exists and can be "tapped" indefinitely without diminishing its value.

The reflector concentrates the source flux onto the receiver of surface area, A_r, where it is absorbed, thus raising the receiver material temperature to T_h, as indicated. Generally the higher the concentration ratio, $C = A_p/A_r$, the higher will be T_h.

Solar-Thermal plants are classified according to the degree and type of concentration. Line focus occurs in the CPC collector ($5 < C < 10$) and the parabolic trough collector ($10 < C < 50$), whereas point focus occurs in circular paraboloids ($50 < C < 100$) and in central receivers ($100 < C < 500$) [e.g., Stine and Harrigan 1985]. Flat plate collectors [e.g., Lund 1989; Lund 1986] have $C = 1$ and therefore are not utilized in power generation. Although higher system efficiencies can be achieved for large C, higher capital costs also occur because of tracking requirements; further discussion of economics may be found in Kreider [1979].

The working fluid of the engine passes through the reciever and removes the useful heat rate, $\dot{Q}_h$, to the heat engine operating between temperatures T_{hi} and T_{ci}, as indicated in Figure 1. However, some rate of energy, $\dot{Q}_\ell$, is lost from the receiver so that not all of $\dot{q}_s A_p$ can be extracted as $\dot{Q}_h$. This is the reason that an optimum receiver temperature, T_h, exists for maximum power output, $\dot{W}$.

The heat rejected is $\dot{Q}_c$, which occurs to a thermal reservoir at T_c for terrestrial plants, whereas it occurs by radiation for space-based plants with $\dot{Q}_c = \dot{q}_c A_c$, as shown.

Although there are many technical design variations for diverse solar-thermal power plants, they all have elements in common with the schematic representation in Fig. 1; as such, it serves as a basis for the following analysis.

3. MODEL EQUATIONS

Let η_o be the optical efficiency associated with the
concentration and absorption processes of the solar flux onto the
receiver, then the rate of energy absorption is

$$\dot{Q}_a = \eta_o \dot{q}_s A_p \tag{3.1}$$

For each type of collection system there are unique technological
challenges for improving η_o, which includes complex processes of
radiation reflectance, transmittance, and absorptance. An energy
balance on the receiver yields the rate of useful energy
collected.

$$\dot{Q}_h = \dot{Q}_a - \dot{Q}_\ell = \eta_o A_p \dot{q}_s - A_r \dot{q}_\ell \tag{3.2}$$

where A_r is the exposed surface area of the receiver. The loss
heat flux from the receiver to the ambient is usually expressed
as the linearized form for low and intermediate temperature
collectors

$$\dot{q}_\ell = U_\ell (T_h - T_c) \tag{3.3}$$

where U_ℓ is the overall unit surface conductance accounting for
convective and linearized radiative effects, and the environment
is taken as T_c. In High temperature receivers, on the other
hand, losses are dominated by radiation so that the flux may be
written as

$$\dot{q}_\ell = \epsilon_r \sigma \, T_h^4 \tag{3.4}$$

where ϵ_r is the effective receiver emittance [Gordon 1988], σ is
the Stefan-Boltzmann constant, and where T_c^4 was neglected
relative to T_h^4.   A more complete loss model was utilized
previously [Howell and Bannerot 1977] however, (3.3) and (3.4)
lead to simpler analytical results, and (3.4) applies almost
exactly for space-based systems.

An important quantity for solar collection systems is the, so called, stagnation temperature, T_s, which is the maximum possible temperature reachable when $\dot{Q}_h = 0$ (e.g., during loss of flow of the working fluid). In this case, the energy absorbed exactly equals that lost, so that from (3.2) and (3.3) for low temperature collectors

$$A_r U_\ell (T_s - T_c) = \eta_o A_p \dot{q}_s \qquad (3.5)$$

and from (3.4) for high temperature collectors

$$\epsilon_r \sigma A_r T_s^4 = \eta_o A_p \dot{q}_s \qquad (3.6)$$

Hence, for either low or high temperature collectors $\dot{q}_s$ in (3.2) may be replaced by the stagnation temperature, T_s:

$$\dot{Q}_h = A_r U_\ell (T_s - T_h) \qquad (3.7)$$

and

$$\dot{Q}_h = \epsilon_r \sigma A_r (T_s^4 - T_h^4) \qquad (3.8)$$

Let $\eta_r = \dot{Q}_h / \dot{q}_s A_p$ be the receiving system efficiency, and let $t = T/T_s$ denote a temperature relative to the stagnation temperature, then the results (3.5) to (3.8) may be summarized as

$$\eta_r = \frac{\eta_o}{\left[1 - t_c \right]^{(4-n)/3}} \left[1 - t_h^n \right] \qquad (3.9)$$

where $n = 1$ for low and intermediate temperature receivers and $n = 4$ for high temperature receivers. Now, with

$$\eta_s = \dot{W}/\dot{q}_s A_p = \eta_r \eta_e$$

we have the overall system efficiency, or fraction of available
solar power converted to mechanical power:

$$\eta_s = \eta_o \frac{1}{\left[1-t_c\right]^{(4-n)/3}} \left[1 - t_h^n\right] \left[1 - \sqrt{t_c/t_h}\right] \qquad (3.10)$$

Since the collector efficiency and endoreversible efficiency have
opposite trends with receiver temperature, it is seen that (3.10)
possesses an optimum temperature for maximizing the power
generation system efficiency.

4. TERRESTRIAL SYSTEMS

In terrestrial systems, T_C refers to low temperature reservoir
which remains fixed, regardless of the magnitude of the rate of

heat rejection, $\dot{Q}_c$. Thus, t_c in (3.10) is a constant, and η_s can
be optimized with respect to t_h. Setting $d\eta_s/dt_h = 0$ results in
the maximum system efficiency for maximum possible power
generation:

$$\eta_s^* = \eta_o \frac{\left[1 - t_h^{*\,n}\right]^2}{\left[1 - t_c\right]^{(4-n)/3}\left[1 + (2n-1)t_h^{*\,n}\right]} \qquad (4.1)$$

where the optimum ratio t_h^* is obtained implicitly from

$$t_c = \frac{4n^2 t_h^{*\,2n+1}}{\left[1 + (2n-1)t_h^{*\,n}\right]^2} \qquad (4.2)$$

A similar result was obtained previously [Gordon 1988]. For a given concentration ratio or concentrator design, $C = A_p/A_r$, T_S may be obtained from (3.5) or (3.6):

$$C = \frac{U_\ell T_S}{\eta_o \dot{q}_S}(1 - t_c) = \frac{U_\ell T_c}{\eta_o \dot{q}_S}\left[\frac{1}{t_c} - 1\right] \qquad (4.3)$$

or

$$C = \frac{\epsilon_r \sigma T_S^4}{\eta_o \dot{q}_S} = \frac{\epsilon_r \sigma T_c^4}{\eta_o \dot{q}_S}\left[\frac{1}{t_c^4}\right] \qquad (4.4)$$

Thus, the intermediate parameter, t_c, may be determined from

$$t_c = \begin{cases} (1 + c_1)^{-1} , & n = 1 \\ (c_4)^{-1/4} , & n = 4 \end{cases} \qquad (4.5)$$

where the scaled concentration ratios are

$$c_1 = \frac{\eta_o \dot{q}_S}{U_\ell T_c}\, C , \qquad c_4 = \frac{\eta_o \dot{q}_S}{\epsilon_r \sigma T_c^4}\, C \qquad (4.6,7)$$

The primary results of the above optimization can now be summarized in terms of these ratios, as indicated in Fig. 2 (where c_4 is 1000 times greater than c_1). With the optimum system efficiency in ratio with the optical efficiency, these functions do not depend on any parameters (except the order of

the receiver heat loss, n), and may be regarded as general design
curves. Nevertheless, specific numerical examples are included
in the following.

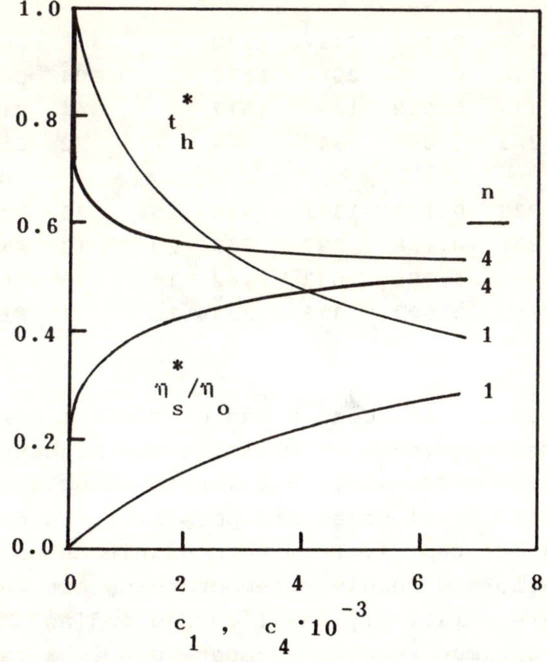

Figure 2. Optimum Efficiency and Temperature Ratios
for Terrestrial Systems

Although it is C that is a design parameter such that t_c is
obtained from (4.3) or (4.4), with specification of T_c, and
thence t_h^* obtained from (4.2) and utilized in (4.1), it is seen
that (4.1) to (4.4) are parametric equations in t_h^*. Thus,
selecting a sequence of t_h^* - values, efficiencies and
corresponding concentration ratios were calculated as in Table 1,
using $\eta_o = 0.85$ and $T_c = 300K$. Only the last column for
calculating the concentration ratio required the following

additional values: $\dot{q}_s = 800$ W/m^2, $\epsilon_r = 0.5$, and U_ℓ in W/m^{2}K, which
defines the type of concentration technology, as indicated.

Table 1. Optimum Conditions for Terrestrial Systems

t_h^*	n	t_c	η_s^*	$T_s(K)$	$T_h(K)$	U_ℓ	C	Type System
0.55	4	0.110	0.428	2727	1500		2306	Non exist.
0.58	4	0.148	0.373	2027	1176		704	Central rec.
0.60	4	0.177	0.338	1695	1017		344	Circ. parab.
0.65	4	0.262	0.255	1145	744		72	Circ. parab.
0.50	1	0.222	0.182	1351	675	25	38	Parab. trough
0.60	1	0.338	0.128	887	532	20	18	Parab. trough
0.70	1	0.475	0.086	632	442	15	8	CPC
0.90	1	0.808	0.023	371	334	10	1	Flat plate

It should be noted for the conditions of Table 1 how practical
limitations on the magnitude of C limits the highest practically
achievable efficiency for maximum power to about 40%, even though
higher theoretical efficiencies are possible; and how the
efficiency drops off rapidly from this magnitude as the
concentration ratio and receiver temperatures are reduced.
However, there are relatively small corresponding changes in t_h,
which means the optimum receiver temperature is a rather broad
maximum, as noted previously [Gordon 1988]. Presently, both
central receiver plants [Hillesland et al. 1988] and parabolic
trough plants [Jensen et al. 1989] are in evaluation; although
the latter converts only about half the power of the former for
the same projected area, as seen in Table 1, it is sufficiently
less costly to produce so that the size of the plant, A_p, can be
correspondingly increased.   There are many design considerations
in a solar-thermal power plant; nevertheless, the preceeding
results are significant in showing the upper limit for maximum
power conversion.

5. SPACE-BASED SYSTEMS
The space-based solar-thermal systems, also described as
solar-dynamic systems [e.g. Labus et al. 1989], differ
fundamentally from the preceeding in that T_c is now the

temperature of the space radiator, which depends on $\dot{Q}_c$, and not a

thermal reservoir. Although there can be significant spatial
temperature variations in the radiator material [e.g. Edwards,
1981], we here take T_C to be a suitably averaged value for the
radiator such that the heat rejection rate is

$$\dot{Q}_c = \epsilon_c \sigma T_c^4 A_c \tag{5.1}$$

However, from (1.4) and (3.8) we also have

$$\frac{\dot{Q}_c}{\dot{Q}_h} = \sqrt{\frac{T_c}{T_h}} = \frac{A_c \epsilon_c \sigma T_c^4}{A_r \epsilon_r \sigma \left[T_s^4 - T_h^4 \right]} \tag{5.2}$$

such that for maximum reversible power output we have a
relationship between T_C and T_h. This constraint may also be
stated as

$$t_c = \left\{ \frac{R^2 \left[1 - t_h^4 \right]^2}{t_h} \right\}^{1/7} \tag{5.3}$$

where the emissivity-area ratio is

$$R = \frac{\epsilon_r A_r}{\epsilon_c A_c}$$

Now, (3.10) still applies for the system efficiency, so that with
(5.3) and $n = 4$ in (3.10) we have

$$\eta_s = \eta_o \left[1 - s \right] \left[1 - R^{1/7} \left\{ 1/s - 1 \right\}^{1/7} \right] \tag{5.4}$$

where $s = t_h^4$. Here, again, we have competing behaviors with t_h,
so that setting $d\eta_s/ds = 0$ we obtain the optimum receiver
temperature relationship, (5.5):

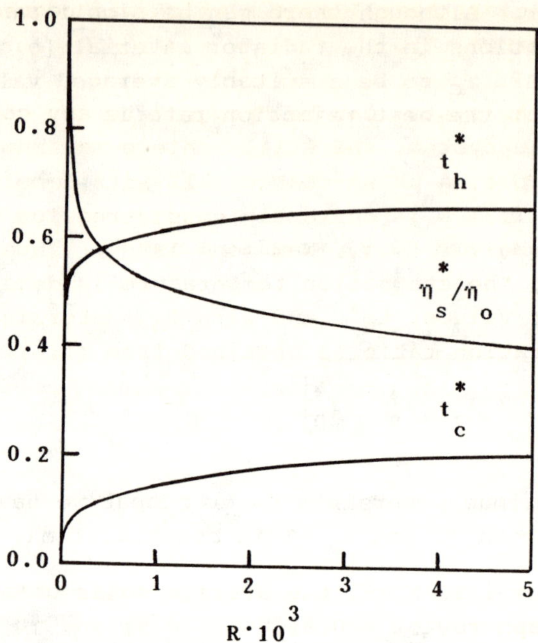

Figure 3. Optimum Efficiency and Temperature Ratios
for Space-Based Systems.

$$R = \frac{s^{*8}}{(1 - s^{*})(1/7 + s^{*})^{7}} = \frac{t_{h}^{*\,32}}{\left[1 - t_{h}^{*\,4}\right]\left[\frac{1}{7} + t_{h}^{*\,4}\right]^{7}} \qquad (5.5)$$

Thus, it is seen that the optimum temperature ratios, t_{h}^{*} and
t_{c}^{*}, and the maximum-power system efficiency are uniquely
determined for a specified emissivity-area ratio, R. The
behaviors of these efficiency and temperature ratios are shown in
Fig. 3.

The remarkable result is, however, that while optimum ratios are
determined, the actual temperatures are not. In this sense, the
solar-dynamic system is seen to thermally _float_, by contrast to

the terrestrial solar-thermal system which was pegged at the
reservoir temperature, T_c.

This means that the system can be "pegged" to any convenient
temperature in the system, and still achieve maximum power
output. For example, a phase-change salt with a melting
temperature T_m = 1121 K is currently considered for thermal
storage in the receiver [e.g., Namkoong 1989]. Thus, setting
T_h = T_m = 1121 K, the stagnation temperature is determined for
various parameter values, t_h^*, and with T_s determined the
required concentration ratio is obtained from (3.6):

$$C = \frac{\epsilon_r \sigma T_s^4}{\eta_o \dot{q}_s} \tag{5.6}$$

where we take $\dot{q}_s$ = 1353 W/m^2, the average solar constant outside
the earth's atmosphere, ϵ_c = 0.85, ϵ_r = 0.5, and η_o = 0.85.
Sample calculations for these conditions are shown in Table 2.

Table 2. Optimum Conditions for Space-Based Systems

t_h^*	1/R	η_s^*	t_c^*	T_s(K)	T_c(K)	C	A_c/A_p
0.55	7180	0.554	0.084	2040	171	425	9.94
0.58	2380	0.495	0.113	1930	219	344	4.07
0.60	1219	0.456	0.136	1870	254	300	2.39
0.65	282	0.365	0.200	1720	345	218	0.76
0.70	83	0.284	0.275	1600	440	162	0.30

It is seen that up to 50% power conversion is possible at a
radiator area of approximately four times the projected
concentrator area; the attainment of this degree of thermal
conversion in space depends strongly on the cost of placing the
power plant in orbit. Current estimates are that the receiver
accounts for a significant fraction of the weight so that a

larger radiator may result in an economic optimization, as well
as the above maximum power optimization.

Quite a number of approximations have been made in generating the
above results, and much more detailed analyses of components of
actual systems is required to ascertain operating characteristics
and power output. Nevertheless, it is the power of thermodynamics, with suitable approximations, to illustrate the system
behavior in simple, comprehensive terms, thus pointing to optimum
operating conditions and providing limits to power conversion.
The advent of finite-time thermodynamics brings the theoretical
limits closer to achievable conditions in practice.

REFERENCES

Andresen, R., P. Salamon and R.S. Berry. 1984. Thermodynamics in
finite time. Phys Today Sept.: 62-70.

Band, Y.B., O. Kafri and P. Salamon. 1982a. Finite time
thermodynamics: Optimal expansion of a heated working fluid.
J. Appl. Phys. 53: 8-28.

Band, Y.B., O. Kafri and P. Salamon. 1982b. Optimization of a
model external combustion engine. J. Appl. Phys. 53: 29-33.

Bejan, A. 1988. Theory of heat transfer-irreversible power
plants. Int. J. Heat Mass Transfer 31:1211-1219.

Bejan, A. 1987a. Second law aspects of solar energy conversion.
In Solar energy utilization, ed. H. Yuncu and E. Paykoc, 145-187.
NATO Adv. Study Inst., Nijhoff Pub., Netherlands.

Bejan, A. 1987b. Unification of three different theories
concerning the ideal conversion of enclosed radiation. J. of Sol
Energy Eng. 109: 46-51.

Bejan, A., W.D. Kearney, and F. Kreith. 1981. Second law analysis
and synthesis of solar collector systems. J. Sol. Energy Eng.13:
681-687.

Callen, H.B. 1985. Thermodynamics and an introduction to
thermostatistics, 2nd ed., New York: J. Wiley.

Chen, J., and Z. Yan. 1988. Optimal performance of an
endoreversible-combined refrigeration cycle. J. Appl. Phys. 63:
4795-4798.

Curzon, F.L., and B. Ahlborn. Efficiency of a carnot engine at
maximum power output. Am J. Phys. 43: 22-24.

De Vos, A. 1988. Thermodynamics of Radiation Energy-Conversion in One and Three Physical Dimensions. _J. Phys Chem S 49_: 725-730.

Duffie, J.A., and W.A. Beckman. 1980. _Solar engineering of thermal processes_. New York: J. Wiley.

Edwards, D.K. 1981. _Radiation heat transfer notes_. New York: Hemisphere.

Gordon, J.M. 1988. On optimized solar-driven heat engines, _Solar Energy_ _40_: 457-461.

Hillesland, Jr., T. and E.R. Weber. 1988. Solar central receiver advancement for utility application, _Proc. 10-th ASME Solar Energy Conf._, L. M. Murphy and T.R. Mancini, ed., Denver, CO: 153-160.

Howell, J.R., and R.B. Bannerot. 1977. Optimum solar collector operation for maximizing cycle work output. _Solar E. 19_: 149-153.

Jensen, C., H. Price and D. Kearney. 1989. The SEGS power plants: performance, _Proc. 11-th ASME Solar Energy Conf._, A.H. Fanney and K.O. Lund, ed., San Diego, CA: 97-102.

Kreider, J.F. 1979. _Medium and high temperature processes._ New York: Academic Press.

Kreith, F, and J.F. Kreider. 1978. _Principles of solar engineering_. New York: McGraw-Hill.

Labus, T.L., R.R. Secunde and R.G. Lovely. 1989. Solar dynamic power for space station Freedom, _Space Power_ _8_:97-114.

Lund, K.O. 1989. General thermal analysis of serpentine-flow flat-plate solar collector absorbers. _Solar Energy_ _42_: 133-142.

Lund, K.O. 1986. General thermal analysis of parallel-flow flate-plate solar collector absorbers. _Solar Energy_ _36_: 443-450.

Mozurkewich, M., and R. S. Berry. 1982. Optimal paths for thermodynamic systems: The ideal Otto cycle. _J. Appl. Phys. 53_: 34-42.

Namkoong, D. 1989. Thermal energy storage flight experiments, _Proc. 11-th ASME Solar Energy Conf._, A.H. Fanney and K.O. Lund, ed., San Diego, CA: 19-24.

Reynolds, W.C., and H.C. Perkins. 1977. _Engineering thermodynamics_, New York: McGraw-Hill.

Salamon, P., Y. B. Band and O. Kafri. 1982. Maximum power from a cycling working fluid. _J. Appl. Phys. 53_: 197-202.

Salamon, P., and A. Nitzan. 1981. Finite time optimizations of a

Newton's law Carnot cycle. _J. Chem. Phys._ 74: 3546-3560.
Stine, W.B., and R.W. Harrigan. 1985. _Solar energy fundamentals and design_. New York: J. Wiley.
Wu., C. 1988. Power optimization of a finite-time Carnot heat engine. _Energy_ 13: 681-687.
Yan, Z, and J. Chen. 1989. An optimal endoreversible three-heat-source refrigerator. _J. Appl. Phys. 65_: 1-4.

Non-Lorentz Cycles in Nonequilibrium Thermodynamics

Mary Jo Ondrechen
Department of Chemistry
Northeastern University
Boston, MA

ABSTRACT

Treatments of non-Lorentz systems operating in finite time
are reviewed. Three systems are described: A reversible engine
driven by an exothermic chemical reaction; A chemical synthesis
which is preheated to accelerate the rate of reaction, and; A
Newton's law engine working between finite heat reservoirs. Two
features of non-Lorentz systems affect the operation of processes
in finite time: The finite size, and hence finite heat capacity,
of heat sources and sinks, and; Finite rate of heat generation,
when the heat source is a chemical reaction. Both of the
features impose limits on the efficiency and power production,
and must be taken into account in the establishment of optimum
operating procedures for such systems.

INTRODUCTION

In many classical treatments of reversible engines, it is
assumed that there exist two infinitely large, isothermal
reservoirs. Heat is removed from the hot reservoir to drive the
engine, yet the temperature of this hypothetical reservoir never
decreases. As the engine produces work, waste heat is deposited
into the hypothetical cold reservoir, the temperature of which
never increases. Cyclical engines working between such
reservoirs are called Lorentz cycles.

Likewise, in most of the treatments presented in the past
fifteen years of the thermodynamics of finite-time processes, it
is assumed that the sources and sinks of heat are isothermal.
Progress in the study of processes operating with time or rate
constraints and with irreversibilities has been the subject of
recent reviews (Andresen 1984a, Andresen 1984b, Orlov 1985).

Model systems have been developed which incorporate such effects
as friction, heat loss, inertial effects and finite rate of heat
conductance. Energy and entropy flow in these systems has been
investigated, with particular emphasis on the derivation of
bounds on process variables. Process paths that yield extremal
values for work or heat have been calculated for a variety of
model finite-time processes. Less attention has been given to
cycles with constraints on the source of heat.

In the present review we focus on processes operating with
constraints on time or rate, with irreversibilities present, and
with the special feature that there are physical and/or rate
constraints on the source of the heat which drives the process.
Sometimes it is reasonable to treat the working substance as a
weak perturbation on the reservoirs, and therefore assume
isothermal reservoirs. However, it is often the case in practice
that heat reservoirs are highly non-isothermal (Ondrechen 1981,
d'Isep 1982). The combustion products in internal combustion
engines, for example, have finite heat capacities and during the
power stroke are cooled to temperatures well below the initial
post-reaction temperature. Although many real energy conversion
cycles employ non-isothermal heat reservoirs, only a few
treatments of the thermodynamics of such non-Lorentz cycles
operating in finite time have been given to date.

Two features of non-Lorentz cycles are stressed here: 1) The
finite heat capacity of any heat reservoir of finite size, and;
2) The finite rate of heat generation in systems driven by a
chemical reaction. Both of these factors affect the maximum
efficiency obtainable by a process operating with a time or rate
constraint. Furthermore, the path required to achieve some
desirable extremum, such as maximum power, is dependent upon
these features of the source of heat for the process.

Three model processes are described below. The first is an
exothermic chemical reaction which generates heat at a finite
rate and drives a reversible heat engine. The second is a system
in which chemical work is done; the reaction mixture is preheated
to accelerate an Arrhenius-law chemical synthesis. Finally, we
examine a Newton's Law cycle operating between finite-sized heat
reservoirs.

The principal goals of this work are to establish the
general, natural bounds on the efficiency, power, and chemical
efficiency for irreversible, finite-time processes, and to
establish general operating principles for systems which serve as
models for real energy conversion processes.

ENGINES DRIVEN BY HEAT GENERATED AT A FINITE RATE
Consider a reversible engine driven by heat generated by an
exothermic chemical reaction, depicted schematically in Figure 1.
The reactants **R** are passed through a flow tube of finite length l

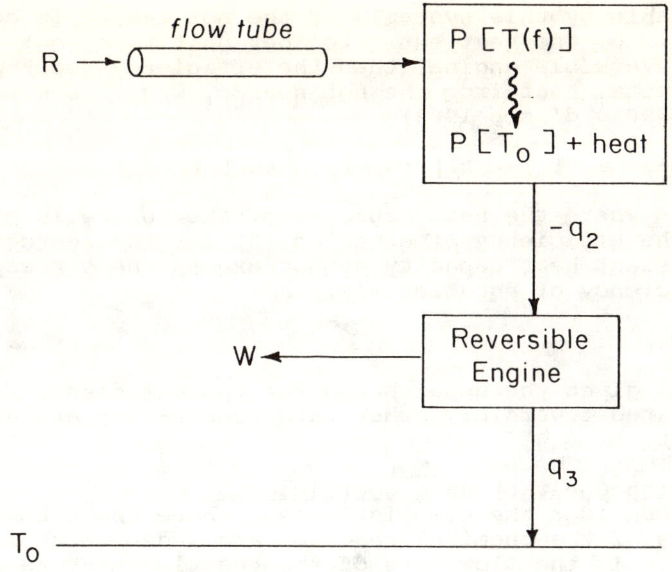

Figure 1. A reversible heat engine driven by an
exothermic chemical reaction.

at a flow rate f to produce products **P**. The reaction begins when
the feed mixture enters the flow tube and stops when the mixture
of products and unreacted starting material exits from the end of
the tube. The product mixture emerges at temperature T(f), and
serves as the hot reservoir for a reversible engine. The product
mixture is cooled to T_o and the heat per unit time q_2 extracted
upon cooling is transferred to the engine at a variable
temperature which lies between T(f) and T_o. The reversible
engine works between this finite hot reservoir initially at T(f)
and a very large cold reservoir at T_o. The engine produces power
w and waste heat per unit time q_3.

Since heat is extracted from the hot source to drive the
engine, the finite-sized source clearly does not stay at the
temperature T(f) and therefore the corresponding Carnot
efficiency given by:

$$\eta_C = 1 - T_o/T(f) \tag{1}$$

is unobtainable by this system. If the hot source is cooled down
to T_o and all of the heat extracted per unit time (q_2) is used to
drive the reversible engine, then the efficiency (work produced
divided by total heat from the hot source, w/q_2) is given by
(Ondrechen 1981, d'Isep 1982):

$$\eta_f = 1 - T_o[T(f)-T_o]^{-1}\ln[T(f)/T_o] \tag{2}$$

for the case where the heat capacity of the source is constant.
Note that the efficiency of equation (2) for the source with
finite, constant heat capacity cannot exceed the corresponding
Carnot efficiency of equation (1):

$$\eta_f \leq \eta_C \tag{3}$$

Equation (2) gives the upper bound for the efficiency for the
engine operated reversibly, when rate constraints are not taken
into account.

<u>Reactions with Constant Rate Coefficients</u>
     First consider the simplistic case where the rate
coefficients of the chemical reaction are independent of
temperature. If the flow rate of the reacting mixture is very
slow, the reaction goes to completion in the flow tube, the
maximum heat per mole of starting material is produced, and the
product mixture emerges at maximum temperature. Thus at slow
flow rate, the efficiency is high (see equation 2), but the heat
transferred to the engine per unit time is slow. Hence the power
produced by the system is low for slow flow rates. On the other
hand, when the flow rate is fast, the reaction does not have time
to go to completion inside the flow tube and the reaction mixture
emerges at a temperature below its maximum. Therefore, both
efficiency and power production fall off at very fast flow rates.

     For reactions following first-order, first-order reversible,
second order and second-order bilinear kinetics, the maximum
power for the system is achieved at a finite, positive flow rate
(Ondrechen 1980a). The maximum fuel efficiency (work produced
divided by reactants consumed, w/f) and the minimum entropy
production are both attained in the uninteresting limit of zero
flow rate.

     As an example, consider the case where the reaction $R \rightarrow P$
obeys first-order kinetics (<u>i.e.</u> $d[R]/dt = -k_1[R]$). The extent
of reaction, defined as the product concentration divided by the
initial reactant concentration $[P(t)]/[R(0)]$, may be expressed as
a function of the dwell time t' in the flow tube as:

$$\epsilon(t') = 1 - \exp(-k_1 t') \tag{4}$$

If the product mixture is assumed to have constant heat capacity
C and the inlet temperature of the feed mixture is assumed to be
T_O (the same as the cold reservoir), the power produced may be
written as:

$$w(t') = CT_O S l D_m \{u\epsilon(t') - \ln[1+u\epsilon(t')]\}/t' \qquad (5)$$

where S is the area of the reaction tube, D_m is the molar density
of the reaction mixture, and u is the molar heat of reaction in
reduced units, given by:

$$u = Q_m/CT_O \qquad (6)$$

Approximate expressions for the optimum dwell time, corresponding
to maximum power, may be obtained for reactions following first-
or higher-order kinetics (Ondrechen 1980a).

Arrhenius Law Chemical Reactions
     We now turn attention to the more realistic situation where
the rate coefficient for the chemical reaction, which generates
the heat to drive the engine, obeys an Arrhenius law, given by:

$$k = A \exp(-E_a/T) \qquad (7)$$

where k is the rate coefficient, E_a is the activation energy, and
T is the temperature of the reaction mixture. The gas constant
is absorbed into the activation energy.

     Again we use the model system of Figure 1. At any point x
inside the reaction tube, the time t it takes for the reacting
mixture to travel from the inlet point (x=0) to x is assumed to
be directly proportional to x, for fixed flow rate. Since the
temperature increases as the mixture travels through the tube,
the rate coefficient k is dependent upon t. In this case, the
extent of reaction becomes:

$$\epsilon_A(t) = 1 - \exp[-\int k(\tau)d\tau] \qquad (8)$$

at time t, where the integration is performed from 0 to t. Of
course, in an exothermic, Arrhenius law reaction, a kind of
cooperative effect is created; the heat of reaction increases the
rate of reaction, which in turn produces heat at a faster rate.
As in the previous system, the total dwell time of the reaction
mixture inside the reaction tube is given by:

$$t' = S l D_m/f \qquad (9)$$

and is inversely proportional to the flow rate f.

     For the engine driven by an Arrhenius law reaction, the
power production may be written as:

$$w(t') = Q_m S1D_m \; \epsilon_A(t')$$
$$\cdot \{1 - T_o[T(t') - T_o]^{-1} \ln[T(t')/T_o]\} \qquad (10)$$

(Ondrechen 1980b). As in the previous (constant coefficient)
system, the power function reaches its peak at some finite,
positive dwell time, and goes to zero as the dwell time
approaches zero or infinity. However, in the present (Arrhenius
law) case, the power is strongly dependent on the activation
energy E_a, except in the limit of very long time. Small changes
in the activation energy cause significant shifts both in the
value of the maximum power and in the dwell time required to
achieve that maximum (Ondrechen 1980b).

In the present system, the temperature profile inside the
flow tube is also very sensitive to the flow rate or dwell time.
Figure 2 illustrates the temperature, in K, of the reacting
mixture as a function of the distance traversed inside the tube,
measured in units of the tube length 1, for a reversible engine
driven by an exothermic Arrhenius law reaction. In this figure
we take E_a = 5.0 kcal/mole, Q_m/CT_o = 1, and T_o = 300 K. It is
assumed that **R** and **P** have the same heat capacity C, which is also
assumed to be independent of temperature within the temperature
range of interest here. The temperature profile is plotted for
three different values for the dwell time: 500, 1000 and 2000,
measured in units of $S1D_m$ and therefore equal to f^{-1}. The dwell
time t = 1000 is very close to that corresponding to maximum
power for the chosen activation energy E_a = 5.0 kcal/mole. The
value chosen for the molar heat of reaction (Q_m/CT_o = 1) results
in a maximum obtainable temperature of 600 K. When the flow rate
is adjusted so that the dwell time corresponds to maximum power,
the reacting mixture is just approaching its maximum obtainable
temperature as it exits the flow tube. This is the case for t =
1000 in the present example. When the dwell time is longer than
that corresponding to maximum power, the maximum temperature is
attained before the mixture emerges from the end of the tube, and
therefore time is wasted. (See the t=2000 curve in Figure 2.)
Furthermore, this maximum temperature is achieved in a sudden
burst inside the tube, where the slope of the temperature profile
is sharply positive. At fast flow rates, when the dwell time is
less than that corresponding to maximum power, such a burst does
not occur within the tube.   The maximum temperature is not
achieved, as in the t=500 curve of Figure 2.

In the present system, maximum fuel efficiency, defined as
w/f (or wt, with t in units of $S1D_m$), is achieved at zero flow
rate, or infinite time. Operation of the system at maximum power
of course requires some sacrifice of fuel efficiency. For the
example shown in Figure 2 at dwell time of about 1000, which
corresponds to peak power, the fuel efficiency is about 90% of
that obtainable at infinite time.

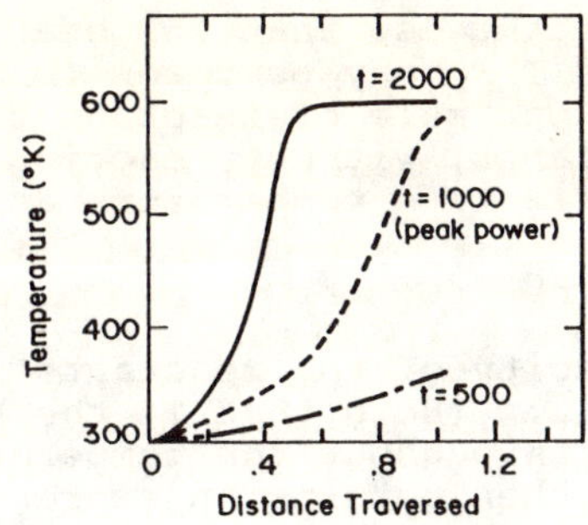

Figure 2. Temperature inside the reaction tube as a function
of distance for an exothermic Arrhenius law reaction.

The systems described in this section illustrate how rate
limitations in the source of the heat that drives an engine limit
the power production of the system. Likewise, operation of the
system at non-zero power imposes limits on the fuel efficiency.
In the next section we examine another system which exhibits a
trade-off between speed and efficiency.

CHEMICAL SYNTHESES
In this section a system doing chemical work in finite time
(Ondrechen 1980b, Termonia 1981) is discussed. The particular
system considered here is an Arrhenius law chemical synthesis
which is preheated in order to produce products at a suitable
rate.

Suppose that the product **P** is synthesized from a reactant **R**
by a reaction of arbitrary order in **R** and in any reaction
intermediate. Let the rate coefficient obey an Arrhenius law,
given in equation (7). It is assumed that the activation energy
E_a is independent of temperature. Suppose further that the
reaction mixture initially at temperature T_o is preheated to a
higher temperature T in order to increase the rate of reaction.
The amount of product made per unit of heat supplied by outside
sources, called the chemical efficiency, is defined as:

$$\xi = q^{-1}(dP/dt) \tag{11}$$

where q is the amount of heat supplied to the reaction mixture
per unit time.

We now wish to find the temperature T_{opt} to which the
reaction mixture should be preheated in order to maximize the
chemical efficiency. Let T' be a temperature close enough to the

optimum operating temperature so that the heat capacity of the
reactants between T' and T_{opt} may be taken to be constant. Let
Q' be the heat required per mole of reactant to increase the
temperature of the reactants, without reaction, from T_o to T'.
We may then define an effective temperature as:

$$T_{eff} = T' - Q'/C \qquad (12)$$

where C is the heat capacity of the reactants in the temperature
range between T' and T_{opt}. (Note that if the heat capacity of
the reactants is linear throughout the temperature range T_o to
T', then $T_{eff}=T_o$.) One then obtains an expression for T_{opt} as:

$$T_{opt} = [E_a + (E_a^2 - 4T_{eff}E_a)^{1/2}]/2 \qquad (13)$$

which corresponds to a local maximum in the chemical efficiency
(Ondrechen 1980b). Note that in general, $E_a \gg T_{eff}$. The
chemical efficiency also has a singularity at approximately T_{eff},
but this corresponds to no preheating; presumably the rate of
reaction is unacceptably slow at that low temperature. Upon
preheating to T_{opt}, the rate of reaction is approximately
$\exp(E_a/T_o - 1)$ times faster than at T_o. Hence preheating becomes
increasingly advantageous for reactions with higher energy
barriers. An energy investment is required to obtain products at
a suitable rate.

In the systems discussed in the previous section and in this
section, rate restrictions are introduced by the kinetics of a
chemical reaction. In the next section, rate restrictions enter
the problem in the form of finite rates of heat conductance.

THE NON-LORENTZ NEWTON'S LAW CYCLE
In this section we examine a heat engine operating in a
cycle with a finite heat source and with a time constraint on the
engine cycle. The system consists of three subsystems: a heat
source with finite heat capacity; a heat engine operating
cyclically, and; a heat sink which may be finite or infinite.
The system is assumed to be endoreversible, _i.e._ the internal
degrees of freedom are in equilibrium within each of the three
subsystems; the only irreversibilities occur at the boundaries
between the subsystems. The system is also assumed to be a
Newton's law system, meaning that all irreversibilities except
for heat resistances are neglected. Thus frictional losses,
inertial effects and heat leakage are neglected.

In one engine cycle, the working substance is placed in
contact with the hot heat source at time t=0. Then at a later
time t_s, called the switch time, that contact is broken and the
working fluid is placed in contact with the cold heat sink until
$t=t_f$, for a total of two contacts per cycle. It is assumed that
the switch in contact occurs instantaneously. The theorem that
the optimal cycle consists of only two contacts was proved

recently (Kuznetsov 1985). The fixed time t_f is allotted for one
engine cycle, so that the working fluid returns to its initial
state after time t_f has elapsed. Naturally, the heat source and
sink do not return to their initial states.

The heat transfer between the reservoirs and the working
substance is assumed to obey a linear law, written as:

$$q_j = dQ_j/dt = K_j(t)[T_j(t) - T(t)] \tag{14}$$

where $q_j = dQ_j/dt$ is the heat flux from the jth reservoir to the
working substance, $T_j(t)$ is the temperature of the jth reservoir,
$T(t)$ is the temperature of the working substance, and $K_j(t)$ is
the thermal conductivity for heat transfer between the jth
reservoir and the working substance. $K_j(t)$ equals the constant
K_j whenever the working fluid is in contact with the jth
reservoir; at other times $K_j(t)$ equals zero and the heat flux for
the jth reservoir is zero. We label the hot reservoir 1 and the
cold reservoir 2.

We shall first assume that the hot reservoir has finite,
constant heat capacity while the cold reservoir has infinite heat
capacity. Then we shall consider the case where both reservoirs
have finite, constant heat capacity.

Infinite Cold Reservoir
We now wish to calculate the path which the system must
follow to produce maximum power. The total work W produced in a
single cycle is given by:

$$W = \int (q_1 + q_2)\, dt \tag{15}$$

where the integral runs from t=0 to $t=t_f$. For the case where the
cold reservoir is infinite in size, there are two constraint
equations. First of all, the entropy change of the working
substance must be zero for one full cycle, as:

$$\int [(q_1 + q_2)/T(t)]\, dt = 0 \tag{16}$$

where the integral is performed from t=0 to $t=t_f$, and where q_1
and q_2 are given by equation (14), except that $T_2(t) = T_2$ is a
constant for this case where the cold reservoir is infinite. A
second constraint is that energy flowing from the hot reservoir
must be conserved, as:

$$C_1(dT_1/dt) - K_1(t)[T(t)-T_1(t)] = 0 \tag{17}$$

where C_1 is the heat capacity of the hot source, assumed now to
be constant.

The problem now is to find the maximum work W per cycle
given by equation (15), subject to the integral constraint of
equation (16) and the differential constraint of equation (17).
This is obtained employing a modified Lagrangian (Salamon 1982).

The path for the working substance which produces maximum
power has been found to be (Ondrechen 1983) an exponential decay
in the temperature during contact with the hot reservoir (_i.e._ t
between 0 and t_s), as:

$$T(t) = uT_1(0) \ exp[-(K_1/C_1)(1-u)t] \qquad (18)$$

where u is a constant which determines the initial temperature of
the working fluid and which may be optimized (Ondrechen 1983).
The cycle producing maximum power is depicted in Figure 3, which

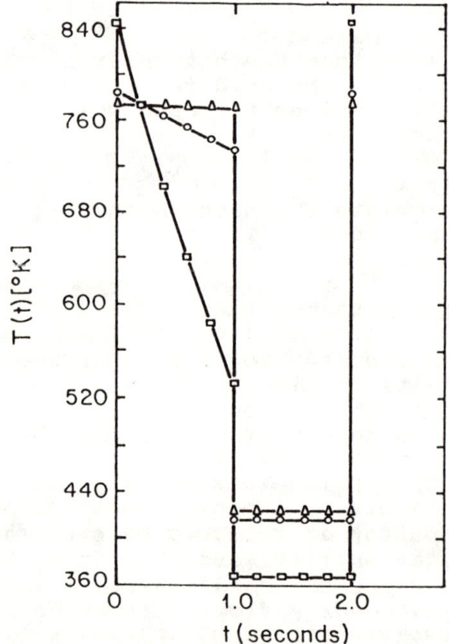

Figure 3. Temperature of the working substance as a function of
time for a non-Lorentz Newton's law cycle with finite source
and infinite sink. Three different values for the source heat
capacity are shown: 1.0 R (□); 10.0 R (o); 100.0 R (△).

148

shows working fluid temperature as a function of time. For this
example, $K_1 = K_2 = 3.0$ R s^{-1}, $T_1(0) = 1000$ K, $T_2 = 300$ K, $t_f =$
2.0 s, and the parameters u and t_s are set at their optimum
values. Three different values for the reservoir heat capacity
are shown: $C_1 = 1.0$ R, 10.0 R, and 100 R, where R is the gas
constant.

Note that, because all irreversibilities except heat
resistances have been neglected in the present model, the
temperature of the working substance is allowed to jump
adiabatically at the points in time when contact is switched (t_s
and t_f). When the working fluid is in contact with the infinite,
isothermal cold reservoir, the maximum power path is an isotherm.
Similarly, when the hot reservoir heat capacity is large ($C_1 =$
100 R), the optimum path for the working fluid approaches an
isotherm during the time when the working fluid and the hot
reservoir are in contact. In this limit of large C_1, the
efficiency, W/Q_1 approaches that for a Newton's law cycle working
between infinite reservoirs, given by (Curzon 1975):

$$W/Q_1 = 1 - (T_2/T_1)^{1/2} \qquad\qquad (19)$$

For smaller C_1, the efficiency corresponding to maximum power is
actually slightly larger than that of equation (19). As C_1
becomes smaller, the working substance sweeps through a much
larger range of temperatures in the maximum power cycle, hence a
small improvement in efficiency is realized.

<u>Both Reservoirs Finite</u>
        When both the hot and cold reservoirs are finite, the
maximum power cycle may be calculated by the same method, except
that there is a second differential constraint, just like
equation (17), for the cold reservoir, which now has finite,
constant heat capacity C_2.

        Figure 4 shows the working fluid temperature as a function
of time for the maximum power cycle for $C_1 = C_2 = 2.0$ R, $T_2(0) =$
300 K, and where the other parameters K_1, K_2, $T_1(0)$ and t_f are
the same as those for Figure 3. When the heat sink is finite,
the low-temperature isotherm is replaced by an exponential rise
in the temperature in the working fluid.

SUMMARY
        Finite heat sources are prevalent in real energy conversion
processes, particularly in systems driven by chemical reactions.
A heat source which is finite in size has finite heat capacity,
and therefore will not stay at constant temperature when heat is
added or extracted. In addition, the generation of heat at a
finite rate will affect the operation of an engine driven by that
heat.

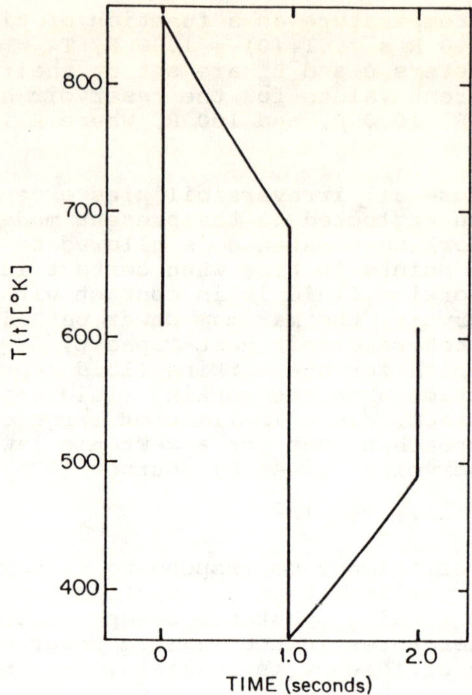

FIgure 4. Temperature of the working substance as a function of
time for a Newton's law cycle with finite source and finite sink.

     For example, for a reversible engine driven by an exothermic
Arrhenius law reaction, maximum power and maximum fuel efficiency
cannot be achieved simultaneously. Some sacrifice in fuel
efficiency is required to produce peak power. Maximum fuel
efficiency requires infinitely slow operation, meaning zero
power. Furthermore, when an exothermic Arrhenius law reaction
takes place in a flow tube, a sudden burst in temperature occurs
inside the tube, provided the flow rate is not too fast. Peak
power for the system shown in Figure 1 is achieved when this
burst occurs just before the mixture exits the tube. For
reactions with high barriers, the reaction front is highly
localized and very sensitive to flow rate. Therefore the system

should not be operated too closely to peak power: a small
fluctuation could displace the burst outside the tube, resulting
in a dramatic drop in power.

In the chemical synthesis accelerated by preheating, we
found a local maximum in the chemical efficiency for a specified
heat input. Heat is expended in order to obtain the desired
product at a reasonable rate. For reactions with high energy
barriers, the rate enhancement obtained from this heat input is a
very sensitive function of the energy barrier.

Finally, we showed that the non-Lorentz Newton's law cycle
operates in a fashion qualitatively different from the
corresponding Lorentz cycle. For smaller values of the source
heat capacity, the working fluid temperature sweeps through an
increasingly wide range of temperatures. Isothermal branches are
eliminated in the maximum power cycle when the heat reservoirs
are finite. Although the maximum work per cycle is a strictly
increasing function of the source heat capacity C_1, the
efficiency is actually slightly higher for smaller C_1's.

Finite heat sources are commonplace features of real energy
conversion systems but are features generally missing from most
of the models studied to date. The present work demonstrates
their importance in the optimal operation of energy conversion
devices.

ACKNOWLEDGMENTS
Acknowledgment is made to National Science Foundation grants
CHE-8820340 and CHE-8700787 and to the Alfred P. Sloan Foundation
for partial support of this work. The author thanks all of the
collaborators with whom she has worked over the years in the area
of nonequilibrium thermodynamics: B. Andresen, Y.B. Band, R.S.
Berry, S. Gozashti, O. Kafri, P. Salamon and M.H. Rubin.

REFERENCES
Andresen, B., R.S. Berry, M.J. Ondrechen, and P. Salamon. 1984a.
    Thermodynamics of processes in finite time. Accounts of
    Chemical Research 17: 266-271.
Andresen, B., P. Salamon, and R.S. Berry. 1984b. Physics Today:
    62-70.
Curzon, F.L. and R. Ahlborn. 1975. Am J. Phys. 43: 22.
d'Isep, F. and L. Sertorio. 1982. Nuovo Cimento B 67: 41.
Kuznetsov, A.G., A.V. Rudenko, and A.M. Tsirlin. 1985. Automation
    and Remote Control 6 (English): 693-705; Avtomatika i
    Telemekhanika 6 (Russian): 20-32.
Ondrechen, M.J., R.S. Berry, and B. Andresen. 1980a.
    Thermodynamics in finite time: A chemically-driven engine.
    J. Chem. Phys. 72: 5118.
Ondrechen, M.J., B. Andresen, and R.S. Berry. 1980b.
    Thermodynamics in finite time: Processes with temperature-
    dependent chemical reactions. J. Chem. Phys. 73: 5838-5843.

Ondrechen, M.J., B. Andresen, M. Mozurkewich, and R.S. Berry. 1981. Maximum work from a finite reservoir by sequential Carnot cycles. <u>Am. J. Phys.</u> <u>49</u>: 681-684.

Ondrechen, M.J., M.H. Rubin, and Y.B. Band. 1983. The generalized Carnot cycle: A working fluid operating in finite time between finite heat sources and sinks. <u>J. Chem. Phys.</u> <u>78</u>: 4721-4727.

Orlov, V.N. and A.V. Rudenko. 1985. <u>Automation and Remote Control</u> <u>5</u> (English): 549-577; <u>Avtomatika i Telemekhanika</u> <u>5</u> (Russian): 7-14.

Salamon, P., Y.B. Band, and O. Kafri. 1982. <u>J. Appl. Phys.</u> <u>53</u>: 197.

Termonia, Y. and J. Ross. 1981. <u>J. Chem. Phys.</u> <u>74</u>: 2339.

Thermodynamics and Economics

B.Å. Månsson
Theoretical Ecology (ATÖ)
Research Center Jülich
Jülich, Federal Republic of Germany

ABSTRACT

Economics, as the social science most concerned with the use and distribution of natural resources, must start to make use of the knowledge at hand in the natural sciences about such resources. In this, thermodynamics is an essential part. In a physicists terminology, human economic activity may be described as a dissipative system which flourishes by transforming and exchanging resources, goods and services. All this involves complex networks of flows of energy and materials. This implies that thermodynamics, the physical theory of energy and materials flows, must have implications for economics. On another level, thermodynamics has been recognized as a physical theory of value, with value concepts similar to those of economic theory. I here discuss some general aspects of the significance of non-equilibrium thermodynamics for economics. The role of *exergy*, probably the most important of the physical measures of value, is elucidated. Two examples of integration of thermo-dynamics with economic theory are reviewed. First, a simple model of a steady-state production system is used to illustrate the effects of thermo-dynamic process constraints. Second, the framework of a simple macro-economic growth model is used to illustrate how some thermodynamic limitations may be integrated in macroeconomic theory.

1. INTRODUCTION

The history of thermodynamics began with an economic question, how much mechanical work a given quantity of coal (or heat derived from burning coal) may yield [Carnot 1824]. Before that, there where attempts to derive an economic theory from a 'physical' basis when the physiocrats (see, e.g., [Weulersee 1931]) tried to use agricultural production as a basis for economic value. (In that setting, the physical limits may be described in terms of finite area, alternatively in terms of finite amounts of solar radiation.) Given the state of the physical sciences at that time, it is not surprising that these efforts failed.

Thermodynamics imposes some limits on what is possible in a process, and thereby implies which physical resources are scarce. In this role, it is obviously relevant for 'the science of best use of scarce resources,' as economics is often defined. However, in the interregnum after the physiocrats, thermodynamics suffered the same fate as almost all other parts of the natural sciences, i.e., it was ignored by economic theory. A few exceptions, mainly concerned with spurious energy theories of value, are noted by Georgescu-Roegen [1986]. Only in the late 1960s were the connections between the physical sciences and economics again taken seriously. To be fair, many of the numerous laws of the physical sciences are of small economical significance. Also, some prominent economists, e.g., Koopmans [1977], have suggested that economists should study the effects of the more important constraints that nature imposes on a production system.

There is one notable exception in the tragicomic history of interaction between economics and the natural sciences. In the bio-economic fields (agriculture, forestry, and fishery) it has been impossible to ignore all biological aspects. However, these fields have been peripheral to the main body of economic theory. Also, even in this context, academic economists have most frequently used *convenient* knowledge from, e.g. ecology, in favour of accurate and pertinent. On the other hand, this is an area of rapid development at present, especially in the fields of 'environmental economics' and 'ecological economics'. Also, the literature on the consequences of incorporating some knowledge from the natural sciences in economic models is fast expanding. Many basic issues for an integration effort were examined by Ayres [1978]. Berry, Salamon and Heal [1978], Berry and Andresen [1982], Lesourd [1985], and Islam [1985], examined consequences of the thermodynamic laws within the framework of microeconomic production theory. The present subject, thermodynamics and economics, was also discussed by, e.g., Bryant [1982] and Ayres and Nair [1984].

Economic processes in a society always involve *matter, energy, entropy,* and *information.* The aim of many economic activities is to achieve a certain *structure.* The recently improved knowledge of non-equilibrium thermodynamic processes, in particular that involving structure formation, could therefore be relevant. Nevertheless economists presume that they can deal with all this at any level, whether it be company, regional, national, or international, without even basic knowledge of thermodynamics. One result of this hubris is that influential but erroneous or obscure descriptions of resource use, resource conversions, and technological options, have inhibited the recognition of available possibilities.

It must also be remembered that whereas the physiocrats did not have access to any useful natural science, the situation today is totally different. This is notably so in the case of thermodynamics, in particular if *statistical mechanics* and *information theory* are included.

From the beginning information theory was close to physical theory [Brillouin 1956], with entropy as a common concept, and today it works and provides useful concepts for many fields of science. Undoubtedly, information theory has much more to give than is possible to discuss here.

Most questions in the social sciences and the humanities involve human communication: how persons, groups or cultures express themselves, how messages are conveyed in the society, how they are interpreted or could be interpreted, interaction between persons or groups through exchange of information. If new concepts are formed within information theory, they may be relevant also to these questions. Therefore, information theory is not only a basis for thermodynamics, it also carries a potential to bridge the gap between the 'two cultures' [Snow 1964], on one hand the natural sciences, mathematics, and technology, on the other the social sciences and the humanities.

Spiritual life needs matter. A psyche may influence and even control material events, but it also needs matter and energy for its very existence. In particular, it needs non-equilibrium matter/energy. We can go on: the material system nurturing the psyche needs more: it needs structure, internally correlated as well as with reference to the outside world. The actions of these physical systems must also be coordinated.

On the surface of our Earth, the contrast between the hot sun and the cold space is the prime prerequisite for emergence and maintenance of structures, some of which takes highly intricate forms as in the social or material infrastructures of human societies. On a large scale, our planet may be loosely described as a generalized heat engine, using energy to build and maintain structures. Obviously, then, thermodynamics is relevant for 'the economics of the spaceship earth' [Boulding 1966]. Also, all economic activity is embedded in the natural flows of energy and materials on the surface of the earth. It is therefore incongruous that economics is still essentially separated from the physical, geophysical and biological sciences, despite numerous excellent efforts made to bridge such gaps [Clark 1976, Ayres 1978, Berndt 1978, Dasgupta and Heal 1979, Daly 1980, Chapman and Roberts 1983, Hall et al. 1986, Cleveland 1987, Faber et al. 1987, Martinez-Alier 1987].

Our civilization is global: the technology and administrative practices of the industrialized societies are spreading all over our world. This involves radical changes, affects all societies and all ecosystems, even the biosphere itself. On the other hand, there is no global culture with due respect and responsibility for the heritage that we have in common with our fellow living beings, the biosphere. To be viable, such a culture must have coherent (trans-disciplinary) scientific knowledge and scientific inquiry as essential cognitive elements; in this, thermodynamics and economics are necessary (but not sufficient).

The foregoing discussion now begs the question: given that part of the theory of thermodynamics should be used in economic theory, which facets of it are most relevant? Depending on the context, one or more of the following six points is likely to be part of the answer.

1. Thermodynamics can be used to identify and quantify important constraints for a wide variety of systems. The economically most significant physical constraints are the laws of mass conservation (matter is conserved in any chemical process) and the first (energy is conserved) and second (entropy increases in all closed systems) law of thermodynamics.

2. Thermodynamics identifies and provides essential relationships between important process variables, e.g., in industrial production processes. This is obviously important for the part of economic theory which deals with production functions.

The examples reviewed here are mainly concerned with the first two points.

3. Thermodynamics provides clear and quantitative measures of efficiency, in particular it elucidates how close to the best possible level a certain process is or could be.

4. Thermodynamics can be used to evaluate (quantify) losses due to irreversibility. In economic terms these may be described as costs.

The first four points are contained in the 'engineering economics' tradition initiated by Carnot. Two currently active branches in this tradition are reviewed in this volume, *thermoeconomics* and *finite time thermodynamics*.

5. Thermodynamic theory is necessary to establish criteria for doing meaningful comparisons between different kinds of energy, as ought to be done regularly in energy economics and, e.g., net energy analysis [Thomas 1977, Gilliland 1978] or energy policy discussions.

6. The modern theory of thermodynamics, with its intimate relationship with information theory, has established an evolutionary perspective in the physical sciences, with *synergetics*, *structure formation*, and *evolution* as central concepts. Some parts of this perspective have already been applied in the social sciences [Weidlich and Haag 1983, Haag 1989], but the possibilities are by no means exhausted. The perspective itself is great and inspiring, despite the fact that the concepts and ideas of information theory are at present sufficient only for the lowest—but all the more fundamental—steps of the evolutionary staircase.

In the concept of evolution there is the notion of *emergence*, in the sense that entirely new features appear. The first step in the evolution process is the step from nowhere and never into space-time and from nothingness into existence. Implicit in this is often a notion of *value*: 'higher' systems—systems of higher value—emerge out of 'lower' systems, and value is in some sense being created. Part of the information carried by an evolving physical system is information on how the system can be maintained or reproduced. In society, science, technology and cultural traditions, play similar roles. The most important economic system, the market economy, is by its very nature evolutionary. The evolution is then primarily a process of reallocation of resources between industries, often caused by technical innovation. This leads to structural changes and keeps the system in *economic* non-equilibrium (not to be confused with thermodynamic non-equilibrium, regardless of some formal analogies).

Economics and values

The question, 'what gives a commodity value?' has different answers in economics, depending on pespective. Often, value or worth of a commodity is derived from some kind of price. In an older tradition, which may be related to the physiocrats, stored-up or *embodied* labour is used as measure of value [Smith 1776, Ricardo 1817, Marx 1867]. Surprisingly, the value concepts used in most of mathematical economic theory are this unsophisticated, see, e.g., [Debreu 1959]. This is closely connected to the use of extremely reduced representations of human beings, organizations, etc. (It also makes the connection with *physical* values so immediate in the sense defined below.)

Value in an economic sense may be an instrument for increasing 'value' in a philosophical sense. In this context, it should be remembered that these values are never totally immaterial. Moral and economic values may in fact be closely tied to very substantial flows of natural resources and to the way the resource-handling systems are embedded in their societal, biotic and abiotic environments. In the end what economic value is, or should be, is closely connected to some *physically* describable service given to human beings.

2. THERMODYNAMICS AND PHYSICAL VALUE

A theory of value based on physics usually involves defining 'values' which are not determined by valuing processes of human actors. A physical object, e.g., an amount of energy, may then be valueable because it can achieve a change in a physical system, i.e., it has instrumental value (or value in use).

Many physically based theories of value use some form of embodiment hypothesis, according to which the value of a commodity derives from how much of some physical quantity has been spent (directly of indirectly) to produce it. The most well-known among these are the numerous energy theories of value (for discussions see, e.g., [Cottrell 1953, Odum 1971, Huettner 1976, IFIAS 1978, Georgescu-Roegen 1979, Costanza 1980, Roberts 1982, Judson 1989]). If the results of these efforts are anything to go by, it is *malapropos* to try to reduce all aspects of value, including economic and philosophical, to a *single* physical standard, such as energy or entropy. Nonetheless, physical concepts which have physical value in the above sense are important for building a general (physico-)economic theory. Foremost amongst these is the *exergy* concept, often denoted 'available' or 'useful' energy [Gibbs 1873, Gouy 1889].

2.1 Exergy

The thermodynamic *exergy*, the amount of mechanical work which may be (reversibly) extracted from a system, has proven to be quite fundamental [Eriksson et al. 1987], also far outside the engineering context in which it was born [Carnot 1824]. It is so constructed that it combines energy and entropy, it takes into account both the first and the second laws of thermodynamics [Gibbs 1873, Maxwell 1875, Rant 1956, Ahern 1980, Moran 1982, Spiegler 1983, Kotas 1986]. Exergy, in contrast to energy, can be consumed; it must be spent in any irreversible process.

2.2 The Cosmic Creation of Exergy

The idea that exergy is an important physical resource and the fact that it is being destroyed naturally leads to the question 'how was it created?'

Our present picture of the universe is that it came into being, together with space-time itself, out of nothing. Like there is no question of 'outside' the universe in space, there is also no question of what was before the start of the universe, the Big Bang. One feature of this picture is that radiation and high temperature matter (together with antimatter) was 'bought' by the universe, during a phase of rapid expansion, inflation, at the expense of a negative energy per unit volume of the growing space.

If the matter/antimatter energy is considered more valuable than the field energy of volume itself, then the inflationary phase also involves creation of value. There may be other phases in the very early universe which for similar reasons can be viewed as value-creating. Eriksson et al. [1982] focused on a process which occured later, after the antimatter had annihilated and the excess matter from an earlier imbalance was present in the form of nucleons. During the era of seconds to hours after the Big Bang, the nuclear exergy was formed which much later came into use as the fuel of the stars. In particular this is the fuel of the Sun, and hence of life on Earth and of human activities. Some of the fuel used by human societies is not taken from the Sun immediately, but has been accumulated on the Earth in the chemical non-equilibrium between the oxygen of the atmosphere and the hydrogen and hydrocarbons of the fossil fuels beneath the surface of the Earth.

An obvious value of the cosmic nuclear exergy lies in the fact that it is being made available by the stars. They transform it into radiation with an equivalent temperature of a few thousand degrees, available to water molecules and to the rich variety of carbon compounds.

2.3 Value of Exergy and Structure

Exergy may be described as a general-purpose 'fuel' which can be used for carrying out various physical processes. From the point of view of thermodynamics exergy is therefore a measure of physical value in the sense above, it can be used to produce changes in physical systems.

The transition from one form of exergy into another, chemical into structural, is discussed in another contribution to this volume. It may be viewed as a process in which value is created—if structural exergy is viewed as more valuable than mere chemical exergy, i.e., if the value concept is further extended to include such things as valuation of the provision of a directing framework for further processes. Thus, if exergy in the form of chemical non-equilibrium is considered as a fuel or a raw material, a means to achieve structure, then self-organization is a production process! However, exergy is only the 'fuel' for structuring matter or for maintaining structure.

A major demand for exergy supply in an economy comes from the requirement that the lost exergy be replaced. A car which loses kinetic energy due to friction, gets it replaced from the exergy contained in the fuel. A house that leaks heat to its environment in winter, gets this heat replaced from electricity or fuel via the heating system. An exergy-rich material which is used for items that are worn out or dispersed through corrosion is replaced by the exergy-demanding processing industry.

In processing industries exergy is used to improve the quality of the processed material. The simplest aspect of the processing of a raw material may be the chemical one, the change in composition. The next aspect is a structural one, e.g., how a solid material is composed of fibers of various size, and within what limits parameters describing this are allowed to vary. Next is the shape of components, etc. In fact, we normally attribute a much higher economic value to structural information, reproducing a prescribed structure, than we do to exergy itself. At room temperature, the 'natural' unit of exergy connected to one bit of information, $kT \ln 2$, is approximately 3×10^{-21} J. For most processes the exergy used to transfer one bit of information is much larger. The exergy content in the shape of a tea cup, for example, is extremely small. The dissipation in the actual production process is clearly many orders of magnitude larger. Even if the exergy of a structure is very small, it can still be quite valuable. Self-organizing systems are interesting and in this sense valuable not because of their exergy content, but because of their content of structure. Thus, as with other thermodynamic concepts, it is futile to try to reduce *all* aspects of value to one standard, exergy. Exergy is necessary to build structure, but structure is more than exergy and has values which are not mirrored by exergy.

3. THERMODYNAMIC SYSTEMS AND ECONOMIC SYSTEMS

An important obstacle when introducing thermodynamics into economics is that the theories operate with different system sizes and aggregation principles. Thermodynamic theory is most effective on small length scales, below the size of a living cell, and on very large scale, of the order of the size of the Earth. Economic theory deals mainly with systems on intermediate spatial scales, from human beings up to a country or a group of countries.

The theories meet best on the scale of industrial production processes, with thermodynamics providing, e.g., the theories of heat engines and chemical reactions, and economic theory concepts such as capital and production function. This is the home ground of 'engineering economics'. The extensive literature dealing with production functions takes many kinds of 'technical constraints' into account, sometimes even what may be identified as thermodynamic constraints. Many such technical constraints are, however, not determined using any laws of physics.

Georgescu-Roegen [1971, 1976, 1979, 1986] has argued in favour of using *entropy* as the basic physical variable in economic analysis. His most serious error lies in ignoring the fact that electromagnetic radiation can carry entropy. This means that the Earth can (and of course does) export its entropy production to the cosmos. Most of this entropy production occurs when the radiation from the sun strikes the surface of the Earth, and this without producing anything else. Only a minor part of the solar exergy is utilized. It is therefore perfectly conceivable (thermodynamically) to have an Earth with the same in- and outflows of entropy as the present, but with the entropy of the planet itself being considerably lower than today. This also means that the dissipation of materials, which Georgescu-Roegen finds so worrying, is of very little significance. Similarly, an economic analysis solely in terms of *energy* is also misleading, since it ignores the important second law. An appropriate physical concept for a model in which the laws of thermodynamics are to be fulfilled is exergy.

3.1 Exergy and the Spaceship Earth Economy

When coining the phrase 'Economics of the Spaceship Earth', Boulding [1966] pointed out that the closed Earth of the future requires different economic principles, by which man may find his place in a cyclical ecological system without unlimited reservoirs for anything. It is a task for economists to ponder what those principles might be, and also to consider how such a system might be made to work. In this, knowledge from the physical sciences must be used.

To make any *economic* model physically meaningful, a minimal requirement is that the laws of thermodynamics are not broken. Note that mass conservation is more difficult (though in principle possible, see, e.g., [Eriksson 1984]) to include in a non-simplistic way, especially in aggregated models.

For a truly *macro*economic model, which encompasses the entire economy of the Earth, thermodynamics provides some hard facts. Since the availability of exergy is a fundamental limiting factor for any production system, an ultimate limiting factor for the spaceship Earth economy is the total solar exergy flow impingning on the Earth, approximately $1.7 \cdot 10^{17}$ W, supplemented by the deposits available in the Earth's crust. Exergy is thus what Koopmans [1973, p. 250] sought in vain for, an "Achilles heel of resource supply, a specific resource that is in limited supply, essential to life and welfare, used dissipatively, and with no substitute in greater supply".

4. THERMODYNAMIC FLOWS AND ECONOMIC VALUES

I have stated that exergy can be given an 'economic' interpretation as a measure of physical value. I shall here pursue this line of reasoning a little further, by considering an economy which is looked upon during a time interval short enough so that it may be treated as stationary.

The physical problem is formulated as the problem of finding the maximal amount of exergy flow (maximal amount of mechanical power) from a given set of sources and sinks, with stationary flows, restricted to certain intervals. In the next stage, the production system provides a flow of services from stationary exergy flow.

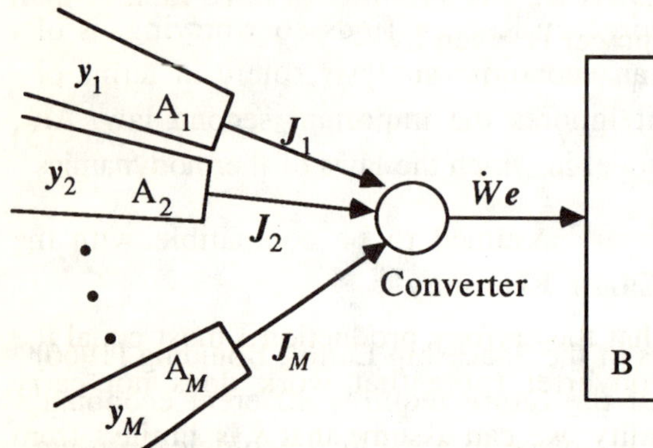

Figure 1. A set of stationary flows to or from distinct systems (parts of the resource base) enter or leave a converter, which delivers mechanical work to a work reservoir B. The work stored in B may then be used as a 'fuel' by production processes to yield services, e.g., in the form of material goods.

Consider a set of separate systems (resources) $A_1,..., A_M$, each of which can yield (or absorb) a stationary flow of extensive quantities (Figure 1). These flows are then vectors $J_1,..., J_M$ in the r-dimensional space of extensive quantities. Assume that $M \geq r+1$, and that each flow carries constant proportions of these quantities, described by unit vectors e_m with non-negative components,

$$J_m = P_m e_m ; e_m \geq 0 ; e_m^2 = 1; \quad (m = 1,..., M), \tag{1}$$

where P_m are flow intensity variables. The (fixed) intensive variables of the flows are assumed to be $y_1,..., y_M$ respectively. The flows are counted positive going *into* the converter. The flow intensities are assumed to be restricted to certain intervals of non-zero widths,

$$\check{P}_m \leq P_m \leq \hat{P}_m \; ; \quad \check{P}_m < \hat{P}_m \; . \tag{2}$$

The source of these restrictions may be, e.g., availability constraints (for flow resources), capacity constraints (for deposited resources), or environmental constraints (for emissions).

The work extracted per unit time (the *power*) into the work reservoir B is assumed to be $\dot{W}e$, where e is a fixed unit vector characteristic for B, thus

$$\sum_{m=1}^{M} J_m = \dot{W}e, \tag{3}$$

which implies the $r-1$ conservation conditions

$$\sum_{m=1}^{M} (e_m - \rho_m \, e)P_m = 0, \tag{4}$$

where $\rho_m = e_m \cdot e$. (The $(M \times r)$-matrix $e_{m\alpha}$ is assumed to have rank r; then $e_{m\alpha} - (e_m \cdot e)e_\alpha$ has rank $r-1$.) The power is given by

$$\dot{W} = \sum_{m=1}^{M} \rho_m P_m \tag{5}$$

The restrictions (2) on the flows are assumed to be compatible with the conservation conditions (4).

Thermodynamics prescribes that the entropy production $\dot{s}$ must equal the net entropy outflow from the converter (note that work does not carry entropy). With little loss of generality, we can assume that $\dot{s}$ is limited from below by a convex function X of the flows. (A higher speed in the conversion processes may lead to higher entropy production.) We get

$$\dot{S} = - \sum_{m=1}^{M} v_m P_m \geq X(\mathbf{P}), \tag{6}$$

where the specific entropy of each flow is $v_m = y_m \cdot e_m > 0$ and $\mathbf{P} = (P_1,..., P_M)$.

The physical optimization problem is then to maximize the linear form (5) subject to the constraints (2), (4) and (6). Since X is convex, the problem has a solution on the boundary of the volume determined by the constraints.

The solution is found [Eriksson et al. 1987] by the Lagrange multiplier method, with the Lagrangian

$$L_{\text{phys}} = \sum_{m=1}^{M} [\rho_m P_m + \breve{\mu}_m(P_m - \breve{P}_m) + \hat{\mu}_m(\hat{P}_m - P_m) +$$

$$+ s \cdot (e_m - \rho_m e) P_m - \tau v_m P_m] - \tau X(\mathbf{P}), \tag{7}$$

where the multipliers satisfies $\breve{\mu}_m$, $\hat{\mu}_m$, $\tau \geq 0$ and $s \cdot e = 0$. Here, the optimal multiplier τ^* (* indicates optimal value) can be interpreted as a reference temperature.

In the stationary economic model, the converter can now be interpreted as driving a combined production and distribution system [Eriksson 1984]. The extracted power $\dot{W}e$ is used to provide a flow of (reasonably composed and distributed) *services* [Ayres 1978]. We assume the service mix to be of certain fixed proportions. The total service flow C is assumed to be variable but limited by the power $\dot{W}$,

$$C \leq \frac{1}{\lambda} \dot{W} \quad (\lambda = \text{constant} > 0). \tag{8}$$

To evaluate the benefits derived from the services, it is customary in economic theory to introduce a *utility flow $U(C)$*. Its first derivative, the 'marginal utility', is assumed to be positive (more service is better) but decreasing,

$$U'(C) > 0; \quad U''(C) < 0. \tag{9}$$

To maximize the utility flow U we can now use an economic Lagrangian which contains the physical Lagrangian (7),

$$L_{\text{phys}} = U(C) - \gamma \lambda C + \gamma L_{\text{phys}}, \tag{10}$$

where γ is a Lagrange parameter, and where ≥ 0 and

$$\gamma(-\lambda C + L_{\text{phys}}) = 0. \tag{11}$$

The service production is said to be *technically efficient*, when (8) holds with equality. Since $L_{\text{phys}} = \dot{W}$, (11) is then satisfied independently of γ. If we assume this to be the case and also assume that $\tau^* \neq 0$ (and connected assumptions about binding conditions and non-degeneracy, see [Eriksson et al. 1987], Ch. 2.4), i.e., that the second law constraint is effective, we can obtain the marginal utility of the various resource flows,

$$\frac{\partial U}{\partial P_m} = \frac{dU}{dC}\frac{1}{\lambda}\frac{\partial \dot{W}}{\partial P_m} = \gamma\left(\kappa_m^* - \tau^* \frac{\partial X(\mathbf{P})}{\partial P_m}\right). \tag{12}$$

Here γ can be identified as the marginal value of exergy, i.e., the utility increase per added unit of exergy. We have here introduced a generalized Carnot factor $\kappa_m^* \equiv (s^* + e^* - \tau^* y_m)\cdot e_m$, which is the exergy content of the mth flow, and the last term in the bracket is the marginal exergy loss (through entropy production) per unit flow.

The service flow may also be limited directly by the flows of energy and materials, e.g., by a constraint

$$aC \le \sum_{m=1}^{M} P_m l_m , \tag{13}$$

where a expresses the requirement of energy and materials, and where l_m expresses their technical availability in various flows. Then the above discussion has to be slightly modified, so that (13) is be taken into account through an extra term in the Lagrangian with a corresponding vector of Lagrange multipliers. The result of this after slight redefinitions is that the marginal utility of the mth flow, besides the terms (12) based on exergy, also obtains a term

$$p\cdot l_m , \tag{14}$$

where p is readily interpreted as an energy/materials *price* vector, reflecting scarcity in the restriction (13), i.e., those components of p vanish which correspond to strict inequality (no scarcity) in (13).

> Different limits hold for the use of energy, materials, water, and land. Many of these will manifest themselves in the form of rising costs and diminishing returns, rather than in the form of any sudden loss of a resource base. But ultimate limits there are, ...
>
> *Our Common Future* [World Commission on Environment and Development 1987, in the chapter "Towards Sustainable Development", p. 45.

5. THERMODYNAMICS AND SUSTAINABILITY

In the second example of usage of non-equilibrium thermodynamics in economic theory, thermodynamic constraints are introduced in a simple macroeconomic model, constructed for analysis of optimal sustainable growth

[Månsson 1986]. A basic idea is to use the 'ultimate limits' set up by thermodynamics on a Spaceship Earth Economy to develop some insight into the
operation of a sustainable economy. (There are of course many other kinds of
ultimate limits, not included here.) Physical constraints are taken into account
through the use of exergy as a resource measure.

The model is constructed in the spirit of Boulding [1966] and Meadows et
al. [1972] (and perhaps Georgescu-Roegen [1971], but I use exergy instead of
entropy as the central physical concept). The economy, which is not a steady-
state economy, is based on one essential, inexhaustible but limited, flow
resource. There are many similar models in the economic literature, though
most of these study exhaustible natural resources, see, e.g., [RES 1974,
Peterson and Fischer 1977, and Smith and Krutilla 1982].

5.1 The Production Function

The state variables of the economic system are the stock of fixed *capital*, K,
measured in terms of exergy, and a '*technology* stock', T, measured in some
suitable dimensionless unit. Being first approximations, these are obviously
quite unsophisticated concepts in comparison with some the ones in, say,
economic growth theory (see, e.g., [Hicks 1939, 1965, 1973, Hoselitz 1960,
Hahn 1971, Solow 1988, Reid 1989]). With technology (technological knowledge) I mean not only a simple understanding of 'how things work', since
theoretical knowledge must be incorporated in labor skills or machine designs
to influence efficiency. Some technology is assumed to be embodied in the
physical capital, reflecting the view that knowledge must have a material basis.
Capital may also include parts of nature, since the environment possesses some
of the characteristics of a capital good [Faber et al. 1987]. The *resource* intake
into the economy, R, is measured in exergy units.

The production system is described in terms of the three factors of
production, capital, technology, and resource intake. The production process
transforms the resource inflow into consumption and investment goods. All
factors of production are presumed fully utilized. The production processes
are described by means of a (dimensionless) gross exergy efficiency, E, which
is a function of technology, physical capital and resource intake. The product
flow (production function) can be written on the form

$$P(R,T,K) = R \cdot E(T,F) \tag{15}$$

where F is the capital per unit resource flow, $F \equiv K/R$ (i.e., twice as many
units of capital can handle twice as much resource flow with the same thermodynamic efficiency). The idea that the quality of the capital depends on the

technology is taken into account by means of the efficiency function.

We use here a physically defined efficiency concept, relating the exergy in- and out-flows of the production process. This efficiency is sometimes called 'effectiveness' or 'second-law efficiency', see, e.g., [Berndt 1978] and [Gaggioli 1980, 1983]. According to thermodynamics

$$0 < E(T,F) \leq 1. \tag{16}$$

The efficiency function is assumed to be smooth and to satisfy $0 < E(0,0)$, $0 < E_T < \infty$, and $0 < E_F < \infty$. (Standard notation for partial derivatives is used throughout, e.g., $\partial W/\partial V = W_V$, and ´ denotes differentiation with respect to the argument of a function of one variable.) Note that to achieve an efficiency close to unity, large stocks of capital and technology are necessary. The minimum saving requirement, to compensate for the depreciation, is then a high proportion of the total output.

Conditions on the second order partial derivatives of the efficiency function would enhance mathematical tractability and enable certain existence proofs, but there is really no valid physical foundation for such assumptions.

The elasticity of production, $e = 1 + TE_T/E > 1 \ \forall \ T,F > 0$, i.e. there are increasing returns to scale. The assumptions regarding the efficiency function do not guarantee that the elasticities of substitution are non-negative everywhere.

5.2 Dynamics and Control

The development of the system is controlled by three variables, the investment rates, I into capital and J into technology, together with the natural resource intake R (all measured in exergy units). Although it is concievable that capital or technology may be used as resource input, it is convenient to disallow this, i.e., to postulate that

$$I \geq 0 \text{ and } J \geq 0. \tag{17}$$

Stocks of natural resources are not considered. Thus, the resource intake must satisfy the inequalities

$$0 \leq R \leq \hat{R}(K), \tag{18}$$

$$R \leq R^*. \tag{19}$$

The limit in (18) reflects an assumption that the resource intake is constrained by the amount of capital available for harvesting the inflow. R^* is a fixed upper limit for the resource intake, e.g., the total solar inflow in a solar-based economy. Part of the resource inflow can be harvested directly without using any capital, i.e., $\hat{R}(0) > 0$, and further that $\hat{R}'(K) > 0$, and that there exists a finite capital stock size K^* defined through $\hat{R}(K^*)=R^*$.

To maintain a certain level of knowledge in a society costs a certain amount of resource input, since the existing level of knowledge, and thereby technology, will deteriorate if nothing is invested in education, training, research and development (cf. [Schuler 1979]). The technology can only increase if more than this amount is invested. There is of course also deterioration of capital, and the embodied technology will deteriorate with the corresponding capital.

I assume proportional depreciation of capital and technology, with constant depreciation rates β_K and β_T, respectively. With γ denoting the (constant) technology increase per invested unit of exergy we have the equations of motion for the state variables

$$\dot{K} = -\beta_K K + I, \tag{20}$$

$$\dot{T} = -\beta_T T + \gamma J + h(\beta_K K, I), \tag{21}$$

where the function h is introduced to take the embodiment of technology in capital into account. Generally, we assume that $h(\beta_K K=I) = 0$, $h(\beta_K K<I) > 0$, and $h(\beta_K K>I) < 0$.

We see immediately that a capital stock greater than R^*/β_K is impossible to maintain; similarly, there is an upper limit for technology, $T < \gamma R^*/\beta_T$.

5.3 Consumption and Utility

The production can be used either for investments or consumption. Assuming no wastage, the consumption rate is

$$C = P - I - J. \tag{22}$$

Whereas the production system could work with zero capital and technology, there is no guarantee that the production would suffice to cover the minimum requirement to sustain the (constant) population under such circumstances.

A conventional assumption in neo-classical economic theory is to use a utility function which is a function of the aggregated consumption only, ignoring, e.g., considerations about the distribution among individuals. Such an approach allows us to concentrate on the effects of physical constraints in a reasonable and simple economic model. The primary disadvantage is that it is dangerously close to a simplistic exergy theory of value, since the consumption is measured in terms of exergy. As usual, the utility rate $U(C)$ is assumed to satisfy the conditions (9).

5.4 Sustainability, Operation Space, and Stationary States

The physical constraints prescribes that this kind of economy must operate in the volume V in the K-T-R-space (Figure 2) where the capital, technology, and resource intake are non-negative, and where the constraints (18) and (19) are fulfilled. Since $P(0,0,0) = \hat{R}(0){\cdot}E(0,0) > 0$ there is always some production. However, this may be less than the minimum necessary consumption.

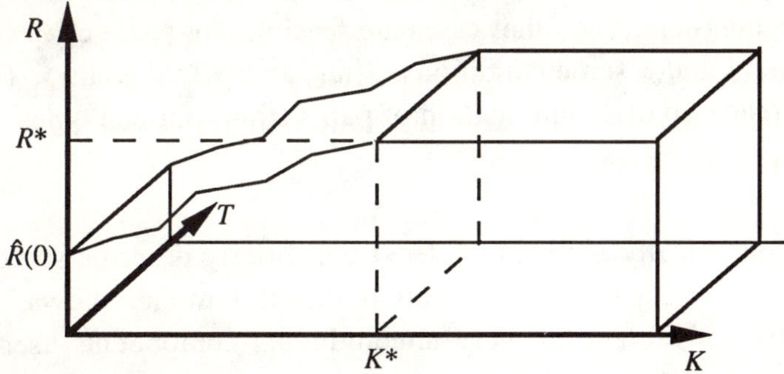

Figure 2. The operation volume V in K-T-R-space. The outer boundaries on the K and T axes are due to the depreciation. Strictly speaking, this also means that the entire volume is not available for a sustainable development.

5.5 The Optimization Problem

The objective of the optimization is to maximize the total utility from the present time, $t = 0$, to some time, t_f, which is either some chosen finite planning horizon or infinity. The economy is assumed to operate within a Ramsey [1928] type of ethics, with no discounting of the utility. Labor and population are treated as constants. Assuming a unique stationary state in which the consumption is larger than in all other stationary states, with consumption C^*, and using it as the final state of the optimal path, the quantity to be maximized is

$$W = \int_0^{t_f} [U(C(\tau)) - U(C^*)] \, d\tau. \tag{23}$$

Some properties of optimal paths can be derived from the necessary conditions of Pontryagin's maximum principle (see, e.g., [Intriligator 1971]). The problem of maximizing W, given the equations of motion (20) and (21) and the constraints (17)-(19), is described by an Hamiltonian function with four time-dependent multipliers which can be interpreted as shadow prices on investments and resource intake, respectively, and two costate variables which can be interpreted as shadow prices on capital and technology. The first order necessary conditions of the maximum principle then yield the equations of motion on the optimal path and conditions on the multipliers. The second order necessary condition can be used to show that for some efficiency functions all optimal paths involve using the maximum allowed resource intake.

Four variables involving the multipliers can each be either zero or positive, giving in all 16 possible combinations (cases). Two of these cases are always internally inconsistent. The other cases are feasible, for some choices of efficiency function and under some conditions. Thus, an optimal solution (if it exists) is generally made up of a combination of paths from fourteen types.

5.6 Results

Primarily for the economically inclined reader, I will briefly describe some of the results. Perhaps the most important result is that this model shows how some mathematically convenient but very unconvincing components used in earlier growth theory models are made unnecessary when physical constraints are taken into account. One example is the assumption that production can grow without limit through (exponentially growing!) technological progress.

The solution of the optimization problem [Månsson 1986] yields a number of properties of the optimal path segments whose union makes up the optimal path leading from any inital state to the final state. The necessary conditions are in fact quite stringent. It is easy to prove that the final state in the optimization cannot lie in the interior of the permitted volume V, in fact, in most cases the resource intake is as large as possible given the constraints (18) and (19). In fact, the optimal resource intake is often quite well specified for any given K and T.

Of the fourteen feasible alternatives, four involve a collapsing economy, where the consumption rate decreases with time. The other ten cases all involve at least one kind of investment.

There are two kinds of optimal path segments: exterior paths where the resource intake is maximal, and interior paths, with no equality in (18) and (19). For both interior and exterior paths there exist conditions for balanced growth, where the developments of the stocks of capital and technology are coupled. If the stocks are unbalanced, there should be no investment at all in the excessive stock.

On all interior optimal paths the elasticities of substitution between resource intake and capital as well as between resource intake and technology are zero; also the marginal productivity of R and the shadow price of the resource intake is zero, i.e., R can be considered as a free good! Optimal paths are excluded from those subvolumes of V where the marginal product of R is negative.

As usual in this kind of model, the general interest rate, r, is positive if the consumption increases with time. It is noteworthy that on optimal paths which involve investments in both stocks, all prices must fulfil the 'Hotelling Rule' (see, e.g., [Dasgupta and Heal 1979, p. 156]).

6. CONCLUDING REMARKS

My primary purpose with this rather superficial survey has been to show that thermodynamics can be useful for economics. This is but a part of a much larger conviction on my part, namely that economics, as the social science most concerned with the use and distribution of natural resources, must start to make use of the knowledge at hand in the natural sciences about such resources. In this, thermodynamics is an essential part.

I am also willing to argue in favour of other parts of the natural sciences, especially information theory (as mentioned above) and ecology. Ecology has a special significance, since it can not only provide hard facts which must be taken into account, but also contains many useful concepts and models, in particular those describing evolution in complex dynamic systems. The connection to economic game theory are obvious. Here, I must caution that economics deals with systems which have essential features which are not present in any ecosystem, e.g, economical actors are in many respects qualitatively different from ecological.

Economics is regarded as a science modelled after physics (in particular classical mechanics). The first example supports this view, showing how closely analogous essential concepts (value, price, cost) are in the two sciences.

In the second example, the model does not include stocks of exhaustible natural resources, e.g., mineral deposits, whereas the production system of today is heavily dependent upon such stocks. Thus, since the model applies to the era after these resources have been exhausted, it is at best only a early step on the road towards 'physico-economics'. Before the exhaustion date (which is not necessarily finite for all of them), the objective is to utilize exhaustible resource stocks optimally, with a 'backstop technology' such as solar energy [RES 1974, Dasgupta and Heal 1979, Spulber 1980].

The examples demonstrate how easy it is to fit knowledge from physics into the existing set of models in economic theory.

ACKNOWLEDGEMENT

I would like to express my sincere gratitude to professor Karl-Erik Eriksson, (Institute for Physical Resource Theory, Göteborg, Sweden) who provided inspiration and guidance for the work reviewed here and who introduced me to some of the fundamental ideas; also to professor Jacqueline McGlade (Theoretical Ecology Working Group, Research Center Jülich, FRG) for her excellent comments to the first draft. Of course, all remaining imperfections are of my own creation.

REFERENCES

Ahern, J.E., 1980, *The Exergy Method of Energy Systems Analysis*, John Wiley & Sons, New York.

Ayres, R.U., 1978, *Resources, Environment, & Economics*, John Wiley & Sons, Inc., New York.

Ayres, R.U., Nair, I., 1984, Thermodynamics and Economics, *Physics Today*, November, 62-71.

Berndt, E.R., 1978, Aggregate Energy, Efficiency and Productivity Measurement, *Annual Review of Energy*, **3**, 225-273.

Berry, R.S., Salamon, P., Heal, G.M., 1978, On a relation between economic and thermodynamic optima, *Resources and Energy*, **1**, 125-137.

Berry, R.S., Andresen, B., 1982, Thermodynamic Constraints in Economic Analysis, in: W.C. Schieve and P.M. Allen, eds., *Self-organization and dissipative structures: Applications in the physical and social sciences*, University of Texas Press, Austin, Texas.

Boulding, K.E., 1966, The Economics of the Coming Spaceship Earth, in: H. Jarret, ed., *Environmental Quality in a Growing Economy*, John Hopkins Press, Baltimore.

Brillouin, L., 1956, *Science and Information Theory*, Academic Press, New York; see also second edition 1962.

Bryant, J., 1982, A thermodynamic approach to economics, *Energy Economics*, **4**, 36-50.

Carnot, N.L.S., 1824, *Réflections sur la puissance motrice du feu et sur les machines propres a développer cette puissance*, Bachelier, Paris; R. Fox, ed., Libraire Philosophique J. Vrin, Paris 1978.

Chapman, P.F., Roberts, F., 1983, *Metal Resources and Energy*, Butterworths, London.

Clark, C.W., 1976, *Mathematical Bioeconomics*, John Wiley & Sons, Inc., New York.

Cleveland, C.J., 1987, Biophysical economics: Historical perspective and current research trends, *Ecological Modelling* **38**, 47-73.

Costanza, R., 1980, Embodied Energy and Economic Valuation, *Science*, **210**, 1219-1224.

Cottrell, F., 1953, *Energy and Society*, McGraw-Hill, New York.

Daly, H.E., 1980, ed., *Economics, Ecology, Ethics—Essays toward a Steady-State Economy*, W.H. Freeman and Company, San Francisco.

Dasgupta, P.S., Heal, G.M., 1979, *Economic Theory and Exhaustible Resources*, Cambridge University Press, Cambridge.

Debreu, G., 1959, *Theory of Value*, Yale University Press, New Haven.

Eriksson, K.-E., 1984, 'Thermodynamical Aspects on Ecology/Economics', in: A.-M. Jansson, ed., *Integration of Economy and Ecology: An Outlook for the Eighties* , Wallenberg Symposium, Stockholm, pp 39-45.

Eriksson, K.-E., Islam, S., and Skagerstam, B.-S., 1982, "A model for the cosmic creation of nuclear exergy," *Nature* **296**, 540. An error in this paper has been corrected in [Eriksson et al. 1987].

Eriksson, K.-E., Lindgren, K., Månsson, B.Å., 1987, *Structure, Context, Complexity, Organization— Physical Aspects of Information and Value*, World Scientific, Singapore.

Faber, M., Niemes, H., Stephan, G., 1987, *Entropy, Environment and Resources—An Essay in Physico-Economics*, Springer, Berlin.

Gaggioli, R.A., 1980, ed., *Thermodynamics: Second Law Analysis* , ACS Symposium Series 122, American Chemical Society, Washington.

Gaggioli, R.A., 1983, ed., *Efficiency and Costing* , ACS Symposium Series 235, American Chemical Society, Washington.

Georgescu-Roegen, N., 1971, *The Entropy Law and the Economic Process*, Harvard University Press, Cambridge, MA.

Georgescu-Roegen, N., 1976, *Energy and Economic Myths*, Pergamon, New York, NY.

Georgescu-Roegen, N., 1979, Energy analysis and economic valuation, *Southern Econ.J.*, **45**, 1023-1058.

Georgescu-Roegen, N., 1986, The entropy law and the economic process in retrospect, *East.Econ.J.* **12**, 3-23.

Gibbs, J.W., 1873, *Collected Works*, Vol. I, p. 53, Yale University Press, New Haven 1948; Originally published in Trans. Conn. Acad., Vol. II, pp. 382-404.

Gilliland, M.W., 1978, ed., *Energy Analysis: A New Public Policy Tool*, AAAS selected symposia series #9, Westview Press, Boulder, Colorado.

Gouy, M., 1889, Sur l'énergie utilisable, *J. de Phys., 2e série*, **8**, 501-518.

Haag, G., 1989, *Dynamic Decision Theory*, Kluwer, Dordrecht.

Hahn, F.H., 1971, *Readings in the Theory of Growth*, Macmillan, London.

Hall, C.A.S., Cleveland, C.J., Kaufmann, R., 1986, *Energy and resource quality—The ecology of the economic process*, John Wiley & Sons, New York.

Hicks, J., 1939, *Value and Capital*, Oxford University Press, Oxford.

Hicks, J., 1965, *Capital and Growth*, Oxford University Press, New York.

Hicks, J., 1973, *Capital and Time*, Oxford University Press, Oxford.

Hoselitz, B.F., 1960, ed., *Theories of economic growth*, Free Press, Glencoe, Ill..

Huettner, D.A., 1976, Net Energy Analysis: An Economic Assesment, *Science*, **192**, 101-104.

IFIAS, 1978, Energy analysis and economics, *Resources and Energy*, **1**, 151-204.

Intriligator, M.D., 1971, *Mathematical Optimization and Economic Theory*, Prentice-Hall, Inc., Englewood Cliffs, N.J.

Islam, S., 1985, Effect of an essential input on isoquants and substitution elasticities, *Energy Economics*, **7**, 194-196.

Judson, D.H., 1989, The convergence of neo-Ricardian and embodied energy theories of value and price, *Ecological Economics*, **1**, 261-281.

Koopmans, T.C., 1973, Some Observations on 'Optimal' Economic Growth and Exhaustible Resources, in: H.C. Bos et al., eds., *Economic Structure and Development* (Essays in Honour of Jan Tinbergen, North-Holland, Amsterdam, 239-255.

Koopmans, T.C., 1977, Concepts of Optimality and Their Uses, *American Economic Review*, **67**, 261-274.

Kotas, T.J., 1986, *The Exergy Method of Thermal Plant Analysis*, Butterworth, London.

Lesourd, J.-B., 1985, Energy and resources as production factors in process industries, *Energy Economics*, **7**, 138-144.

Månsson, B.Å., 1986, Optimal development with flow-based production, *Resources and Energy*, **8**, 109-131.

Martinez-Alier, J., 1987, *Ecological Economics—Energy, Environment, and Society*, Basil Blackwell, Oxford.

Marx, K., 1867, *Das Kapital*, vol. 1.

Maxwell, J.C., 1875, *The theory of heat*, 4th ed., Longmans, Green, and Co., London.

Meadows, D.H., Meadows, D.L., Randers, R. and Behrens III, W.W., 1972, *The Limits to Growth*, Universe Books, New York, NY.

Moran, M.J., 1982, *Availability Analysis: A Guide to Efficient Energy Use*, Prentice-Hall, Englewood Cliffs, New Jersey.

Odum, H.T., 1971, *Environment, Power, and Society*, Wiley-Interscience, New York, NY.

Peterson, F.M., Fisher, A.C., 1977, The Exploitation of Extractive Resources. A Survey, *Economic Journal*, **87**, 681-721.

Ramsey, F.P., 1928, A Mathematical Theory of Saving, *Economic Journal*, **38**, 543-559.

Rant, Z., 1956, Exergie, ein neues Wort für 'technische Arbeitsfähigkeit', *Forschung auf dem Gebiete des Ingenieurwesens* **22**(1), 36.

Reid, G.C., 1989, *Classical economic growth—an analysis in the tradition of Adam Smith*, Basil Blackwell, Oxford.

Review of Economic Studies (RES), 1974, Symposium articles.

Ricardo, D., *The Principles of Political Economy and Taxation*, London 1817.

Roberts, P., 1982, Energy and value, *Energy Policy*, **10**(September), 171-180.

Schuler, R.E., 1979, The Long Run Limits to Growth: Renewable Resources, Endogenous Population, and Technological Change, *Journal of Economic Theory*, **21**, 166-185.

Smith, A., 1776, *An Inquiry into the Nature and Causes of the Wealth of Nations*, Penguin Books Ltd, Harmondsworth, Middlesex, England 1979.

Smith, V.K. and J.V. Krutilla, eds., 1982, *Explorations in Natural Resource Economics*, John Hopkins University Press, Baltimore, Md.

Snow, C.P., 1964, *The Two Cultures and A Second Look*, Cambridge University Press, Cambridge.

Solow, R.M., 1988, *Growth theory: and exposition*, Oxford University Press, New York.

Spiegler, K.S., 1983, *Principles of Energetics*, Springer-Verlag, Berlin.

Spulber, D.F., 1980, Research and development of a backstop energy technology in a growing economy, *Energy Economics*, **2**, 199-207.

Thomas, J.A.G., 1977, ed., *Energy analysis*, Westview Press, Boulder, Colorado.

Weidlich, W., Haag, G., 1983, *Quantitative Sociology*, Springer, Berlin.

Weulersee, G., 1931, *Les physiocrates*, G. Doin & Cie, Paris.

World Commission on Environment and Development, 1987, *Our Common Future*, Oxford University Press, Oxford.

Equipartition of Entropy Production:
A Design and Optimization Criterion
in Chemical Engineering

Daniel Tondeur

Laboratoire des Sciences du Génie Chimique
CNRS-ENSIC-1, 54000 Nancy, France

ABSTRACT

The present text deals with the optimal design or operation of heat and mass transfer processes and develops the following conjecture : *for a given duty, the best configuration of the process is that in which the entropy production rate is most uniformly distributed*. This principle is first analyzed in detail on the simple example of tubular heat exchangers, and within the framework of linear irreversible thermodynamics. A main result is established, which states that the total entropy production is minimal when the local production is uniformly distributed (equipartition). Corollaries then result, which relate the entropy production and the variance of its distribution to economic factors such as the duty, the exchange area, the fluid flow-rates, and the temperature changes.

The equipartition principle is then extended to multiple independent variables (time and space), multicomponent transfer, and non-linear but concave flux vs force relationship. Chemical Engineering examples are discussed, where the equipartition property has been applied implicitly or explicitly : design of distillation plates, cyclic distillation, optimal state of feed, and flow-sheets in chromatographic separations.

Finally, a generalization of the equipartition principle is proposed, for systems with a distributed design variable (such as the size of the various elements of a system). *The optimal distribution of investment is such that the investment in each element (properly amortized) is equal to the cost of irreversible energy degradation in this element*. This is equivalent to saying that the ratio of these two quantities is uniformly distributed over the system, and reduces to equipartition of entropy production when the cost factors are constant over the whole system.

1. INTRODUCTION

The operating costs of industrial chemical or physical processes are usually to a large extent energy dependent. Saving operating costs thus implies reducing energy degradation in the process, in other words minimizing the irreversibilities, that is the entropy production.

In a "classical" approach, this objective is approached by "slowing down" all the processes, which implies either to reduce the production rate (the "duty" of the process) or to make additional investments (in heat transfer area, distillation plates, or volume of reactor for example) if the duty is to be maintained.

The approach adopted here is one of overall economics under finite constraint ; in other words, we account simultaneously for all cost elements (operating and investment), **and** we impose a finite duty on the process. We want to investigate whether some configurations, or some operating modes are more economical than others to achieve the same specified job. It will be shown that not only the total value of the entropy production, but also its **distribution** along the process plays an essential role in this respect.

The simplest and most familiar illustration of this problem is the comparison between one-pass tubular heat exchangers, operating counter-currently or co-currently. The better performance of the former can be related to the more uniform driving force for heat transfer along the tube, which in turn determines the uniformity of the entropy production. We attempt here to generalize this property, to give it a fundamental thermodynamic background, and to apply it to less trivial situations.

This approach is clearly related to the general search for extremum principles, and in that respect, should find synergy with much other material of the present book. Our emphasis is on engineering, and more precisely, on the optimal economic choice of process configurations under finite-time or finite duty constraints, rather than on evolution of given systems.

We shall first focus our attention on the simple tubular heat exchanger mentionned above, and develop a detailed approach for that case, in the framework of linear irreversible thermodynamics. The rest of the paper will be devoted to extending the approach, or to find its limits, for other less trivial processes, for multicomponent systems, for non-linear situations. The presentation made here is largely inspired from (Tondeur 1987), but some new results and developments are presented.

2. SIMPLE CONFIGURATIONS : TUBULAR HEAT EXCHANGERS AT STEADY-STATE

Figure 1 shows the classical one-pass tubular heat exchangers. It is well known experimentally that the counter-current flow of the two fluids (1a) leads to "better performances" than the co-current flow (1b), and classical heat transfer calculations confirm this. If this pragmatic property has some character of generality, one may wonder whether there exists some deep underlying thermodynamic principle. What we propose here is that this underlying principle is that suggested in the introduction : the equipartition of the irreversibilities.

In this paper, we shall try to clarify the meaning of terms like "better performances" and "equipartition of irreversibilities". For the time being, consider that a process is the more irreversible, the farther it takes place from equilibrium. Looking at the temperature profiles along the heat exchangers of Figure 1, we understand that thermal equilibrium would require the temperatures of the two fluids to be equal at all points along the exchanger. In fact, there is everywhere a temperature difference, which acts as a driving force. What we observe is that this temperature difference is distributed differently in the two configurations : it is roughly evenly distributed in the counter-current case (1a) and quite unevenly in the co-current case (1b). Admitting that the irreversibility is measured somehow by the deviation from equilibrium, we may state that the counter-current exchanger is closer to "equipartion of irreversibilities" than the co-current exchanger.

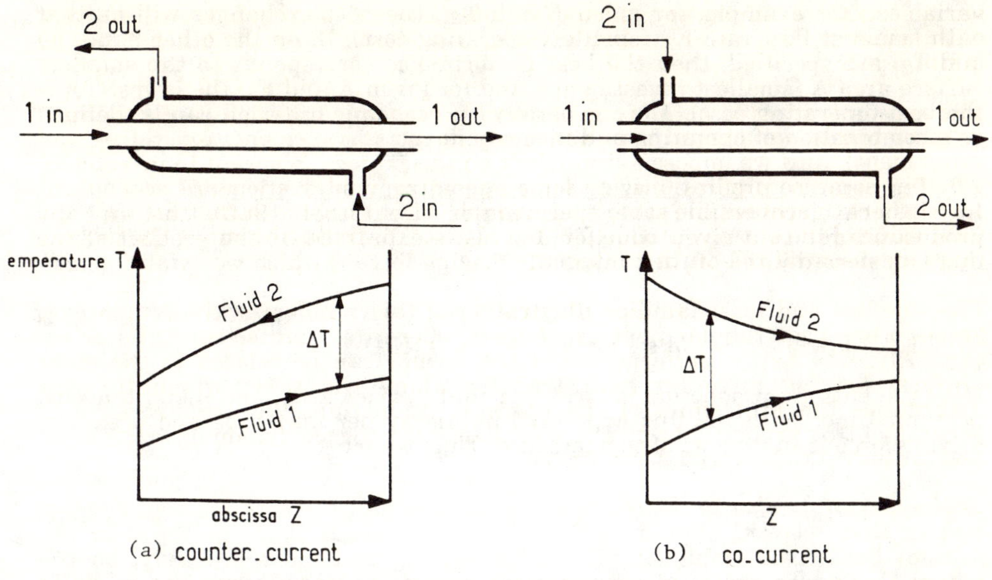

Figure 1. Configurations of one-pass tubular heat exchangers, and corresponding temperature distributions.

In order to set that first observation on more solid ground, we shall now discuss the three following points : in what sense can one say that counter-current flow is better than co-current flow ; what relation is there between the temperature-difference profile and entropy production ; what relation is there between entropy production and performance of the exchangers.

<u>2.1 Performance and duty of heat exchangers</u>
For the sake of simple reasoning, let us assume the two exchangers are geometrically and structurally identical, except for the relative flow direction of the two fluids, and in some cases for the length. In addition we assume for the time being that the overall heat transfer coefficient is the same, and independent of the flow-rates and temperatures, and constant along the tube. We consider steady-state operations.

Let us now specify the **duty** of the exchangers. A realistic way to do this is to specify all the characteristics of the cold fluid (fluid 1) : its flow-rate, F_1, its inlet temperature, T_{1i}, and its outlet temperature, T_{1o}. The amount of heat to be transferred per unit time from the hot to the cold fluid is then specified and equal to $F_1 C_{p1}(T_{1o}-T_{1i})$. This is our **"finite duty constraint"**.

The "free" variables are then the surface area, A, (or the length) of the exchangers, the flow-rate, F_2, and the inlet temperature, T_{2i}, of the hot fluid. The performance of the exchangers will be measured in terms of these variables. For example, for given A and T_{2i}, the best exchanger will be that with smallest flow-rate F_2 (smallest operating cost). If, on the other hand, F_2 and T_{2i} are specified, then the best performance corresponds to the smallest surface area A (smallest investment). And for given A and F_2, the lowest T_{2i} is the best (operating cost). More generally an economic criterion can be defined as a combination of operating and investment costs.

2.2 Temperature profiles, driving force and entropy production

In classical irreversible thermodynamics (De Groot 1962), the entropy production due to a given transfer process is expressed as the product of the flux transferred j and of the conjugate driving force f, which we symbolize as

$$\underset{(J\ K^{-1}\ s^{-1}\ m^{-2})}{\sigma} \quad = \quad \underset{(J\ s^{-1}\ m^{-2})}{j} \quad \bullet \quad \underset{(K^{-1})}{f} \tag{2.1}$$

Here the entropy production is local, per unit surface area, and instantaneous, per unit time. With the flux expressed in energy per unit time and area, the driving force is in inverse of temperature. Thus we let

$$f = \left(\frac{1}{T_1} - \frac{1}{T_2}\right) \tag{2.2}$$

where T_1 and T_2 are local temperatures at any abscissa along the tube. To Equations 2.1 and 2.2, we associate the classical **linear transfer law**

$$j = L \bullet f \tag{2.3}$$

where L is a phenomenological heat transfer coefficient (in units $J\ K\ s^{-1}\ m^{-2}$), which we assume constant in time and along the exchanger. The local entropy production thus appears to vary as the square of the driving force :

$$\sigma = L f^2 \tag{2.4}$$

The total entropy production is obtained by integration over the time and the surface area A of the exchanger. We assume steady-state and consider the total entropy production per unit time P :

$$P = \int_A \sigma\, dA = L \int_A f^2 dA \tag{2.5}$$

Furthermore, we consider that the duty of the exchanger is specified, as discussed above, by fixing the flow-rate and the temperature change of the cold fluid, so that the total amount of heat transferred is specified. We thus write

$$J_{(\text{specified})} = \int_A j\, dA = L \int_A f\, dA = F_1 Cp_1 (T_{1o} - T_{1i}) \tag{2.6}$$

We define the average driving force $\overline{f}$, over the surface area A, by

$$\overline{f} = \frac{1}{A} \int f dA \tag{2.7}$$

so that (2.6) and (2.7) yield

$$J_{(specified)} = LA\,\overline{f} \tag{2.8}$$

It can be seen that for given surface area A and constant L, specifying J amounts to specifying the **average** driving force $\overline{f}$. The question then is whether, for a given $\overline{f}$, there exists a **distribution** of f which minimizes the total entropy production P.

<u>2.3 Minimizing the entropy production</u>
The question above is of course a simple variational problem, of minimizing an integral (Eq. 2.5) subject to a constraint (Eq. 2.6). The solution is furnished by the corresponding Euler equation, in terms of the variable f :

$$\frac{\partial}{\partial f}[f^2 + \lambda f] = 0 \tag{2.9}$$

where λ is a Lagrange multiplier (a constant). Equation 2.9 is satisfied by $f = -\lambda/2$, that is, by a constant value of f. In addition, we have :

$$\frac{\partial^2}{\partial f^2}[f^2 + \lambda f] > 0 \tag{2.10}$$

implying that the extremum is a minimum.

Thus, with a constant transfer coefficient, we conclude the following :

the distribution of driving force which minimizes the entropy production, under the constraint of a specified duty, is a uniform distribution.

Clearly, this also implies (from Eq. 2.3) that the transfer flux j is also uniformly distributed, and so is the entropy production (from Eq. 2.4).

However, in **real** tubular heat exchangers, the driving force **is not constant** along the tube, eventhough it is more uniform in the counter-current case than in the co-current case. The question then is whether two configurations can be compared quantitatively by their deviation from uniformity of driving force or entropy production.

The present analysis is consistent with more classical ones, such as the exergy analysis as presented for example in (Le Goff 1980). However, we expect the present approach to be somewhat more general and, possibly, closer to the roots of thermodynamics.

<u>2.4 Characterizing the deviation from equipartition</u>
Let us designate by P_0 the overall entropy production of a fictitious exchanger, of given surface area A, achieving a specified duty J_0, and such that equipartition of entropy production is exactly verified. In this case, f is constant, equal to its mean $\overline{f}$, and imposed by the specified duty :

$$\overline{f} = J_0/LA \tag{2.11}$$

From Eq. 2.5, we then have :

$$P_0 = L \int f^2 dA = L(\overline{f})^2 = J_0 \overline{f} \qquad (2.12)$$

Consider now a "real" exchanger, of the same surface area A and transfer coefficient L, and achieving the same duty J_0. From Equation 2.11, the average driving force is then $\overline{f}$. Its overall entropy production P is then given by Equation 2.5, but with f variable.

Let us express the deviation of the real exchanger from the equipartition case, by the difference $P-P_0$ of the entropy productions, written explicitly using Equations 2.5 and 2.12 :

$$P-P_0 = L \int f^2 dA - L(\overline{f})^2 = LA\left[\overline{f^2} - (\overline{f})^2 \right] \qquad (2.13)$$

The quantity $\overline{f^2}$ is the mean of f^2 over the surface area A, defined similar to Equation 2.7. The difference in the bracket on the right of Equation 2.13 is the difference between the mean of f square and the square of f mean : this is nothing but the mean quadratic deviation from the mean, in other words, **the variance** $s^2(f)$ of the distribution f, a positive quantity. Thus :

$$P-P_0 = LAs^2(f) = LA \, \overline{(f-\overline{f})^2} > 0 \qquad (2.14)$$

and the equipartition property may be restated as follows :

Any deviation from a uniform distribution leads to an excess of entropy production, and the "dissipativeness" of a configuration may be measured by the magnitude of the variance s^2 of the distribution of the driving force.

Note that this principle may as well be expressed in terms of the flux j, and thus an equivalent expression of Equation 2.14 is :

$$\frac{P-P_0}{A} = Ls^2(f) = \frac{1}{L} s^2(j) \qquad (2.15)$$

where $s^2(j)$ is the variance of the distribution of the transfer flux.

2.5 Entropy production, its distribution and engineering performances

Configurations that minimize $s^2(f)$, or, $s^2(j)$, also minimize entropy production. It may seem intuitive that less dissipative configurations should somehow be better, not only from a thermodynamic point of view, but also from an engineering point of view. However we shall see that it is worthwhile to try to clarify the relationships between these points of view.

For that purpose, let us consider a first set of constraints, and compare two exchangers a and b, operating at steady state and of the same exchange area A, the same transfer coefficient L, and the same total entropy production P, but different distributions of driving force, characterized by $s^2(f)$, such that :

$$s_a^2 < s_b^2 \qquad (2.16)$$

From Equation 2.14, this is equivalent to :

$$\frac{P_a\text{-}P_{oa}}{LA} < \frac{P_b\text{-}P_{ob}}{LA} \tag{2.17}$$

and since by hypothesis, $P_a = P_b$:

$$P_{oa} > P_{ob} \tag{2.18}$$

P_{oa} and P_{ob} are the entropy productions of fictitious exchangers satisfying equipartition ($s^2=0$) and achieving respectively the same duties as exchangers a and b, with the same L and A (these exchangers might use different flow-rates or temperatures of the hot fluid for example). P_o is given by Equation 2.12, so that we may write :

$$(\bar{f}_a)^2 > (\bar{f}_b)^2 \tag{2.19}$$

which entails :

$$|\bar{f}_a| > |\bar{f}_b| \tag{2.20}$$

and from Equation 2.8 :

$$|J_a| > |J_b| \tag{2.21}$$

In practice, this means that for example the cold fluid will be heated more, or a larger flow-rate may be treated, in exchanger a. The final outcome of this calculation is that, **although the two exchangers have the same total entropy production, the one with smallest s^2 will achieve the largest duty.** This result is interesting in that it shows that the distribution of entropy production may play a more important role than the total production itself.

A similar conclusion may be reached in considering exchangers having the same duty, the same total entropy production, but different exchange areas A, and different s^2. Starting again from Equation 2.16, we obtain :

$$\frac{P_{oa}}{A_a} > \frac{P_{ob}}{A_b}$$

thus

$$\frac{(\bar{f}_a)^2}{A_a} > \frac{(\bar{f}_b)^2}{A_b} \tag{2.22}$$

and the duty is :

$$J = LA_a\bar{f}_a = LA_b\bar{f}_b \tag{2.23}$$

Multiplying Equation 2.22 by Equation 2.23, we get :

$$(\bar{f}_a)^3 > (\bar{f}_b)^3 \qquad \text{thus} \qquad \bar{f}_a > \bar{f}_b \tag{2.24}$$

which in turn, from Equation 2.23, entails :

$$A_a < A_b \tag{2.25}$$

Thus the exchanger with the smallest s^2 requires the smallest exchange area.

The reader may investigate other sets of constraints. Here, we should like to discuss an example which requires a slightly different approach. This time, we seek a condition on the hot fluid. We consider again two exchangers a and b, operating at steady-state and at constant pressure, achieving the same duty, and undergoing no exchange with the environment. An overall entropy balance may then be written as

$$\text{exchanger a} \qquad F_a \Delta s_a = -\Delta S + P_a \tag{2.26}$$
$$\text{exchanger b} \qquad F_b \Delta s_b = -\Delta S + P_b \tag{2.27}$$

where F is the flow-rate of the hot fluid, Δs the specific entropy change of that fluid between outlet and inlet, P the total entropy production and ΔS the total entropy change of the cold fluid, which is assumed the same for both exchangers, and determined by the specified duty. The latter is expressed using the specific enthalpy changes of the hot fluid Δh, by :

$$J = F_a \Delta h_a = F_b \Delta h_b \tag{2.28}$$

Assume now that P_a is smaller than P_b. By substracting Equation 2.27 from Equation 2.26, one then obtains :

$$F_a \Delta s_a - F_b \Delta s_b = P_a - P_b < 0 \tag{2.29}$$

Accounting for the fact that Δs is negative, because the hot fluid becomes colder, this entails :

$$F_a | \Delta s_a | > F_b | \Delta s_b | \tag{2.30}$$

Dividing the two sides of Equation 2.28 by the two sides of inequality 2.30, one gets :

$$\frac{\Delta h_a}{\Delta s_a} < \frac{\Delta h_b}{\Delta s_b} \tag{2.31}$$

On a Mollier diagram (enthalpy vs entropy) the terms of this relationship represent the slopes of chords to the constant pressure lines connecting inlet and outlet conditions. The constant pressure lines are convex (i.e. $(\partial^2 h/\partial s^2) > 0$), so that qualitatively, if the inlet conditions of the hot fluid are the same in both exchangers, we have the situation of Figure 2a. Thus inequality 2.31 entails :

$$| \Delta h_a | > | \Delta h_b | \tag{2.32}$$

which , because of Equation 2.28, in turn implies :

$$F_a < F_b \tag{2.33}$$

Hence we can conclude :
The exchanger a, having the smallest entropy production, then requires a smaller flow-rate of hot fluid.

Notice that in the present example, we had to make use of a thermodynamic property of the hot fluid, and this is not unexpected inasmuch as we are seeking a performance index (the flow-rate) relating directly to the use of this fluid. The validity of the conclusion relies on this thermodynamic property (positiveness of $\partial^2 h/\partial s^2$ at constant pressure) which is always verified for pure fluids. For the case of mixture, the situation may well be different.

We might also examine the conclusions that can be reached when a pressure drop is accounted for in the hot fluid. Figure 2b illustrates that Inequality 2.31 then implies Inequalities 2.32 and 2.33 **and/or a smaller pressure drop** for exchanger a. If we consider points 1,2,3,4,5 as possible outlet points for exchanger a, while the outlet conditions of exchanger b are unchanged, we have the following characteristics :

$$
\begin{array}{llll}
1. & |\Delta h_a| < |\Delta h_b| \; ; & F_a > F_b & \text{and } \Delta P_a < \Delta P_b \\
2. & |\Delta h_a| = |\Delta h_b| \; ; & F_a = F_b & \text{and } \Delta P_a < \Delta P_b \\
3. & |\Delta h_a| > |\Delta h_b| \; ; & F_a < F_b & \text{and } \Delta P_a < \Delta P_b \\
4. & |\Delta h_a| > |\Delta h_b| \; ; & F_a < F_b & \text{and } \Delta P_a = \Delta P_b \\
5. & |\Delta h_a| > |\Delta h_b| \; ; & F_a < F_b & \text{and } \Delta P_a > \Delta P_b
\end{array}
\tag{2.34}
$$

3. EXTENSIONS AND LIMITATIONS OF EQUIPARTITION

3.1 Time as a variable. The steady-state of homogeneous systems

In the previous approach, it is easy to see that we can replace the exchange area A of the heat exchanger by some other independent variable, for example time. In that case, we are concerned by the optimal time evolution of an exchanger with a specified duty. Let us illustrate this on the example inspired by (Storck 1989) of an electrochemical cell. We require the electrode to transfer a given amount of electricity Q over a finite but arbitrary time θ, so that :

$$
Q = \int_0^\theta i\,dt = \bar{i}.\theta
\tag{3.1}
$$

where i is the instantaneous electric current, and $\bar{i}$ its average over the interval θ. With the assumption that Ohm's law holds within the cell :

$$
\eta = Ri
$$

where η is the electric potential difference between the electrodes and R an ohmic resistance, the power dissipated irreversibly in the system is :

$$
P = \int_0^\theta \eta i\,dt = R \int_0^\theta i^2 dt
\tag{3.2}
$$

As before, minimizing the power dissipated for specified Q, with θ arbitrary, implies $i = \text{constant} = \bar{i}$ and the deviation from this optimal case is given by an Equation similar to Equation 2.15 :

$$
\frac{P-P_0}{\theta} = Rs^2(i) = \frac{1}{R}s^2(\eta)
\tag{3.3}
$$

Thus, we have the following principle :

The steady-state is the less dissipative regime of a homogeneous reactor performing a specified duty in a finite time.

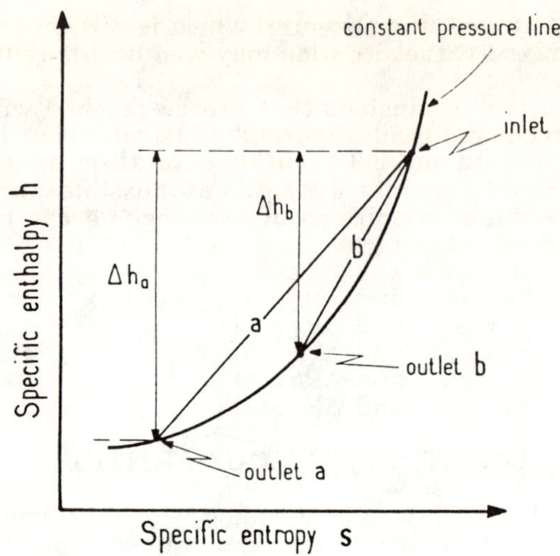

(2a). Illustration of Eqs. 2.31 and 2.32 at constant pressure.

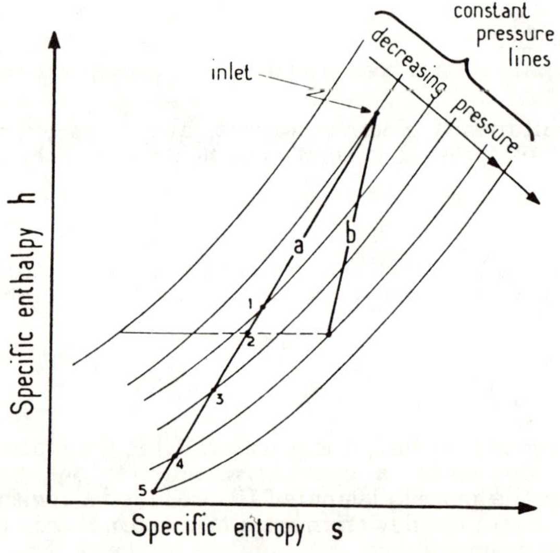

(2b). Illustration of relations 2.34 at variable pressure.

Figure 2. Schematic Mollier diagram for fluid 1.

This result is close to Prigogine's theorem (Prigogine 1947 ; De Groot 1962) concerning the steady-state of open homogeneous systems, subject to constraints on the boundaries.

3.2 Multiple independent variables : time and space

It is easy to see that the previous approaches may be generalized to several space and time variables, including discrete situations. Keeping the electrochemical cell as an example, suppose we have several working electrodes, or equivalently, that the working electrode is divided into N different zones which may work at different potentials. The overall specified duty will then be defined as :

$$Q = \sum_{1}^{N} \int_{0}^{\theta} i dt = \bar{i} N \theta \qquad (3.4)$$

where the time integration is as before, and the discrete summation Σ extends over the N zones. The average $\bar{i}$ so defined is thus taken over time and number of zones (the order of integration summations is indifferent). The total entropy production is expressed as in Equation 3.2, but with the summation over N zones. The rest of the approach follows the same path as before, and if P_0 is the entropy production corresponding to $i = \bar{i}$ at all times and on all zones, we have :

$$\frac{P\text{-}P_0}{N\theta} = Rs^2(i) \qquad (3.5)$$

Hence,
The minimal dissipation for a specified duty corresponds to equipartition of flux, driving force and entropy production along the time and space variables of the process.

3.3 Multicomponent systems

Staying within the framework of Onsager's linear irreversible thermodynamics, the instantaneous and local fluxes and entropy production are expressed as :

$$\mathbf{j} = [\mathbf{L}].\mathbf{f} \qquad (3.6)$$

$$\sigma = \mathbf{f}^T.\mathbf{j} = \mathbf{f}^T.[\mathbf{L}].\mathbf{f} \qquad (3.7)$$

where fluxes and forces are vectors, σ is a scalar, $[\mathbf{L}]$ is the matrix of (constant) phenomenological coefficients, a symmetric matrix (owing to Onsager's relations) with positive diagonal elements. All products of tensorial quantities used here are inner (dot) products (the order of the product is the sum of the order of the factors, minus twice the number of dots). The total entropy production is then the integral of σ over time and space. An overall duty may be specified as :

$$\mathbf{J} = \int \mathbf{j} dV = \int [\mathbf{L}].\mathbf{f} dV = V[\mathbf{L}].\bar{\mathbf{f}} \qquad (3.8)$$

$\bar{\mathbf{f}}$ is the average driving force vector, defined by Equation (3.8) such that its components $\bar{f}_i$ are the averages of the individual driving forces f_i (the

integration over the variable V symbolizes integration over time and space).
Defining P_0 as before as the entropy production for the case where $\mathbf{f}$ is constant
and equal to $\bar{\mathbf{f}}$, we express the "excess entropy production" as :

$$P-P_0 = \int \left[\mathbf{f}^T.[\mathbf{L}].\mathbf{f} - \bar{\mathbf{f}}^T.[\mathbf{L}].\bar{\mathbf{f}} \right] dV \qquad (3.9)$$

To show that this quantity is positive, we notice that $[\mathbf{L}]$ being a positive
symmetric matrix has a unique symmetric "square root" $[\mathbf{R}]$, so that the
quadratic forms in the above integral may be written :

$$\mathbf{f}^T.[\mathbf{L}].\mathbf{f} = \mathbf{f}^T.[\mathbf{R}].[\mathbf{R}].\mathbf{f} = (\mathbf{R}.\mathbf{f})^T.(\mathbf{R}.\mathbf{f}) \qquad (3.10)$$

Letting :

$$\mathbf{e} = [\mathbf{R}].\mathbf{f}$$

the excess entropy production becomes :

$$P-P_0 = \int (\mathbf{e}^T.\mathbf{e} - \overline{\mathbf{e}}^T.\overline{\mathbf{e}})dV = \left(\overline{\mathbf{e}^T.\mathbf{e}} - \overline{\mathbf{e}}^T.\overline{\mathbf{e}} \right)V \qquad (3.11)$$

The latter quantity is positive, by the Cauchy-Schwarz inequality (Bass 1968),
and we have again $P-P_0 > 0$, thus $P_0 = Min(P)$ and the **less dissipative system is
that where all driving forces, and thus entropy productions, are evenly
distributed along time and space.**

Equation (2.15) may be generalized if one defines the covariance of the scalar
driving forces as :

$$cov(f_i,f_j) = \overline{f_i f_j} - \overline{f_i}\ \overline{f_j} \qquad (3.12)$$

(when i=j, we obtain the ordinary variance). Then, by developing $\mathbf{f}^T.[\mathbf{L}].\mathbf{f}$, it is
easy to show that the excess entropy production is :

$$\frac{P-P_0}{V} = \sum_i \sum_j L_{ij} \cdot cov(f_i,f_j) \qquad (3.13)$$

This multicomponent generalisation is formally interesting, but it is not clear
as yet whether it has any practical usefulness. The main reason is that it is
doubtful whether one would want to specify the duty, in a multicomponent
transfer process, by Equation 3.8. In general, one would put a specification on
the flux of one or two components. The approach has then to be revisited. In
addition, going from minimum dissipation to engineering criteria as was done
in the heat transfer case is likely to be more difficult. Probably practical
examples need to be worked out to evaluate this approach.

3.4 Non-linear irreversible thermodynamics (NLIT)
Clearly, an important issue is whether the equipartition theorem breaks down
when the linear phenomenological relations cease to be valid. The following
derivations are from (Latifi et al. 1990). We write quite generally :

$$P = \int j.f dV \qquad J = \int j dV = J_o = V\,\overline{j} \qquad\qquad (3.14)$$

with

$$f = f(j) \qquad\qquad or\ j = j(f)$$

As the only conditions on these relationships, we require :

$$jf > 0 \quad ; \quad f' = \frac{df}{dj} > 0 \quad ; \quad j = 0 \quad for \quad f = 0$$

and we form the Lagrangian :

$$L(j) = P + \lambda(J\text{-}J_o) = \int (jf + \lambda j - \lambda\,\overline{j}\,)dV \qquad\qquad (3.15)$$

and the Euler equation, corresponding to an extremum of P :

$$\frac{\partial L}{\partial j} = \int (jf' + f + \lambda)dV = 0 \qquad (V\ arbitrary) \qquad\qquad (3.16)$$

which entails :

$$jf' + f + \lambda = \frac{\partial}{\partial j}[(f + \lambda)j] = 0 \qquad\qquad (3.17)$$

thus

$$(f + \lambda)\,j = constant \qquad\qquad (3.18)$$

This equation has two types of solutions :

$$a)\quad j = constant \quad ; \quad f = constant$$

$$b)\quad f = \frac{k}{j} - \lambda$$

Solution b is incompatible with the requirements on f and j mentioned above, and is therefore not an acceptable solution.

Again we find that P is stationary when flux and force, thus entropy production, are uniformly distributed. It remains to investigate whether this stationary value is a minimum. This is done by examining the sign of the second derivative :

$$\frac{\partial^2}{\partial j^2}[(f + \lambda)j] = f''j + f' \qquad\qquad (3.19)$$

Since j and f' are positive, this quantity is always positive when f'' > 0 ; in other words, when f is a convex function of j, or j a concave function of f (Fig. 3b). We can therefore extend the result obtained in the framework of Onsager's linear thermodynamics and state :

When the flux j is a linear or a concave function of the driving force f ($\partial^2 j/\partial f^2 < 0$), then equipartition of entropy production corresponds to minimal dissipation.

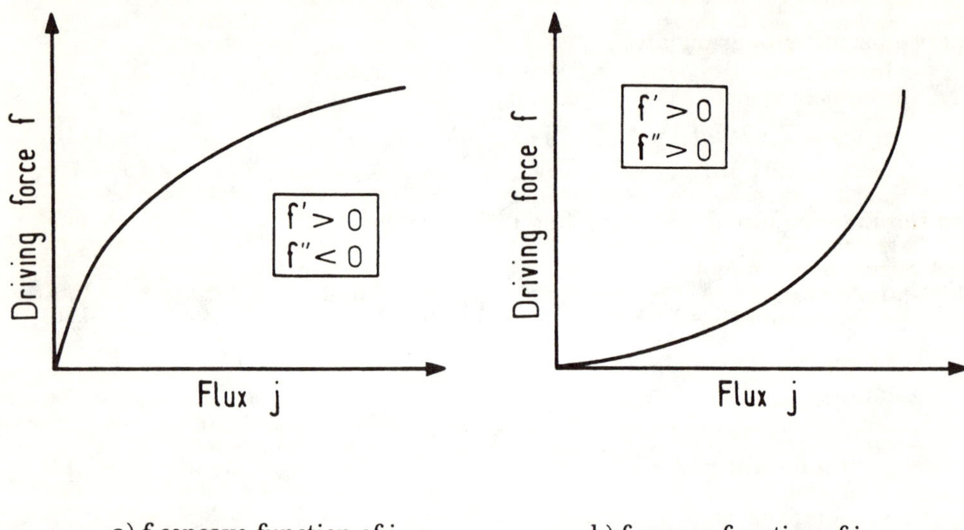

a) f concave function of j   b) f convex function of j

Figure 3. Concavity and convexity of force vs flux relationship.

3.5 Convex flux : the trend toward heterogenity

When f " is negative in Equation 3.19, then the sign of the second derivative may be either positive or negative, and may actually change along the process. Thus no immediate conclusion can be drawn without a numerical examination. What may happen when $\partial^2 L/\partial j^2$ is negative ? Clearly, the entropy production is no longer minimal but maximal. Then :

When the flux vs force curve is "sufficiently" convex, the most dissipative configuration is the uniform one. Non-uniformity is then the "economic" trend.

Such a situation may arise for example in electrochemical cells having a strongly non-ohmic behaviour, as illustrated in (Latifi 1990). A similar statement for the time evolution of homogeneous open systems with boundary constraints was already given by (Prigogine 1947). We have here again a generalization of this result.

Note that the approach of (Prigogine 1947) and (De Groot and Mazur 1962), based on specific expressions of the fluxes and forces, allows an evolutionnary criterion to be determined : the spontaneous relaxation of a system, under constant constraints, tends toward the minimum entropy production. In the case of strongly convex flux curves, this should then correspond to the appearance of some ordered heterogeneity, some dissipative structure.

From the point of view adopted here, that of optimal design of engineering systems, the practical conclusion is opposite to that of the Onsager case : the "good" choice of heat exchanger of Figure 1 would be the co-current configuration. Ramadane and Le Goff (Ramadane 1988) have given an illustration of a similar situation. Their example concerns falling film evaporators. The presence of obstacles on the surface such as turbulence promoters, creates heterogeneities of the flow-rate (of the film thickness). Owing to the shape of the curve of Nusselt number vs Reynolds number (Figure 4), it is on the average more advantageous to have zones of very different Reynolds numbers (Re_1 and Re_2 on the Figure), lying on the two branches of the curve, rather than to have a uniform film with an average Reynolds (Re_m on the Figure), **for the same overall flow-rate**. The average Nusselt number $\overline{Nu}$ is higher in the heterogeneous case, and the heat transfer thus more effective, than in the homogeneous case (where $Nu = Nu_m$).

The curve of Figure 4 presents a laminar branch and a turbulent branch. It is the opposite trends on these two branches and the particular convexity of the curve that are responsible for this situation. Clearly, on the turbulent branch, (the branch with positive slope), the system is rather far from equilibrium, and therefore from linear flux-force relations. However this curve is **not** a flux vs force curve, and it is not immediately obvious in what way this example relates to the previous analysis. Let us merely observe the analogy and conjecture that it relates to the same basic idea.

4. FURTHER EXAMPLES : DISTILLATION PLATES AND ION-EXCHANGE

4.1 Configuration and operation of distillation plates and columns

Lewis has shown (Lewis 1936) that "parallel flow" plates have a higher efficiency than "antiparallel flow" plates. Parallel here means that the direction of the liquid flow on two successive plates is the same, whereas antiparallel means that the flows are in opposite directions (Fig. 5). The efficiency may be as much as doubled in the parallel flow arrangement. Figure 6 shows examples of construction of plates satisfying this arrangement (Hausen 1935 ; Heucke 1987 ; Néel 1983).

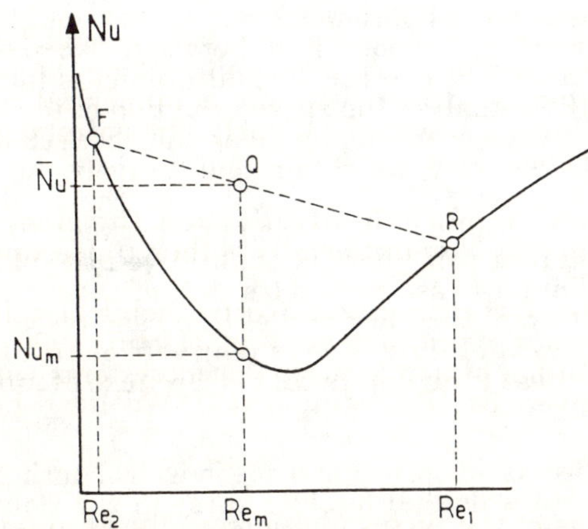

Figure 4. Nusselt number versus Reynolds number for falling film evaporation, showing that heterogeneous flow ($\overline{Nu}$) is more efficient than homogeneous flow (Nu_m).

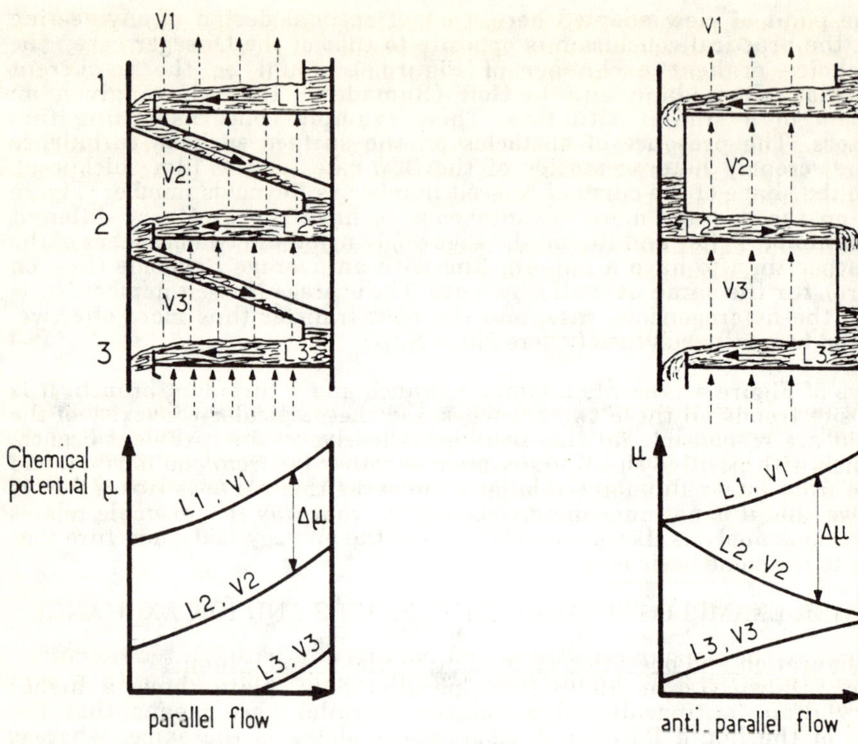

Figure 5. Relative flows and chemical potential profiles in distillation or absorption plates.

However, this performance is achieved only inasmuch as the liquid flow on the plate is close to plug flow **and** the vapour flow between two plates is well seggregated (not mixed). If this is the case, important gradients in composition arise along the liquid path. If we analyze the space evolution of the composition (or chemical potential) of both vapour and liquid along the abscissa of the liquid path, from the downcomer to the weir, we obtain qualitatively the patterns of Figure 5 (bottom). We have represented schematically the chemical potentials, and assumed **local** equilibrium at all points on the plate between the liquid and the vapour leaving. The important point here is qualitative : the pattern of the driving force $\Delta\mu$ of the parallel flow case resembles that of the counter-current heat exchanger, and the pattern of the antiparallel flow case resembles the co-current heat exchanger. **The parallel flow case is more efficient for the same reasons as developed for the heat exchangers that is, because of more uniform driving force distribution $\Delta\mu$.**

Cyclic or periodic distillation is an operating mode in which, during one part of the cycle, the vapour flows upwards and bubbles through the liquid which is stationnary on the plates ; during the other part of the cycle, the liquid flows down from one plate to the next while the gas flow is practically stopped. It has been shown that this operating mode may increase considerably the overall plate efficiency, actually by a factor 2 in the ideal case (McWhirter 1961 ; Gaska 1961). This improvement is not due to the flow pattern of the liquid on the plate or to the gas seggregation. It actually works as well for a perfectly mixed liquid on each plate and perfectly mixed vapour between plates, but allowing for the time evolution of both compositions. More generally, it is known that the cyclic regime of some chemical engineering operations may be more efficient than the steady-state.

In order to understand the origin of this improvement of performances, we shall use a simplified model of both the "parallel flow" plate at steady-state, and the "well-stirred" plate in cyclic operation.

The differential equation expressing mass conservation on a vertical "slice" of liquid phase on the plate at steady state is :

$$V[y_{n+1}(\theta) - y_n(\theta)] = L\frac{dx_n}{d\theta} \tag{4.1}$$

where V and L are the vapour and liquid flow-rates, y and x the vapour and liquid mole fractions, θ being a non-dimensional abscissa increasing on the plate along the liquid path ; n is the plate number, the numbering being from top to bottom. To this differential balance, we adjoin the specification of a **local** gas phase efficiency :

$$E = \frac{y_{n+1}(\theta) - y_n(\theta)}{y_{n+1}(\theta) - y_n{}^*(\theta)} \tag{4.2}$$

where $y_n{}^*$ is the vapour composition in equilibrium with x_n. The boundary condition expressing that the flows are parallel on successive plates (Fig. 5a) is expressed as :

$$x_{n+1}(0) = x_n(1) \tag{4.3}$$

Coming to the cyclic regime, we shall assume that the transfer of liquid from one plate to the plate below is instantaneous, involves no mixing and is accompanied by no mass-transfer between phases. Each plate being considered as a perfect mixer for the liquid, the differential equation governing its evolution in time, during the vapour-flow period, is exactly Equation 4.1, but with L defined as the liquid hold-up on the plate, and θ a normalized time varying from 0 to 1 along the vapour flow period. Similarly, we may define an **instantaneous** plate efficiency by Equation 4.2. The boundary condition 4.3, when written in terms of the normalized time, implies exactly the transfer of the liquid from plate to plate without mixing. Note that the ratio L/V in this case is also numerically the ratio of the overall flow-rates of the two phases in the column.

The parallel-flow plate and the cyclic regime are thus described by the same equations and the same boundary conditions (such is not the case for the antiparallel plate for which the boundary condition would be different). What this implies is that **the time evolution of the compositions, or chemical potentials, of vapour and liquid in the cycling column with well-mixed plates resembles the space evolution of the parallel-flow plate** (Fig. 5a), in other words, resembles the counter-current pattern.

Can one combine the advantages of both cyclic regime and parallel-flow ? This is theoretically possible, but one should be aware that the cyclic regime improves equipartition of driving forces, thus reversibility, only when the spatial configuration does not allow a good approach to equipartition. But this question remains to be investigated in more depth.

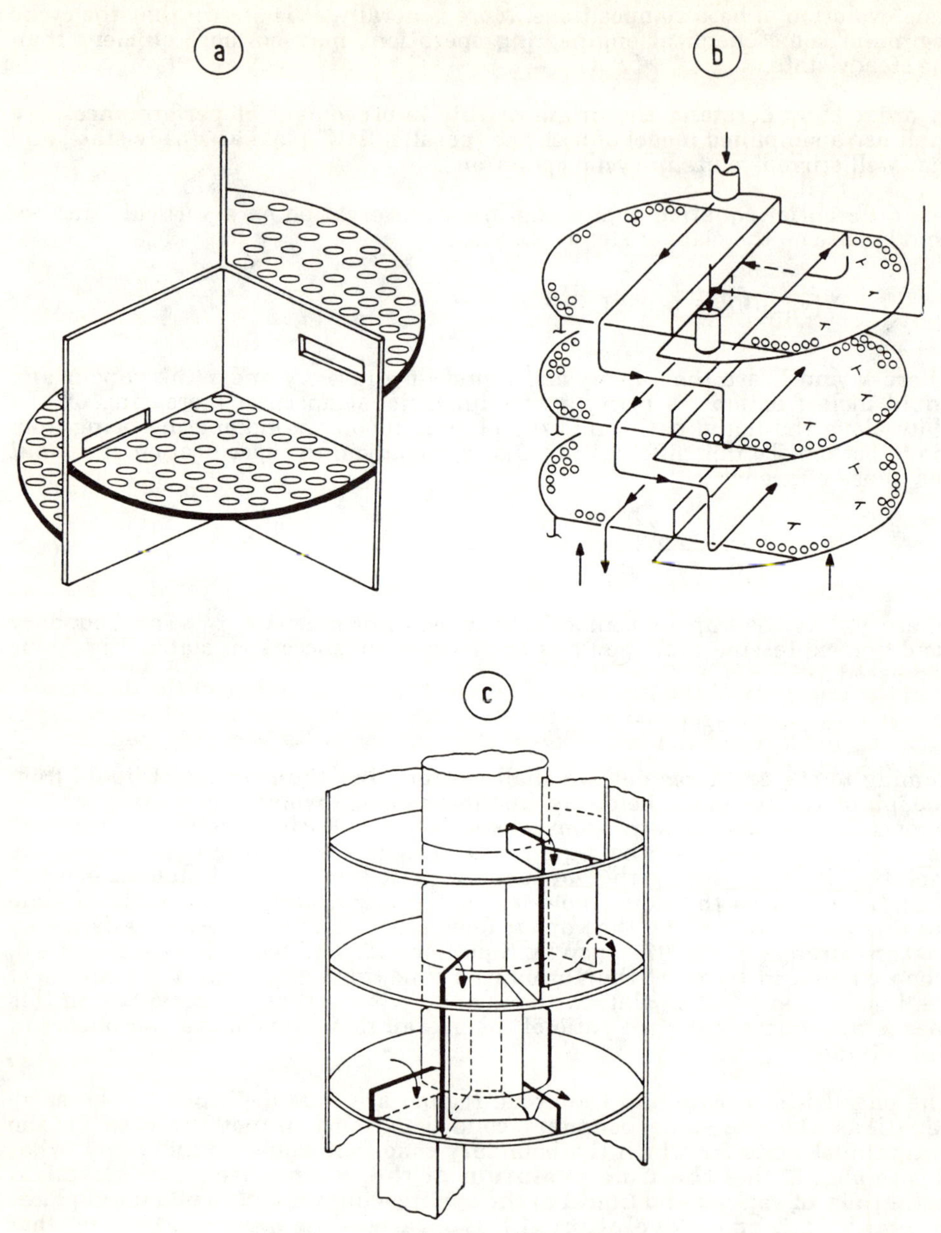

Figure 6. Industrial arrangements for parallel-flow plates
a : Neel 1983 ; b : Fur zer 1978 ; c : Hausen 1935.

Various configurations have been considered to effect cyclic operation and to overcome the limiting phenomena which are the mixing of the liquids of successive plates during downflow, and the control of holdup on each plate. These configurations have been reviewed by (Neel 1983 ; 1990).

We conclude on distillation by the statement that both parallel-flow plates and cyclic distillation draw their efficiency from a better equipartition of driving forces and of entropy production along their process path (space or time respectively).

<u>4.2 An example in sorption processes</u>
Figure 7 represents the effluent concentration of a cation-exchange column, initially in equilibrium with a solution of hydrogen chloride, receiving as feed a solution of sodium chloride for some time, and then again hydrogen chloride. This pictures, in a somewhat schematic way, operations that take place, for example, in a metal recovery process from solutions : metal uptake during a sorption step, followed by regeneration (elution) with acid.

We observe two "fronts" or "waves" (zones where the concentrations change), separated by a "plateau" where the metal is the only cation species. The first front to break through, produced by the sodium uptake, is sharp, whereas the second, produced by the elution of sodium by hydrogen ion, is "dispersive". Figure 8 shows the sodium-hydrogen ion-exchange equilibrium isotherm (solid curve), as the solid-phase ionic fraction y of sodium versus its liquid phase ionic fraction x at equilibrium. On the same graph, we have plotted as dashed lines the so-called "operating curves" of the two fronts, that is the curves which represent the **actual** solid-phase versus liquid-phase ionic fractions along the fronts. The operating line of the sharp sorption front is the straight dashed line (1) in the concavity of the isotherm, whereas the operating line of the dispersive elution front is the dashed curved (2) lying just above the isotherm. Both operating lines rejoin the isotherm at the points representing the plateaus of pure sodium and pure hydrogen.

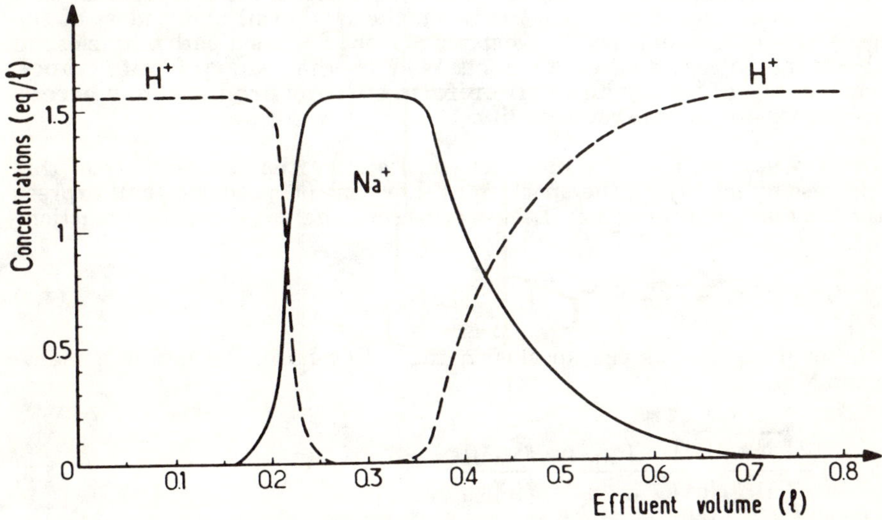

Figure 7. Effluent concentration of ion-exchange column. Polystyrene-sulfonated cation-exchanger, initially equilibrated with HCl 1.5 N, receiving a pulse of NaCl 1.5 N, then HCl.

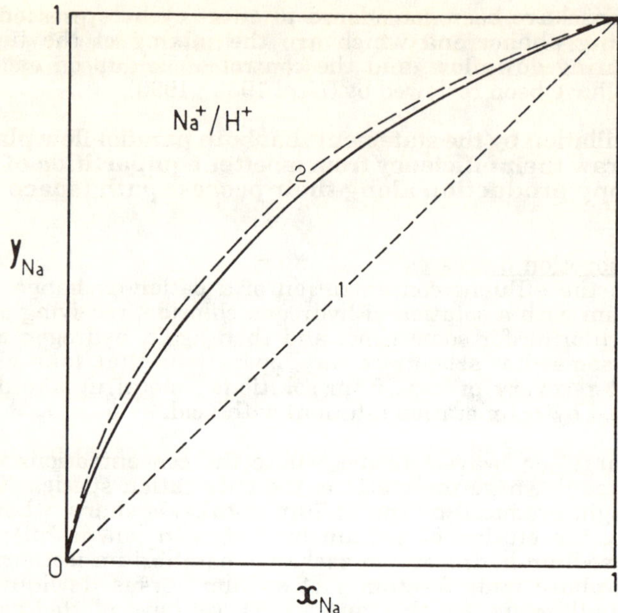

Figure 8. Isotherm for exchange of Na$^+$ against H$^+$ (solid curve) and operating lines (dashed) for sodium uptake (1) and for sodium elution (2).

We notice the important fact that the sharp sorption front corresponds to a situation far from equilibrium (that is, from the isotherm) along most of the exchange path. On the contrary, the dispersive front follows a path very close to the isotherm (i.e. to reversibility) all along. We infer that **sharp fronts (shock waves) are zones of strong and non-uniform entropy production, whereas dispersive zones are close to reversibility**.

As an illustration, let us show how one can improve the reversibility of this simple process by modifying the shock front. For that purpose, we shall express explicitly the entropy production. Let us represent the ion-exchange "reaction" by :

$$H^+ + \overline{Na} \rightleftharpoons Na^+ + \overline{H} \tag{4.4}$$

where the overline means the species in the solid-phase. We define a mass-action ratio M by :

$$M = \frac{[\overline{Na}][H^+]}{[\overline{H}][Na^+]} = \frac{\gamma_{Na}\gamma_H}{\gamma_H\gamma_{Na}} \frac{y_{Na}x_H}{y_H x_{Na}} \tag{4.5}$$

where the brackets symbolize activity, γ an activity coefficient, and x and y are ion-fractions in the liquid and solid respectively. At equilibrium, M is equal to the equilibrium constant K.

194

The entropy production for this ion-exchange process is best expressed as for a chemical reaction, using the affinity A as overall driving force. The affinity is defined from the chemical potentials μ_i and the stoichiometric coefficients ν_i (here, all ν are equal to 1) :

$$A = \Sigma\, \nu_i\mu_i = \Sigma\, \nu_i\mu_i^\circ + RT\, Ln\, M \tag{4.6}$$

Since at equilibrium, the affinity vanishes, we must have :

$$RT\, Ln\, K = -\, \Sigma\, \nu_i\mu_i^\circ \tag{4.7}$$

and therefore :

$$A = RT\, Ln\, \frac{M}{K} \tag{4.8}$$

The local and instantaneous entropy production (here expressed per unit mass of absorbent) may be written as :

$$\sigma = -\,\frac{1}{T} \sum_i J_i.\Delta\mu_i \tag{4.9}$$

where the summation extends over the species sodium and hydrogen, J_i is the flux transferred (moles per second and per unit mass of adsorbent), $\Delta\mu$ is the chemical potential difference between the phases. Since we have a strictly stoichiometric ion-exchange, we have $J_{Na} = -\, J_H$, and it is easy to see that σ becomes :

$$\sigma = \frac{J_{Na}}{T} A \tag{4.10}$$

Assuming a linear mass-transfer law :

$$J_{Na} = kA \tag{4.11}$$

then

$$\sigma = \frac{k}{T} A^2 = kR^2T \left(Ln\, \frac{M}{K} \right)^2 \tag{4.12}$$

The total instantaneous entropy production over the entire front is obtained by integrating this expression over the concentration range.

Let us do this with the assumptions that k, K and T are constant, and that the ratio of activity coefficients appearing in Equation 4.5 is equal to 1. Then, along the operating line of the shock front, since y = x, we have M = 1 so that the total instantaneous entropy production is :

$$\left. \frac{dP}{dt} \right)_I = kR^2T \int_0^1 \left(Ln\, \frac{1}{K} \right)^2 dx = kR^2T(LnK)^2 \tag{4.13}$$

Now we should like to illustrate that an operating mode may be devised which achieves the same saturation of the ion-exchanger in sodium, but in a more reversible manner. This is obtained by dividing the shock front into a succession of "smaller" ones (meaning that the amplitude of the concentration change is smaller). Before describing such an operating mode, let us evaluate what entropy production we are likely to save in taking this trouble. Taking for example $K = 2$, consider Figure 9 in which we have two successive shocks, separated by a plateau at equilibrium at a composition ($x_{Na} = 0.5$, $y = 0.667$). These two shocks are represented by the solid operating lines a and b. Their equations are :

$$y_1 = 1.315\,x_1 \qquad ; \qquad 1.5(1-y_2) = 1-x_2 \qquad\qquad (4.14)$$

To obtain the total instantaneous entropy production (per unit mass of adsorbent), we need to calculate the integrals :

$$\left.\frac{dP}{dt}\right)_{II} = kR^2T\left[\int_0^{0.5} F_1(x)dx + \int_{0.5}^1 F_2(x)dx\right] \qquad\qquad (4.15)$$

where F_i is obtained by substituting the expressions (4.15) for y_i into the ratio $[y(1-x)/x(1-y)]$. After some calculations, we find :

$$\left.\frac{dP}{dt}\right)_{II} = kR^2T\,(0.040 + 0.018) = 0.058\;kR^2T$$

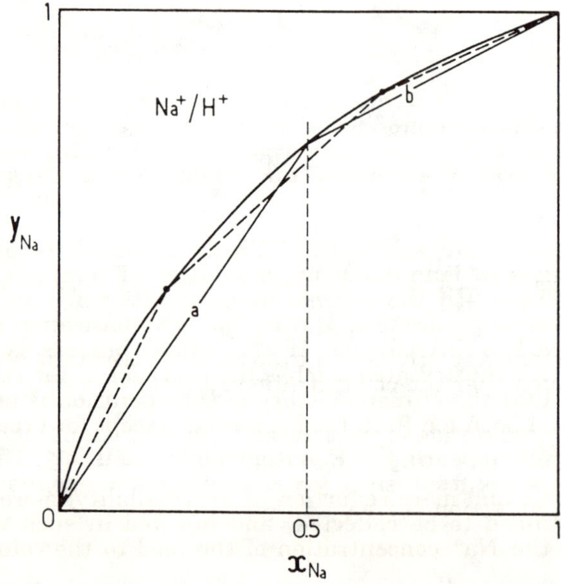

Figure 9. Operating lines for the "2-segmented recycle" process illustrated on Figure 10 (full lines a and b) and for a 3-segmented process such as that of Figure 11c (discontinuous lines).

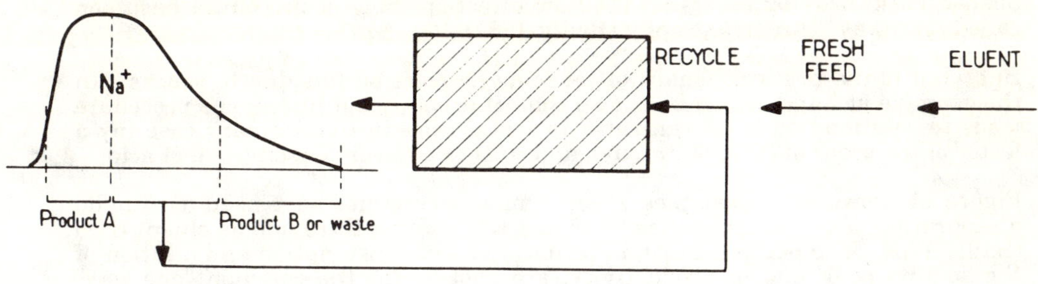

Figure 10. Principle of the segmented recycle process : part of the effluent of the
elution step is "recycled" as a "pre-feed" in the sorption step.

whereas from Equation 4.13, in the case of the full shock :

$$\left.\frac{dP}{dt}\right)_I = 0.48 \ kR^2T$$

**Splitting the shock in two at x = 0.5 thus reduces the entropy production by an
order of magnitude. Further reduction could be obtained by splitting into more,
smaller fronts, closer and closer to the isotherm, thus distributing more evenly
the entropy production along the composition change, and at the same time,
reducing it.**

How do we design a process that realizes this ? Figure 10 shows the principle of
such a process. It consists in manufacturing a solution of composition x_{Na} =
0.5 by collecting the effluent of the regeneration operation for an adequate
period. This effluent is used as "pre-feed" during the next saturation step, then
fresh feed of pure NaCl is introduced. If the whole process is properly
optimized, the second front thus created catches the first just at the column exit
and merges with it, so that the effluent history of the operation is exactly the
same as if pure NaCl had been fed from the beginning (except for breakthrough
times, which are different).

To achieve more splitting, and more reduction of irreversibility, more fractions
of mixtures Na$^+$ + H$^+$ need to be collected, and injected in such way as to
increase progressively the Na$^+$ concentration of the feed to the column. This

amounts to realizing a gradient injection. One efficient way to obtain a gradient such that it "recompresses" to form a full shock at the column outlet is to store the effluent profile of a regeneration. This profile may be stored in part of the column itself, and instead of recycling it fraction by fraction, it may simply be pushed backwords by reversing the flow direction. Such a procedure has been called "two-way" chromatography (Bailly 1981).

But what is the pratical benefit of reducing irreversibilities due to shocks ? In the example at hand, it can be shown that, if properly optimized, the procedure leads to solutions more concentrated in metal than in the ordinary case (by a factor of the order of 30 to 60 %, say), and thus to a lesser consumption of acid.

Figure 11 shows some examples where similar techniques have been used to do a separation of two ionic species (Na^+ and K^+) using a third one as eluent (H^+) (Bailly 1984). The performances, in terms of eluent consumption and dilution of the sodium peak and of productivity, are poorest for the common "one-way" chromatographic experiment (a). Here, a large shock of Na^+/H^+ exchange is created by injection of the concentrated mixture to be separated, but this shock collapses as it moves along the bed. The entropy production is thus initially high, then lower and lower (bad equipartition). The process involving recycling of a single fraction, mixed with fresh feed (b) greatly improves the performance : the Na^+/H^+ shock is of smaller amplitude, because the mixture injected is already diluted, but keeps this amplitude throughout the process. Further improvements are obtained by recycling separately several fractions as shown on Figure 11c. The corresponding operating lines of the shocks, before they merge, are shown as dashed lines on Figure 9. Finally, storing a part of the K^+ elution profile in the bed and reversing the flow, we have "two-way" chromatography (d) (Bailly 1981).

What relation is there between entropy production in shocks and the eluent consumption ? In adsorption or ion-exchange, the Gibbs energy necessary to effect a separation is brought by the eluent (usually a solution of a pure compound), and this energy is transferred to the products by dilution (the separation of potassium and sodium in our example is "compensated" by the mixing of both with hydrogen ion). But the eluent brings also the Gibbs energy which corresponds to the entropy produced (for practically non-thermal processes such as ion-exchange, we have $\Delta G = - T\Delta S$). **Thus all irreversibilities are "paid" in terms of use of eluent, that is, in dilution of the products.**

<u>4.3 The pinch technology : an application of equipartition</u>
The "pinch" analysis of thermal systems, as developed for example by Linhoff (Linhoff 1983) and (Gourlia 1982 ; 1989), consists in plotting the Carnot temperature $(T-T_0)/T_0$ versus enthalpy of all the fluids in a plant that give exergy and of all the fluids that receive exergy (exergy is defined as $H-T_0S$, similarly to Gibbs energy, but with a constant reference temperature T_0, usually that of the environment). This method is commented elsewhere in the present book (Le Goff). In this diagram (Fig. 12), the area between the two resulting lines represents the exergy destroyed, which corresponds to entropy produced, in the heat exchange processes. To reduce these losses amounts to modifying the flow-sheets and the working temperatures so as to bring these two curves closer together, but this is most often achieved at the cost of increased exchange area. The limiting factor in this respect is the so-called "pinch point", that is, the point where the two lines get to touch each other. An infinite exchange area would be required at the corresponding heat exchanger. It is easy to see that the pinch phenomenon may be avoided, or delayed, if the two curves are parallel, in other words when **equipartition** of the exergy losses is realized in the whole plant.

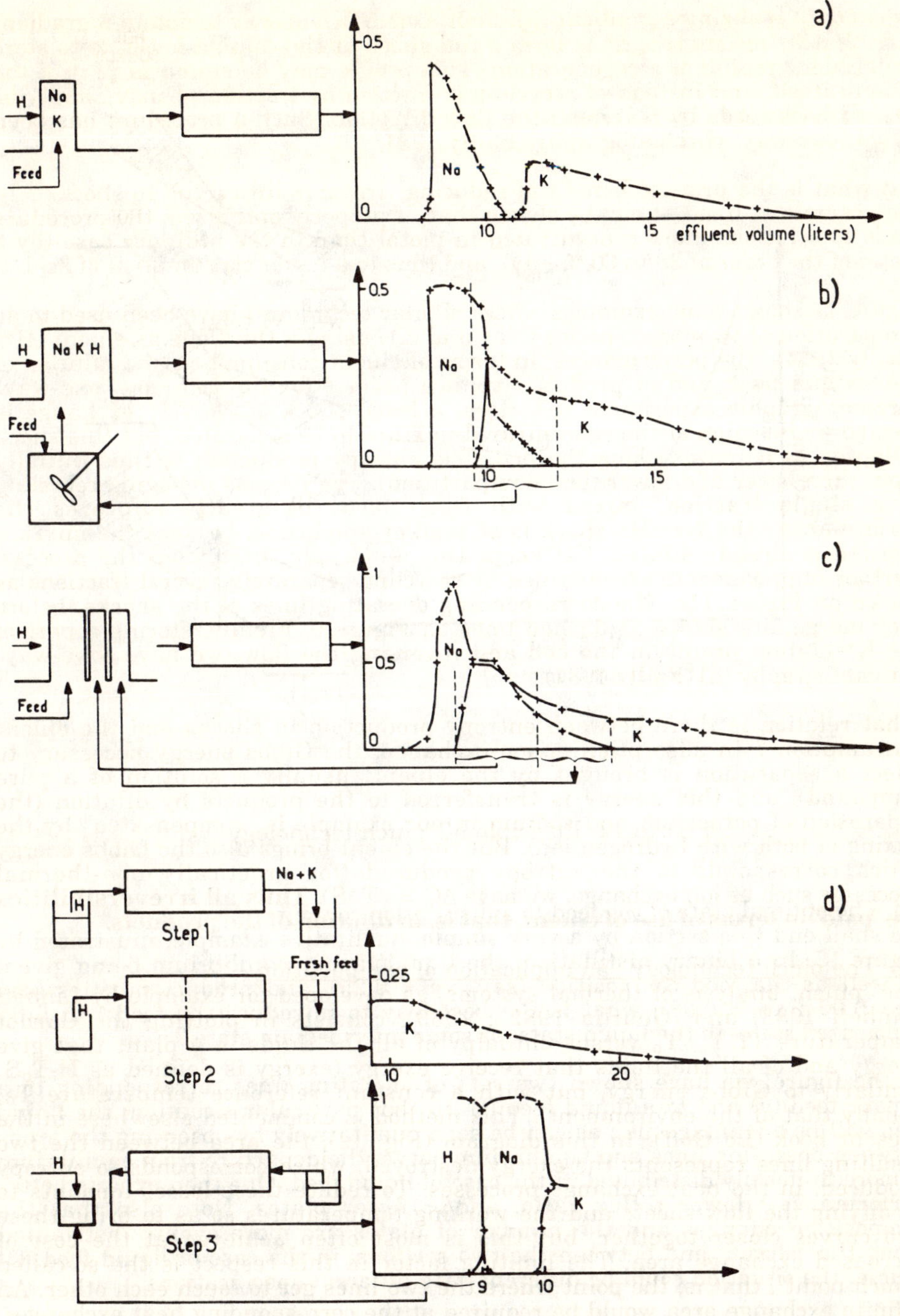

Figure 11. Comparison of other chromatographic processes.
a) Ordinary one-way elution-separation.
b) Elution with mixed recycle.
c) Elution with 2-segemented recycle.
d) "Two-way" chromatography (storage of elution profile and flow reversal).

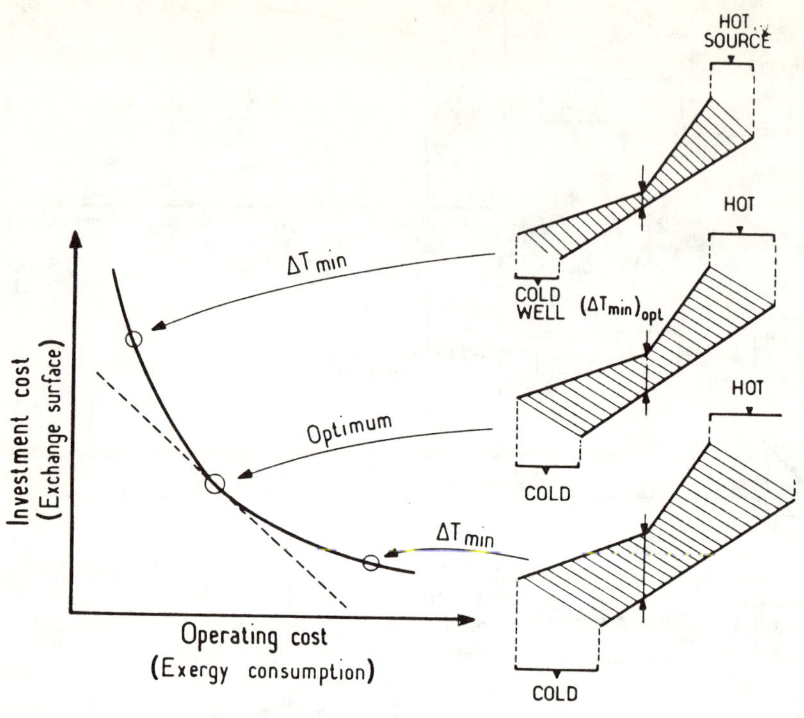

Figure 12. Principle of "pinch technology".

4.4 The optimal state of the feed to a distillation column

We shall end this section by a very simple, qualitative example, illustrated by Figure 13. In a binary distillation, the liquid/vapour equilibrium being given, as well as the feed composition (say, $x_F = 0.25$), and product purities (say $x_B = 0.02$; $x_D = 0.98$), what is the best way to introduce the feed : in fully vapourized state, in the liquid state, or some intermediate state ?

In the figure, we have shown two sets of operating lines corresponding to a vapour feed (v) and to a liquid feed (l). In that particular situation, the liquid feed will be advantageous ; as can be seen qualitatively by observing the areas between operating lines and equilibrium curve, the departure from equilibrium is more uniformly distributed in the case of liquid feed. One then expects better performance : lower reflux ratio, and/or fewer plates. Note also that the composition changes and the number of plates are more evenly distributed along the column, and between the two sections, in the case of liquid feed. Of course, the situation could be different with different specifications.

200

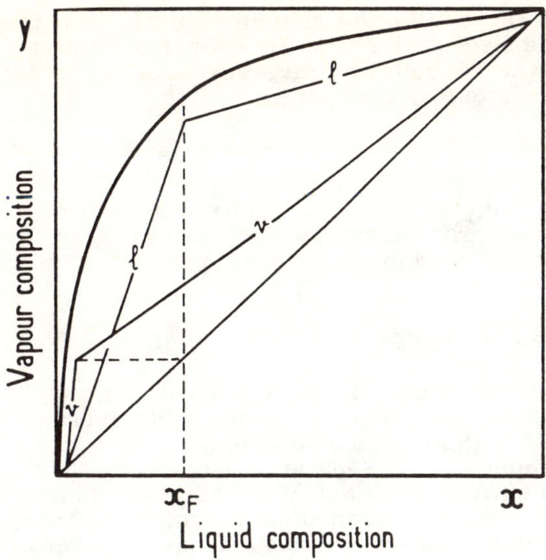

Figure 13. Optimal enthalpic state of the feed to a distillation column.
Operating lines for liquid feed (l) and for vapour feed (v).

5. AN ECONOMIC GENERALIZATION : OPTIMAL ALLOCATION OF INVESTMENT IN A MULTISTAGE SYSTEM

In the previous sections, we considered systems or processes in which the distribution of entropy production was globally determined by the flow configuration. In the tubular heat exchanger, once counter-current flow or co-current flow has been chosen, we no longer control the internal temperature profiles, and we cannot adjust them to approach equipartition. The optimization is thus a discrete binary choice.

On the other hand, if one considers a cascade of connected elements which can be adjusted separately, or if some control variable may be acted upon along the process, it may be possible to approach equipartition of entropy production. The question is then how the profile of the decision variable should be arranged in the process to minimize some cost function.

Examples of such distributed processes are networks of heat exchangers or of distillation columns, the allocation of wash water in a cross-current washing process, and "diabatic" or "diathermal" distillation, which we use here as an illustration, and in which heat exchange is effected all along the column, or at discrete levels of the column (Hausen 1932). Such distributed heat exchange makes it possible to adjust the flow ratio of the two phases, and thus the shape of the operating line and the driving force all along the column, as illustrated in Figure 14. Of course, some expenses need to be made for heat exchange area.

With a certain number of assumptions, the problem of optimal allocation of heat exchange area along the column can be handled simply, and allows us to illustrate the relationship between the thermodynamic property of equipartition and economic optimization.

Let us assume that the distillation column is made of a given number of distinct elements, the size of which can be defined separately, and including some heat transfer area. We define an investment cost C_i for each element i as a linear function of the size A_i of the element :

$$C_i = \alpha_i A_i + \beta_i \tag{5.1}$$

where α is a proportional cost factor and β a fixed cost (these costs include the heat transfer area). Similarly, we assume that the operating cost $\bar{C}_i$ is a linear function of the exergy destroyed in the element i per unit time (say, a year) :

$$\bar{C}_i = \gamma_i(E_i + T_0 P_i) + \delta_i \tag{5.2}$$

where γ_i is a proportional cost factor and δ_i a fixed cost and T_0 is a reference temperature for exergy, generally the ambient temperature. The exergy destroyed is split into a thermodynamic minimum E_i, which corresponds for example to the minimum work of separation, and an irreversible contribution $T_0 P_i$, where P_i is the entropy produced. We shall assume in addition, and this is specific to such processes as diabatic distillation, that each element is small enough that equipartition of entropy production may be approximately achieved inside it, by adjusting the heat exchange flux, thus the ratio of flow-rates of the phases. We require in addition each element to perform a specified duty, expressed on a transfer flux J_{io}.

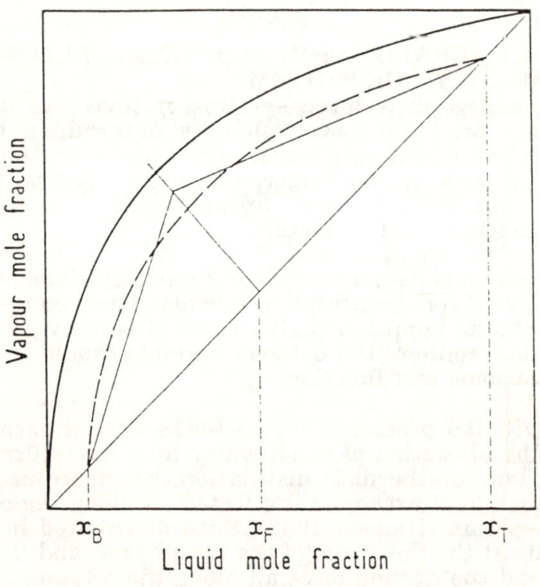

Figure 14. Operating lines for classical (full straight lines) and diabatic distillation (discontinuous curve).

The total cost function C_t which we seek to minimize is then :

$$C_t = \sum_i \left(\overline{C}_i + \tau C_i \right) \tag{5.3}$$

where τ is the yearly amortization rate, and the summation extends to all elements. We use the variational approach, and minimize a Lagrangian as

$$\Omega = \sum_i \left(\overline{C}_i + \tau C_i \right) + \Sigma \lambda_i (J_i - J_{io}) = 0 \tag{5.4}$$

which we thus require to satisfy :

$$\frac{\partial \Omega}{\partial A_i} = 0 \quad ; \quad \frac{\partial^2 \Omega}{\partial A_i^2} > 0 \qquad \text{(all i)} \tag{5.5}$$

In order to explicitate these derivatives, we need to express the costs in terms of the size variables A_i (note that we could minimize as well with respect to $\overline{C}_i$). To do this, we express P_i in Equation 5.3, using the fact that equipartition is assumed, from Equations 2.11 and 2.12, as :

$$P_i = J_i^2 / L_i A_i \tag{5.6}$$

so that Ω becomes :

$$\Omega = \sum_i \left[\tau \alpha_i A_i + \tau \beta_i + \gamma_i E_i + \delta_i + \frac{\gamma_i T_o J_i^2}{L_i A_i} + \lambda_i (J_i - J_{io}) \right]$$

and taking derivatives with respect to A_j gives a system of independent equations :

$$\frac{\partial \Omega}{\partial A_j} = \tau \alpha_j - \frac{\gamma_j T_o J_j^2}{L_j A_j^2} = 0 \qquad \text{(all j)} \tag{5.7}$$

or

$$\boxed{\tau = \frac{\gamma_j T_o P_j}{\alpha_j A_j}} \qquad \text{(all j)} \tag{5.8}$$

and

$$\frac{\partial^2 \Omega}{\partial A_j^2} = \frac{2\tau}{A_j} > 0$$

Equation 5.8 is the essential result of this procedure. We note that, τ being a given constant, depending only on the financial conditions, the right-hand side of Equation 5.8 must be independent of the element j, thus an "invariant" along the process. Let us try to give it a more physico-economical meaning. The product $\tau \alpha_j A_j$ is the amortized proportional part of the investment, whereas the product $\gamma_j T_o P_j$ is the annual cost related to irreversible energy degradation (in terms of exergy destruction). These two quantities should thus be equal in any element (but not the same from one element to another !). We may now restate this result as the

Optimal size distribution theorem : Under the assumptions stated, the optimal size distribution of the elements is that which realizes the equipartition of the ratio $\gamma T_0 P/\alpha A$ on all elements, or also, in which the cost of irreversible energy degradation in any element is equal to the amortized proportional investment cost in that element.

Let us emphasize that the cost parameters α and γ and the temperature levels T need not be the same in all elements. However, in the special case when α and γ are constant, then the theorem above reduces simply to equipartition of the ratio P_i/A_i, in other words **equipartition of the local rate of entropy production**. This establishes the connection with the result of the previous sections.

When the cost of exergy is the same in all elements (that might be the case for example in an isothermal process), but not the investment costs, then it is the ratio $P_i/\alpha_i A_i$ which should be equipartitioned. We then recover a statement by Sieniutycz (Sieniutycz 1984) : "... in the optimal process, the changes in apparatus price are balanced by the changes of thermodynamic irreversibility... for expensive apparata, only very intensive processes can be optimal..." or also a statement by Bejan (Bejan 1982) : "... the heat transfer area should be concentrated in that region where the heat transfer is most intense...".

6. CONCLUSIONS

The generality of the equipartition property, its limitations, as well as its scientific status, needs to be carefully ascertained. Within Onsager's thermodynamics (linear flux-force relationship) and for a single species transferred, the situation is clear : the optimality of equipartition may be demonstrated rigorously, as was done at the beginning of this article. It may therefore be called a theorem, and has several corollaries with practical implications. These results are rigorously extended to concave flux vs force relationships. For multicomponent systems, although the formal result still holds, the practical implications remain to be investigated, probably on specific situations.

Outside of Onsager's thermodynamics, and for non-concave flux vs force relations, situations may arise where non-uniformity becomes optimal. This, of course, may be compared to stability studies. In the neighborhood of thermodynamic equilibrium, stability is generally associated with steady-state and/or uniform systems. On the other hand, far from equilibrium, in strongly non-linear situations, the most stable configurations may correspond to unsteady regimes (possibly periodic) and/or non-uniformity in space (dissipative structures). We may conjecture that equipartition as well as stability considerations are part of a more general "law" or "principle" of economy or evolution, which would govern both natural and industrial systems. The former would tend to minimize their entropy production (thus to equipartition, if they are close to equilibrium). The optimization of the latter would, in fact, implicitly include the equipartition principle.

The idea of an optimum distribution of investment related to the distribution of irreversibility is, in fact, more or less explicit in much of the literature on exergy analysis (Szargut 1980). We have attempted to formalize this idea and to give it a somewhat general and fundamental basis.

I believe that for practical purposes, the equipartition concept should be considered as a conjecture, primarily of qualitative interest, in that it helps finding new routes for process improvement and avoiding misconception in the design of new processes.

BIBLIOGRAPHY

Bailly, M., and D. Tondeur. 1981. Two-way chromatrography. Chem. Eng. Sci. 36 : 455-469.

Bailly, M., and D. Tondeur. 1984. Reversibility and Performances in Productive Chromatography. Chem. Eng. Process. 18 : 293-302.

Bass, J. 1968. Cours de Mathématiques, Vol. 1, Paris : Masson.

Bejan, A. 1982. Entropy generation through heat and fluid flow. New-York : Wiley.

De Groot, S.R., and P. Mazur. 1962. Non-equilibrium Thermodynamics, Amsterdam : North-Holland.

Furzer, I.A. 1978. Can. J. Chem. Eng. 56 : 747-750.

Gaska, R.A., and M.R. Cannon. 1961. Controlled cycling distillation in sieve and screen plate towers. Ind. Eng. Chem. 53 : 629-631.

Gourlia, J.P. 1989. La méthode du pincement ou exploitation des diagrammes température/enthalpie. Rev. Générale de Thermique 327 : 154162.

Hausen, H. 1932. Zeitung Techn. Phys. 6 : 271.

Hausen, H. 1935. Forsch. Ingenieurwesen 6 : 9.

Heucke, C. 1987. Vorteile von parallelen Strömen bei Rektifikation, Absorption und Extraktion. Chem.-Ing.-Tech. 59 : 107-111.

Latifi, A., G. Risson, P. Nabonnand, and A. Storck. 1990. Commande optimale d'un réacteur électrochimique discontinu : application à l'électrodéposition du cuivre. Entropie, in press.

Le Goff, P. (Editor). 1979, 1980, 1982. Energétique Industrielle. Vol. 1, 2 and 3. Paris : Lavoisier

Le Goff, P. 1990. This book.

Lewis, W.K. 1936. Ind. Eng. Chem. 28 : 399.

Linhoff, B., and E. Hindmarsh. 1983. The pinch design method for heat exchanger networks. Chem. Eng. Sci. 38 : 745-763.

McWhirter, J.R., and M.R. Cannon. 1961. Controlled cycling distillation in a packed-plate column. Ind. Eng. Chem. 53 : 632-634.

Neel, L. 1983. Distillation cyclique et distillation hélicoïdale. Doctorate Thesis. Nancy : Institut National Polytechnique de Lorraine.

Neel, L., M. Bailly, and D. Tondeur. 1990. Influence des régimes spatiaux et temporels d'écoulement sur l'efficacité des plateaux de distillation ou d'absorption. Entropie (in press).

Prigogine, I. 1947. Etude thermodynamique des phénomènes irréversibles. Liège : Desoer.

Ramadane, A., P. Le Goff, and B. Clauzade. 1988. In Proceedings of the Symposium Progrès récents dans les échangeurs thermiques. Ed. Centre d'Etudes Nucléaires de Grenoble.

Sieniutycz, S. 1984. A development of the relation between drying energy savings and thermodynamic irreversibility. Chem. Eng. Sci. 39 : 1647-1659.

Storck, A. 1989. Personal communication.

Szargut, J. 1980. Energy 5 : 709.

Tondeur, D., and E. Kvaalen. 1987. Equipartition of entropy production. An optimality criterion for transfer and separation processes. Ind. and Eng. Chem. Res. 26 : 50-56.

LIST OF SYMBOLS

A   surface area of heat exchanger (eq. 2.5), m^2
       size of element of diabatic distillation column (eq. 5.1 (m^2 or m^3 for example)
A affinity of ion-exchange reaction (eq. 4.8)
$\underline{C}_i$ investment cost of element i (eq. 5.1)
C_i operating cost of element i (eq. 5.1)
C_t total cost (eq. 5.3)
C_p specific heat $(J.kg^{-1}.K^{-1})$
E local efficiency in distillation (eq. 4.2)
       work of separation (in eq. 5.2)
f driving force, defined by eq. 2.2 for heat transfer
$\underline{f}$ vector of driving forces (eq. 3.6)
$\bar{f}$ driving force averaged over the whole system (eq. 2.7)
F flow-rate $(kg.s^{-1})$
h specific enthalpy (eq. 2.28) $(J.kg^{-1})$
       heat transfer coefficient in Nusselt number $(m.s^{-1})$
i current intensity (eq. 3.1) (Amp.)
j local transfer flux (eqs. 2.1 and 2.2) $(J.s^{-1}.m^{-2}$ in heat transfer)
J total flux of energy transfer over the whole system (eq. 2.6) $(J.s^{-1})$
J_o specified value of J
J_i flux of species i transferred in ion-exchange $(mol.s^{-1}.kg^{-1}$ adsorbent)
k mass-transfer coefficient in ion-exchange (eq. 4.11) $(mol^2.J^{-1}.s^{-1}.kg^{-1}$ adsorbent)
K equilibrium constant of ion-exchange (eq. 4.7) (n.d.)
L phenomenological transfer coefficient $(J.K.s^{-1}.m^{-2}$ in heat transfer)
L liquid flow-rate (eq. 4.1 only)
$[L]$ matrix of phenomenological coefficients (eq. 3.6)
M mass-action ratio (eq. 4.5) (n.d.)
Nu   Nusselt number (Fig. 4) = (hd/λ) (n.d.) ; h is a heat transfer coefficient, λ a thermal diffusivity and d the pipe diameter
P total rate of entropy production, defined by eq. 2.5 $(J.K^{-1}.s^{-1})$
P_o total rate of entropy production of an "equipartitioned" system
Q quantity of electricity (eq. 3.1) (Coulomb)
R electric resistance (eq. 3.2) (Ω)
       gas constant (eq. 4.6) $(J.mol^{-1}.K^{-1})$
Re   Reynolds number (Fig. 4) = (ud/v) (n.d.) ; u is flow velocity ; v viscosity
s specific entropy (eqs. 2.26, 2.27) $(J.kg^{-1}.K^{-1})$
s^2 variance of distribution, of driving force or flux, (eq. 2.14) (n.d.)
S entropy (eqs. 2.26, 2.27) $(J.kg^{-1})$
t time (s)
T temperature (K)
T_o ambient temperature, or reference temperature for exergy (eq. 5.2)
V vapour flow-rate in distillation (eq. 4.1) $(mol.s^{-1})$
       volume or space coordinate vector (eq. 3.8)
x liquid mole fraction (eq. 4.1) (n.d.)
y vapour mole fraction (eqs. 4.1, 4.2) (n.d.)

Greek symbols

$\alpha, \underline{\beta}, \gamma, \delta$ cost coefficients in eqs. 5.1, 5.2
$\gamma, \bar{\gamma}$ activity coefficient (eq. 4.5)
η overall electric potential difference (eq. 3.2)
μ_i chemical potential of species i (eq. 4.6) $(J.mol^{-1})$
λ Lagrange multiplier (eq. 3.15) ; thermal diffusivity in Nusselt number $(m^2.s^{-1})$
v_i stoichiometric coefficients in ion-exchange reaction (eq. 4.6) n.d.)

dynamic viscosity in Reynolds number ($m^2.s^{-1}$)

σ local rate of entropy production (eq. 2.1) ($J.K^{-1}.s^{-1}.m^{-2}$)
in ion-exchange, defined by eq. 4.9 ($J.K^{-1}.s^{-1}.kg^{-1}$ adsorbent)

θ time interval (eq. 3.1) (s)
normalized abscissa along distillation plate (eqs. 4.1, 4.2) (n.d.)
normalized time in cyclic distillation

τ amortization rate (eq. 5.3)

Ω Lagrangian functional of cost (eq. 5.4)

Subscripts

a, b   two configurations of exchangers

i, o   inlet, resp. outlet of heat exchanger

i, j   refer to species i or j in multicomponent or multi-element systems (eqs. 3.12, 4.6, 5.1)

n   plate number in distillation (eq. 4.1)

1, 2   cold, resp. hot fluid in heat exchangers

Superscripts or overline

*   equilibrium between phases (eq. 4.2)

–   average over system (eq. 2.7)

–   solid phase composition or activity (eqs. 4.4, 4.5)

GLOSSARY

*** Inner (dot) product** (eqs. 3.6 to 3.11)

*** Exergy** : defined as $Ex = H - T_0 S$, where T_0 is some reference temperature, often ambient temperature. This function, introduced by Gouy (1889), called the "availability function" by Keenan (1932), has been coined the name "exergy" by Rant (1956), and is widely accepted in Europe (Le Goff, 1982 for example). As compared to the Gibbs energy $G = H - TS$, it measures the fraction of thermal energy that may be converted to mechanical energy, using a reversible machine and a cold source at T_0.

*** Entropy production or dissipation** : we use these terms to designate a **rate** of production (that is, entropy per unit time).

*** Duty** : designates a specification on the amount of matter or energy a system is required to transfer or transform.

*** Equipartition or uniform distribution of a quantity** : means that this quantity is constant along the process variable (constant with length in a tubular exchanger at steady-state ; constant in time in the time evolution of a homogeneous system, etc.).

*** Operating lines** : when two phases are contacted along an apparatus, and exchange chemical species or enthalpy, the plot of the composition or enthalpy of one phase versus that of the other phase, at any point of the contactor where these phases "cross" each other, is called an operating line in chemical engineering. The compositions thus plotted are actual compositions, generally not at equilibrium. In many steady-state operations, the resulting operating lines are straight.

*** Fronts, waves, shocks**. These terms designate the moving concentration distributions that occur in fixed-bed adsorption or ion-exchange operations. Fronts or waves may be "sharp" or "dispersive". Sharp waves, sometimes tend to take a constant shape. They are thus called "shocks" by analogy with the phenomena in compressible flow.

Exergy Optimization in a Class of Drying Systems with Granular Solids

Z. Szwast

Institute of Chemical Engineering, Warsaw Technical University
00-645 Warsaw, Poland

ABSTRACT

Because of their large energy consumption, drying processes belong to the most uneconomical unit operations. To decrease the total economic costs (i.e. the sum of investment and operational costs) the use of unconventional, optimally controlled drying processes is recommended,'the parameters of the drying agent being varied with the residence time of solid (Sieniutycz 1973).In such processes less intensive drying usually takes place at the beginning of the process, and more intensive drying is applied at the end of the process. The rigorous principles of rational control are not generally known for drying processes, so there is a need for the formulation and solution of their mathematical optimization. This paper deals with the steady-state multistage fluidized drying processes, the parameters of the drying agent being varied (for unconventional processes) with the stage number.

Unconventional and conventional processes in the crosscurrent and countercurrent cascade are taken into account.

For the crosscurrent cascade, the following two variants of the unconventional processes are considered: the first - the optimal inlet air temperature, humidity and flow rate are varied with stage number, and the second variant - the optimal inlet air temperature and flow rate are varied with stage number, whereas the inlet air humidity is assumed to be equal to the ambient humidity for every stage of the cascase; in this variant the possibility of appearance of the processes with product recycle is taken into account. Also for the conventional processes two variants are considered: the first - the optimal inlet air temperature, humidity, and flow rate are constrained to be the same for every stage, and the second variant - the optimal inlet air temperature and flow rate are constrained to be the same, whereas the inlet air humidity is asumed to be equal to the ambient humidity.

For the countercurrent cascade, the system with heat exchangers for heating of the drying air leaving any stage and entering next one, is considered. For the unconventional processes the optimal inlet air temperature is varied with stage number, whereas for the conventional processes the optimal inlet air temperature is constrained to be the same for every stage. Moreover, for the conventional processes the preliminary dehumidity of the inlet air entering the first dryer of the cascade,as well as the drying air recycle are also taken into account.

The optimization results are obtained for silica gel-air-water system. Thermodynamic equilibrium between the gaseous and solid phases leaving the fluidized stage is assumed.

The economic effectiveness, corresponding to the sum of the
investment and operational costs,is accepted as a performance criterion. To express the cost of the drying air in terms of the drying parameters (air temperature and air humidity) an economic balance of the dryers investigated is made. In this balance the so-
-called exergy tariff of prices is applied for the purpose of quantitative evaluations.In the performance criterion,together with
physiochemical variables,only one parameter λ, connected with the
sum of the investment and gas pumping costs,appeared.The value of
λ for a concrete drying process can be computed. The optimization
results are obtained for the family of drying processes corresponding to various values of the parameter λ.

INTRODUCTION

        Because of their large energy consumption, drying processes
belong to the most uneconomical unit operations. To decrease the
total economic costs (i.e. the sum of investment and operational
costs) the use of unconventional, optimally controlled drying processes is recommended, the parameters of the drying agent being
varied with the residence time of solid (Sieniutycz 1973).In such
processes less intensive drying usually takes place at the beginning of the process, and more intensive drying is applied at the
end of the process, The rigorous principles of rational control
are not generally known for drying processes, so there is a need
for the formulation and solution of their mathematical optimization. This paper deals with both cases of steady-state multistage fluidized drying processes: conventional and unconventional
(optimally controlled). In the last case, temperature, humidity
and flow of inlet gas (or only one of these parameters) are varied with the stage number of the multistage process.Crosscurrent
and countercurrent processes are taken into account. Also considered are the crosscurrent processes with product recycle, and the
countercurrent processes with drying gas recycle. The optimal data for the performance indices (total process costs expressed in
exergy units) for the various modes of drying are calculated and
compared. It should be remembered that, in industry, even a small
percentage decrease in costs is usually important. For the both,
unconventional processes and conventional ones, the properties of
the optimal solid and gas parameters are discussed.

PROBLEM FORMULATION

        A moist solid (silica gel) has to be dried by a gas (air) in
Nth stage of a crosscurrent or countercurrent cascade of fluidi-

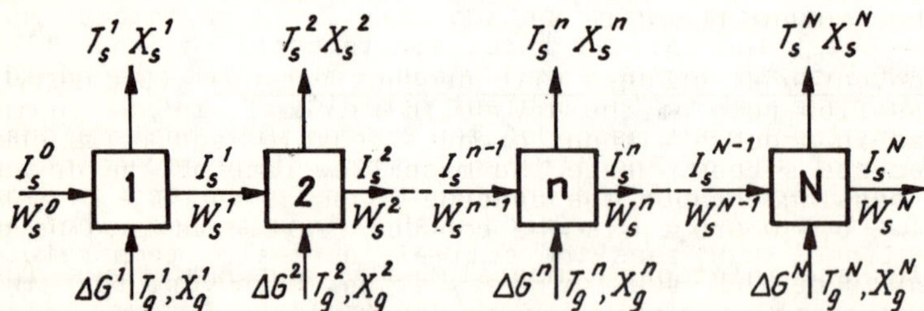

Fig.1.Scheme of the multistage crosscurrent fluidized drying.

zed dryers.The mass flow rate of the dry solid is equal to S.An isobaric process is assumed. The initial state of the solid,defined by moisture content and enthalpy (or temperature) is prescribed. For the final solid state the only requirement pertains to the final moisture content which has to be preassigned.The final solid enthalpy (or temperature) is undetermined. It should assume the value which optimizes the performance index.

The inlet air temperature is not constrained in the present paper. Here we shall assume that the only constrained quantity is the the absolute air humidity which should be sufficiently large to avoid serious troubles connected with the preparation of very dry air. Therefore, we assume that the minimum allowable air humidity is equal to X_*. An upper constraint on the humidity can also be imposed to avoid condensation of gas on the cold solid. However, since abnormally high gas humidities result in high total costs,the initial calculations showed that this constraint was not operative when minimum drying costs were considered and thus it was omitted in the final optimization model presented here. For similar reasons the constraint on the gas temperature is not imposed either (too large absolute differences between the gas temperature and the ambient temperature are always costly and so an unconstrained optimum of costs with respect to the inlet gas temperature is reasonable).

The remaining properties of drying processes considered here depend on mode of process, and they are presented below for each mode.

I. Crosscurrent processes

The drying gas mass flow rate at every stage of the cascade,ΔG^n and the total gas flow rate,G^N, are undetermined.The air from every stage of the cascade is released into the atmosphere.

I.1. Unconventional (controlled) process (Sieniutycz 1982).The decision variables are inlet gas temperatures, humidities and flows (T_g^n, X_g^n and ΔG^n, respectively) for n = 1,...N, fig.1.

I.2. Conventional processes. The optimal values of the decision variables are the same for every stage,i.e, $T_g^n = T_g$, $X_g^n = X_g$, $\Delta G^n = \Delta G$. The two variants of conventional processes are considered.

I.2.1. The decision variables are temperature T_g ,humidity X_g and flow ΔG of the inlet gas.

I.2.2. The decision variables are temperature T_g and flow ΔG of the inlet gas. The humidity of the inlet gas at every stage is equal to the ambient humidity X_o.

I.3. Unconventional process with product recycle (Sieniutycz 1984). The mass flow rate of the dry solid is equal to S in the feed stream (as in process without product recycle) and it is equal to R in the recycle stream. Hence the mass flow rate of the dry solid in stream passing through the cascade is equal to S + R (see fig.2) or $S(1 + r_1)$ where r_1 is the product recycle ratio. The decision variables are inlet gas temperatures T_g^n and gas flows ΔG^n for n=1,...N. The humidity of the inlet gas at every stage is equal to the ambient humidity X_o.

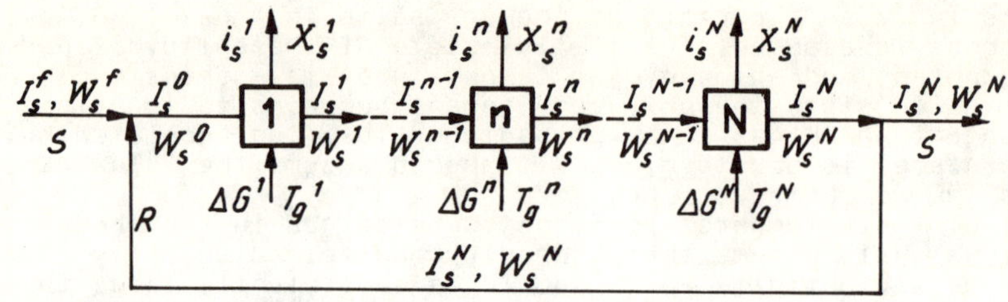

Fig.2.Scheme of the multistage crosscurrent fluidized drying with product recycle.

II. Countercurrent processes

As in the case of crosscurrent processes,every stage of cascade is the fluidized dryer. Starting with stage number n=1 the drying air passes through every stage of the cascade being heated between stage n and n+1 (for n=1,..N-1) from temperature T_{sg}^n to temperature T_g^{n+1} and conserving humidity $X_g^{n+1} = X_{sg}^n$ during heating, where T_{sg}^n and X_{sg}^n represent temperature and air humidity in equilibrium between the gaseous and solid phases leaving the fluidized stage n. Before stage n=1 the drying air is heated to temperature T_g^1. The mass inflow of the fresh drying air,G,is undetermined. The dry solid passes through every stage of the cascade starting with stage number n=N. The general scheme of the multistage countercurrent fluidized drying processes (including processes with drying air recycle) is presented in fig.3.

II.1. Unconventional process without air recycle. The decision variables are inlet gas temperatures T_g^n for n=1,..N and gas inflow G. In the general scheme, fig.3, gas recycle stream G_r is equal to zero (i.e. all air from the last stage is released into the atmosphere) and the absorber for inlet gas preparation is omitted. Thus $T_{sg}^0 = T_o$ (where T_o is the ambient air temperature) and $X_{sg}^0 = X_w = X_o$.

II.2. Conventional process without air recycle. The decision variables are inlet gas temperature T_g ($T_g^n = T_g$ for n=1,..N) and gas inflow G. In the general scheme,fig.3,gas recycle stream is equal to zero,$G_r = 0$. The possibility of preliminary dehumidity of the inlet air,from X_o to X_w, in the isothermal absorber is assumed. The numerical results of the optimization for various values of X_w are presented.

II.3. Conventional process with drying air recycle. The decision variables are inlet gas temperature T_g ($T_g^n = T_g$ for n = 1,...N) and the fresh air inflow G. The absorber for inlet gas preparation is omitted, thus $X_w = X_o$. Temperature and humidity of the drying air entering the first heat exchanger result from the mixing of the

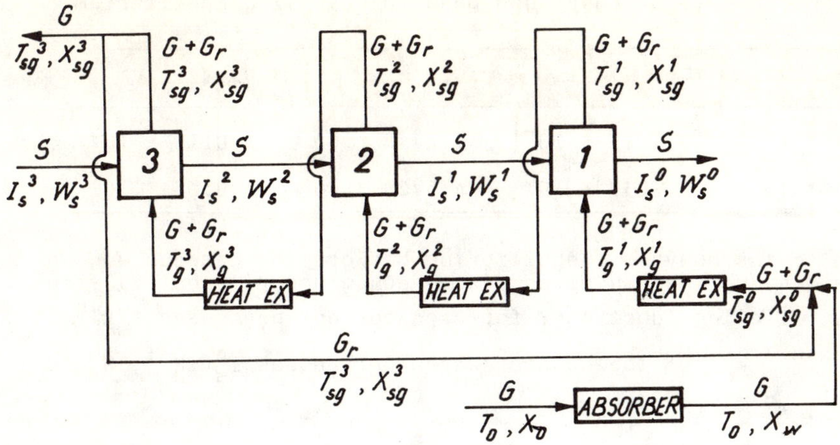

Fig.3.The general scheme of the multistage countercurrent fluidi-
      zed drying.

fresh and recycle air streams. The numerical results of the optimization for various values of the recycle ratio $r_2 = G_r/G$ are presented (Reda 1983 , Szwast 1984).

EQUILIBRIUM DATA

Thermodynamic equilibrium between the gaseous and solid phases
leaving the fluidized stage is commonly observed in beds of heights
such as those used in industry. This means that the simplest (equilibrium-stage) model can be accepted, replacing the more complicated bubble (kinetic) model with an error of the order of 5%.
The gas-solid equilibrium is described by the semiempirical e-
quations

$$i_s(I_s, W_s) = 4.19\left[a_1 W_s - b_1 + (c_1 W_s + d_1)/Y\right]^{-2} \tag{1}$$

$$X_s(I_s, W_s) = \left[a_2 W_s - b_2 + (c_2 W_s + d_2)/Y\right]^{-2} \tag{2}$$

where

$$Y = Y(I_s, W_s) = 0.24 I_s + 32.6 W_s/(0.094 + W_s)$$

which hold true for the silica gel-water-air system considered here.
The coefficients in equations (1) and (2) are given in Table 1.
These equations were found to describe the tubular data (Sieniutycz
1973a) to an accuracy of 10%. They hold in the range $0.1 \leqslant W_s \leqslant 0.2$
kg/kg and $0 \leqslant T_s \leqslant 100^{\circ}C$. The effects of heat of moisture sorption
and drying equilibrium are also taken into account in these equa-
tions.

TABLE 1. Coefficients of Equations Describing Drying Equilibrium

i	a_i	b_i	c_i	d_i
1	$7.1114 * 10^{-2}$	$1.3913 * 10^{-1}$	$1.6448 * 10^{1}$	$5.0111 * 10^{0}$
2	$5.2355 * 10^{0}$	$5.6037 * 10^{0}$	$4.5728 * 10^{2}$	$1.8365 * 10^{2}$

Owing to thermodynamic equilibrium between the gaseous and solid phases leaving the fluidized stage, equations (1) and (2) represent the following functions: for crosscurrent processes $i_s^n(I_s^n, W_s^n)$ and $X_s^n(I_s^n, W_s^n)$ or for countercurrent processes $i_s^n(I_s^{n-1}, W_s^{n-1})$ and $X_s^n(I_s^{n-1}, W_s^{n-1})$ for $n=1,..N$.

ECONOMIC COST ANALYSIS LEADING TO A PERFORMANCE CRITERION

The total economic cost, i.e. the sum of total investment costs and total exploitation costs, was chosen to be the performance criterion of the multistage processes considered here. This criterion involves, amongst others, some special costs (related, for example, to such items as administration, lighting, nonenergetic value inputs, etc.(Sieniutycz 1978)), which are independent of the process itself and, as such, can be neglected in the working form of the performance criterion. Consequently, we formulate the performance criterion as the sum of the investment costs of the cascade and the variable part of the exploitation costs, related to the successive drying modes specified earlier.

<u>Drying mode I.1.</u>

$$F' = \left(\frac{\tilde{z}}{T_{gr}} + \beta\right) \frac{J}{T_u S} + \sum_{n=1}^{N} \frac{e_{el} \Delta P \Delta G^n}{\eta \rho S} + \sum_{n=1}^{N} c_g(T_g^n, X_g^n) \frac{\Delta G^n}{S} \qquad (3)$$

In equation (3) the first term describes the investment costs, the second the gas pumping costs, and the third the cost of preparation of dry hot gas with parameters T_g^n and X_g^n (from the worthless ambient air with T_o and X_o). All costs refer to unit mass of flowing solid.

For the concrete drying problem the quantities $\tilde{z}, T_{gr}, \beta, T_u, S, e_{el}, \Delta P, \eta$ and ρ are known constants. Also, the explicit form of the function $c_g(T_g, X_g)$ is known. In this paper the so-called exergy tariff of prices is used for this purpose. Whereas the total value J of the N-stage cascade can be described by formula (4),

$$J = \sum_{n=I}^{N} (j_o + pA_a^n) = Nj_o + \frac{p}{\rho_v} \sum_{n=1}^{N} \Delta G^n \qquad (4)$$

for industrial fluidized equipment the apparatus price per unit
surface area of sieve bottom,p,is the characteristic measure of e-
quipment value (Sieniutycz 1978). The constant j_o is the value of
the fixed costs of every stage N which appears even if the process is not conducted. For the concrete problem the unit apparatus
price p and the superficial gas velocity v are known constants. We
assume that the total number of stages,N,is fixed in our problem.
    To obtain a more suitable expression for the performance in-
dex we substitute eqn.(4) into eqn.(3) and simplify the result,re-
jecting additive constants. As a result, for constant N,the expression for the performance index can be written in the form:

$$F'' = \sum_{n=1}^{N} \left[c_g(T_g^n , X_g^n) + \lambda_e \right] \theta^n \tag{5}$$

where

$$\theta^n = \Delta G^n / S \tag{6}$$

and λ_e is the so-called economic coefficient of the investment
costs including the cost of gas pumping. On the basis of eqns.(3)
and (4) it is easy to see that λ_e is expressed by the economic and
technical quantities as

$$\lambda_e = \left(\frac{\tilde{z}}{T_{gr}} + \beta \right) \frac{p}{T_u \rho v} + \frac{e_{el} \Delta P}{\eta \rho} \tag{7}$$

i.e.,it is related to both investment and gas pumping costs. All
quantities appearing on the right-hand side of eqn. (7) are given
constants. Therefore λ_e is also known. This constant, λ_e,is a very

suitable parameter for our optimization solution.
    The variable quantities appearing on the right-hand side of e-
quation (5) are exclusively decision variables: dimensionless gas

flows $\theta^n = \Delta G^n / S$, gas temperatures T_g^n, and gas humidities X_g^n,for n=
=1,..N. We seek the the optimal values of these quantities which
make the performance index (5) a minimum. This is,of course,equiva-
lent to the minimization of the total cost (3).

Drying mode I.2.

    Variant I.2.1. Owing to $T_g^n = T_g$, $X_g^n = X_g$ and $\Delta G^n = \Delta G$ (hence $\theta^n =$
$= \theta$), the performance index can be written in the following form
(compare eqn.(5)):

$$F'' = N \left[c_g(T_g , X_g) + \lambda_e \right] \theta \tag{8}$$

    Variant I.2.2. Owing to $T_g^n = T_g$, $\Delta G^n = \Delta G$ and $X_g^n = X_o$ the perfor-
mance index can be written in the following form:

$$F'' = N \left[c_g(T_g) + \lambda_e \right] \theta \tag{9}$$

Drying mode I.3.

Owing to T_g^n and ΔG^n varied with $n=1,..N$ and $X_g^n=X_o$, the performance index can be written in the following form (compare eqn.(5)):

$$F'' = \sum_{n=1}^{N} \left[c_g(T_g^n) + \lambda_e \right] \theta^n \tag{10}$$

where θ^n is defined by eqn.(6), or in the form:

$$F'' = (1 + r_1) \sum_{n=1}^{N} \left[c_g(T_g^n) + \lambda_e \right] g_1^n \tag{11}$$

where

$$g_1^n = \frac{\Delta G^n}{S+R} \tag{12}$$

and product recycle ratio, r_1, is defined as

$$r_1 = R/S \tag{13}$$

It should be undelined that dimensionless gas flow θ^n describes the real gas consumpion in fluidized dryer n per unit final product stream whereas dimensionless gas flow g_1^n describe the real mass balance in this dryer. The recycle ratio, r_1, is a known parameter.

Drying mode II.1.

For countercurrent processes, it is the same mass stream of gas passing through every stage of the cascade, $\Delta G^n=G$. Moreover, the inlet gas for any stage $n=2,..N$ is prepared from the outlet gas for $n=1,..N-1$, respectively. Thus, the unit cost of this preparation (heating) can be expressed as the gas price increment. The performance index can be written in the following form (compare eqn. (5) and fig.3 without gas recycle and without absorber):

$$F'' = \sum_{n=1}^{N} \left[c_g(T_g^n, X_g^n) - c_g(T_{sg}^{n-1}, X_{sg}^{n-1}) + \lambda_e \right] \theta \tag{14}$$

with $X_g^n = X_{sg}^{n-1}$ for $n=1,..N$ and $X_{sg}^0 = X_o$, $T_{sg}^0 = T_o$. The values of X_{sg}^n and T_{sg}^n for $n=1,..N$ result from the energy balance for every stage. The decision variables are T_g^n for $n=1,..N$ and θ.

Note: For crosscurrent processes, θ^n (or θ, respectively) describes the gas consumption in dryer n per unit final product, whereas the total gas comsumption is described by $G^N/S = \sum_{n=1}^{N} \theta^n$ (or $N*\theta$, respectively). For countercurrent processes, $\theta=G/S$ describes dimensionless gas flow for every stage of the cascade, as well as, the total gas consumption per unit final product.

Drying mode II.2.

The optimal values of the inlet gas temperature are the same for every stage, $T_g^n = T_g$. The preliminary dehumidity of inlet air, from X_o to X_w, is assumed. Owing to air price increment in the isothermal absorber, expressed as $c_g(T_o,X_w) - c_g(T_o,X_o)$ the performance index can be written in the following form:

$$F'' = \sum_{n=1}^{N} \left[c_g(T_g,X_g^n) - c_g(T_{sg}^{n-1},X_{sg}^{n-1}) + \lambda_e \right] \theta +$$

$$+ \left[c_g(T_o,X_w) - c_g(T_o,X_o) \right] \theta \tag{15}$$

with $X_{sg}^0 = X_w$ and $T_{sg}^0 = T_o$. The decision variables are T_g and θ, whereas X_w is a known parameter.

Drying mode II.3.

Owing to $\Delta G^n = \Delta G = G + G_r$ and gas recycle ratio, r_2, defined as

$$r_2 = G_r/G \tag{16}$$

and hence

$$\frac{G + G_r}{S} = g_2 = (1 + r_2)\frac{G}{S} = (1 + r_2)\,\theta \tag{17}$$

the performance index can be written in the following form:

$$F'' = \sum_{n=1}^{N} \left[c_g(T_g,X_g^n) - c_g(T_{sg}^{n-1},X_{sg}^{n-1}) + \lambda_e \right] g_2 \tag{18}$$

or

$$F'' = (1 + r_2)\sum_{n=1}^{N} \left[c_g(T_g,X_g^n) - c_g(T_{sg}^{n-1},X_{sg}^{n-1}) + \lambda_e \right] \theta \tag{19}$$

with T_{sg}^0 and X_{sg}^0 resulting from the mixing of the fresh air with X_o,T_o and recycle air with X_{sg}^N,T_{sg}^N. The decision variables are T_g and θ, whereas gas recycle ratio, r_2, is a known parameter.

Note: Dimensionless gas flow $\theta = G/S$ describes total fresh gas consumption per unit final product, whereas dimensionless gas flow $g_2 = (G+G_r)/S$ describes the mass balance for every stage of the cascade.

EXERGY TARIFF

The drying air price c_g can be computed with the help of a so-called exergy tariff, i.e. as the product of the unit exergy,

$b_g(T_g,X_g)$, and the unit exergy price e, which is estimated as e =
= 8 \$/GJ (Sieniutycz 1978). However, the use of the exergy tariff
is not necessary - one may use another scale of prices which is acceptable in the concrete factory.

The exergy of humid air as a function of its temperature T_g
and humididy X_g is described by the formula (Sieniutycz 1978, Szargut 1965):

$$b_g(T_g,X_g) = (c_p + X_g c_w)(T_g - T_o - T_o \ln \frac{T_g}{T_o}) + \frac{RT_o}{M_p} \ln \frac{M_w/M_p + X_o}{M_w/M_p + X_g} +$$

$$+ \frac{RT_o X_g}{M_w} \ln \frac{X_g(M_w/M_p + X_o)}{X_o(M_w/M_p + X_g)} \qquad (20)$$

Here we shall use the convenient simplified form of exergy that is
obtained after Taylor expansion of the exact exergy function around the point (T_o,X_o), neglecting third-order and higher terms. The
gas unit price is then obtained as

$$c_g(T_g,X_g) = \frac{1}{2} e \left[A(T_g - T_o)^2 + B(X_g - X_o)^2 \right] \qquad (21)$$

where

$$A = (c_p + X_o c_w)/T_o \qquad (22)$$

and

$$B = RT_o/(M_w + X_o M_p)X_o \qquad (23)$$

Note: If $X_g = X_o$ the last term in eqn.(21), $B(X_g - X_o)^2$, vanishes - compare the inlet air price in drying mode I.2. variant I.2.2.

MATHEMATICAL MODEL OF OPTIMIZATION

After applying eqn.(21) in the formulas describing the performance indices for the drying modes mentioned above and defining
the exergy coefficient of investment and gas pumping costs

$$\lambda = \lambda_e/e \qquad (24)$$

one can obtain the new form of the performance indices

$$F = F''/e \qquad (25)$$

The performance indices F, expressed in exergy units, are used in
our optimization.

Under the above-mentioned equilibrium-stage assumption the
process state equations are obtained as a result of elementary balances of energy and mass at arbitrary stage n for n=1,..N.

The performance index, state equations, boundary conditions and
algebraic constraint imposed on the minimum allowable air humidity

complete the mathematical model of the process. The mathematical models, for each drying mode considered here, are presented below.

Drying mode I.1.

$$F = \sum_{n=1}^{N} \left[\frac{1}{2} A (T_g^n - T_o)^2 + \frac{1}{2} B (X_g^n - X_o)^2 + \lambda \right] \theta^n \tag{26}$$

State equations (27) and (28) for $n=1,\ldots N$

$$I_s^n - I_s^{n-1} = \left[i_g^n(T_g^n, X_g^n) - i_s^n(I_s^n, W_s^n) \right] \theta^n \tag{27}$$

$$W_s^n - W_s^{n-1} = \left[X_g^n - X_s^n(I_s^n, W_s^n) \right] \theta^n \tag{28}$$

where $i_g^n(T_g^n, X_g^n)$ is the inlet air enthalpy at the nth stage computed as

$$i_g^n(T_g^n, X_g^n) = (c_p + X_g^n c_w)(T_g^n - 273.15) + 2502.7 X_g^n \tag{29}$$

The initial conditions for the coordinates I_s^0, W_s^0 are prescribed for both state equations (27) and (28). For eqn. (28) the final condition for W_s^N is prescribed, whereas the state coordinate I_s^N is undetermined.
   The problem comes down to finding the optimal decision variables θ^n, T_g^n and X_g^n for $n=1,\ldots N$ which maximize the performance index F (eqn. (26)) under the constraints resulting from the state equations (27) and (28), the boundary conditions for I_s^0, W_s^0 and W_s^N and the constraint $X_g^n \geqslant X_*$. The decisions T_g^n and θ^n are not constrained and the total air flow is undetermined.

Drying mode I.2.
   Variant I.2.1.

$$F = N \left[\frac{1}{2} A (T_g - T_o)^2 + \frac{1}{2} B (X_g - X_o)^2 + \lambda \right] \theta \tag{30}$$

$$I_s^n - I_s^{n-1} = \left[i_g(T_g, X_g) - i_s^n(I_s^n, W_s^n) \right] \theta \tag{31}$$

$$W_s^n - W_s^{n-1} = \left[X_g - X_s^n(I_s^n, W_s^n) \right] \theta \tag{32}$$

The coordinates I_s^0, W_s^0 and W_s^N are preascribed, and I_s^N is undetermined. The decision variables are θ, T_g and X_g with constraint $X_g \geqslant X_*$.

   Variant I.2.2.

$$F = N \left[\frac{1}{2} A (T_g - T_o)^2 + \lambda \right] \theta \tag{33}$$

The state equations for n=1,..N are eqns.(31) and (32) for $X_g = X_o$.
The decision variables are θ and T_g.

Drying mode I.3.

The performance index can be written in the form

$$F = \sum_{n=1}^{N} \left[\frac{1}{2} A \, (T_g^n - T_o)^2 + \lambda \right] \theta^n \tag{34}$$

or

$$F = (1 + r_1) \sum_{n=1}^{N} \left[\frac{1}{2} A \, (T_g^n - T_o)^2 + \lambda \right] g_1^n \tag{35}$$

The state equations (36) and (37) for n=1,..N

$$I_s^n - I_s^{n-1} = \left[i_g^n(T_g^n) - i_s^n(I_s^n, W_s^n) \right] \frac{1}{r_1 + 1} \theta^n \tag{36}$$

$$W_s^n - W_s^{n-1} = \left[X_o - X_s^n(I_s^n, W_s^n) \right] \frac{1}{r_1 + 1} \theta^n \tag{37}$$

or (38) and (39) for n=1,..N

$$I_s^n - I_s^{n-1} = \left[i_g^n(T_g) - i_s^n(I_s^n, W_s^n) \right] g_1^n \tag{38}$$

$$W_s^n - W_s^{n-1} = \left[X_o - X_s^n(I_s^n, W_s^n) \right] g_1^n \tag{39}$$

where the inlet air enthalpy is computed as (compare eqn.(29)):

$$i_g^n(T_g^n) = (c_p + X_o c_w)(T_g^n - 273.15) + 2502.7 X_o \tag{40}$$

The state of the solid before the first stage of the cascade is described by following mixing functions:

$$I_s^0 = M_1(I_s^N) = \frac{S I_s^f + R I_s^N}{S + R} = \frac{I_s^f + r_1 I_s^N}{1 + r_1} \tag{41}$$

$$W_s^0 = M_2(I_s^N) = \frac{S W_s^f + R W_s^N}{S + R} = \frac{W_s^f + r_1 W_s^N}{1 + r_1} \tag{42}$$

The coordinates I_s^f, W_s^f and W_s^N are preascribed, and I_s^N is undetermined. The product recycle ratio, r_1, is a known parameter. The decision variables are θ^n and T_g^n.

<u>Drying mode II.1.</u>

The performance index and state equations for $n=1,..N$ can be
written as the formulas (43) and (44),(45) - respectively (compare
eqn.(14) writting eqn.(43), and see fig.3 writting eqns (44),(45))

$$F = \sum_{n=1}^{N} \left[\frac{1}{2} A \, (T_g^n - T_o)^2 + \frac{1}{2} B \, (X_g^n - X_o)^2 \right.$$

$$\left. - \frac{1}{2} A \, (T_{sg}^{n-1} - T_o)^2 - \frac{1}{2} B \, (X_{sg}^{n-1} - X_o)^2 + \lambda \right] \theta \qquad (43)$$

$$I_s^n = I_s^{n-1} - \left[i_g^n(T_g^n, X_g^n) - i_{sg}^n(I_s^{n-1}, W_s^{n-1}) \right] \theta \qquad (44)$$

$$W_s^n = W_s^{n-1} - \left[X_g^n - X_{sg}^n(I_s^{n-1}, W_s^{n-1}) \right] \theta \qquad (45)$$

or in the simplified form (because of $X_g^n = X_{sg}^{n-1}$ for $n=1,..N$ - see information below eqn.(14))

$$F = \sum_{n=1}^{N} \left[\frac{1}{2} A \, (T_g^n - T_o)^2 - \frac{1}{2} A \, (T_{sg}^{n-1} - T_o)^2 + \lambda \right] \theta \qquad (46)$$

$$I_s^n = I_s^{n-1} - \left[i_g^n(T_g^n, X_{sg}^{n-1}) - i_{sg}^n(I_s^{n-1}, W_s^{n-1}) \right] \theta \qquad (47)$$

$$W_s^n = W_s^{n-1} - \left[X_{sg}^{n-1} - X_{sg}^n(I_s^{n-1}, W_s^{n-1}) \right] \theta \qquad (48)$$

with $X_{sg}^0 = X_o$, $T_{sg}^0 = T_o$ (see information below eqn.(14)) and T_{sg}^n for
$n=1,..N$ computed from eqn.(49) describing the outlet air enthalpy

$$i_{sg}^n (T_{sg}^n, X_{sg}^n) = (c_p + X_{sg}^n c_w)(T_{sg}^n - 273.15) + 2502.7 \, X_{sg}^n \qquad (49)$$

The equilibrium outlet air enthalpy and humidity, $i_{sg}^n(I_s^{n-1}, W_s^{n-1})$
and $X_{sg}^n(I_s^{n-1}, W_s^{n-1})$ are computed with using of eqns (1) and (2).
The coordinates I_s^N , W_s^N (the inlet solid parameters) and W_s^0 are
prescribed, whereas the outlet solid enthalpy I_s^0 is undetermined.
The decision variables are T_g^n for $n=1,..N$ and θ.

<u>Drying mode II.2.</u>

The performance index can be expressed by eqn.(46) for $T_g^n = T_g$
and enlarged by the inlet air exergy increment in the isothermal
absorber (the last term in eqn.(50)). Thus

$$F = \sum_{n=1}^{N} \left[\frac{1}{2} A \, (T_g - T_o)^2 - \frac{1}{2} A \, (T_{sg}^{n-1} - T_o)^2 + \lambda \right] \theta +$$

$$+ \frac{1}{2} B \, (X_w - X_o)^2 \theta \qquad (50)$$

The state equations for n=1,..N are eqns (47) and (48) for $T_g^n = T_g$ and with $X_{sg}^0 = X_w$. The boundary conditions are the same as for the drying mode II.2. The air humidity X_w is a known parameter. The decision variables are T_g and θ.

Drying mode II.3.

The performance index can be written in the following form (see eqn.(19) and compare transformation of eqn.(14) to eqn.(46)):

$$F = (1 + r_2) \sum_{n=1}^{N} \left[\tfrac{1}{2} A (T_g - T_o)^2 - \tfrac{1}{2} A (T_{sg}^{n-1} - T_o)^2 + \lambda \right] \theta \quad (51)$$

The state equations for n=1,..N can be written in the form of eqns (52) and (53) (compare eqns (47) and (48) for the same inlet air temperature for every stage i.e. $T_g^n = T_g$,and see definition of variable θ for the countercurrent process with the drying air recycle, eqn.(17)):

$$I_s^n = I_s^{n-1} - \left[i_g^n(T_g, X_{sg}^{n-1}) - i_{sg}^n(I_s^{n-1}, W_s^{n-1}) \right] (1 + r_2) \theta \quad (52)$$

$$W_s^n = W_s^{n-1} = \left[X_{sg}^{n-1} - X_{sg}^n(I_s^{n-1}, W_s^{n-1}) \right] (1 + r_2) \theta \quad (53)$$

with X_{sg}^0 and T_{sg}^0 resulting from the mixing of the fresh and recycle air. For the air humidity and enthalpy we can write

$$X_{sg}^0 = \frac{GX_o + G_r X_{sg}^N}{G + G_r} = \frac{X_o + r_2 X_{sg}^N}{1 + r_2} \quad (54)$$

$$i_{sg}^0 = \frac{Gi_g(T_o, X_o) + G_r i_{sg}^N(T_{sg}^N, X_{sg}^N)}{G + G_r} = \frac{i_o + r_2 i_{sg}^N}{1 + r_2} \quad (55)$$

Then, T_{sg}^0 can be computed from eqn.(56)

$$i_{sg}^0(T_{sg}^0, X_{sg}^0) = (c_p + X_{sg}^0 c_w)(T_{sg}^0 - 273.15) + 2502.7 X_{sg}^0 \quad (56)$$

The coordinates I_s^N , W_s^N (the inlet solid parameters) and W_s^0 are preascribed, and the outlet solid enthalpy I_s^0 is undetermined. The drying air recycle ratio, r_2, is a known parameter. The decision variables are T_g and θ.

COMPUTATIONAL PROCEDURE

Below, the computational procedure for the drying mode I.1. only is presented in details. The procedures for the other modes

222

are mentioned at the end of this chapter.

For the drying mode I.1., the mathematical model of the process, eqns (26) - (28), is linear with respect to the unconstrained decisions θ^n. This fact enables the use of the special discrete optimization algorithm of the maximum principle with a constant discrete Hamiltonian (Szwast 1978, Sieniutycz 1983).

The Hamiltonian function for $n=1,..N$ (see Appendix):

$$H^{n-1}(I_s^n, W_s^n, z_1^{n-1}, z_2^{n-1}, T_g^n, X_g^n) = \tfrac{1}{2} A (T_g^n - T_o)^2 + \tfrac{1}{2} B (X_g^n - X_o)^2 + \lambda +$$

$$+ z_1^{n-1} \left[(c_p + X_g^n c_w)(T_g^n - 273.15) + 2502.7 \, X_g^n - i_s^n(I_s^n, W_s^n) \right] +$$

$$+ z_2^{n-1} \left[X_g^n - X_s^n(I_s^n, W_s^n) \right] = 0 \tag{57}$$

Since the total gas flow is undetermined (free $G^N/S = \sum_{n=1}^{N} \theta^n$), the constant value of the Hamiltonian (57) is equal zero,

The adjoint equations for $n=1,..N$:

$$\frac{z_1^n - z_1^{n-1}}{\theta^n} = - \frac{\partial H^{n-1}}{\partial I_s^n} = z_1^{n-1} \frac{\partial i_s^n(I_s^n, W_s^n)}{\partial I_s^n} + z_2^{n-1} \frac{\partial X_s^n(I_s^n, W_s^n)}{\partial I_s^n} \tag{58}$$

$$\frac{z_2^n - z_2^{n-1}}{\theta^n} = - \frac{\partial H^{n-1}}{\partial W_s^n} = z_1^{n-1} \frac{\partial i_s^n(I_s^n, W_s^n)}{\partial W_s^n} + z_2^{n-1} \frac{\partial X_s^n(I_s^n, W_s^n)}{\partial W_s^n} \tag{59}$$

As I_s^N is undetermined and W_s^N is prescribed we have the following boundary conditions associated with eqns (58) and (59) respectively

$$z_1^N = 0 \tag{60}$$

$$z_2^N - \text{undetermined} \tag{61}$$

Computation of the partial derivatives appearing in iqns (58) and with the help of eqns (1) and (2) and solution of these equations with respect to z_1^{n-1} and z_2^{n-1} yield

$$z_1^{n-1} = \frac{z_1^n - \theta^n(z_1^n \beta_2^n - z_2^n \beta_1^n)}{1 + \theta^n(\alpha_1^n - \beta_2^n) + (\theta^n)^2(\alpha_2^n \beta_1^n - \alpha_1^n \beta_2^n)} \tag{62}$$

$$z_2^{n-1} = \frac{z_1^n \theta^n \alpha_2^n + z_2^n(\alpha_1^n \theta^n + 1)}{1 + \theta^n(\alpha_1^n - \beta_2^n) + (\theta^n)^2(\alpha_2^n \beta_1^n - \alpha_1^n \beta_2^n)} \tag{63}$$

where

$$\alpha_1^n = 2Y^n(c_1 W_s^n + d_1)E_1^n \tag{64}$$

$$\beta_1^n = 2Y^n(c_2 W_s^n + d_2)E_2^n \tag{65}$$

$$\mathcal{A}_2^n = 2Y^n \left[a_1(Y^n)^2 + c_1 Y^n - \frac{3.064(c_1 W_s^n + d_1)}{(0.094 + W_s^n)^2} \right] E_1^n \tag{66}$$

$$\beta_2^n = 2Y^n \left[a_2(Y^n)^2 + c_2 Y^n - \frac{3.064(c_2 W_s^n + d_2)}{(0.094 + W_s^n)^2} \right] E_2^n \tag{67}$$

where

$$Y^n = Y^n(I_s^n, W_s^n) = 0.24 I_s^n + 32.6 W_s^n/(0.094 + W_s^n) \tag{68}$$

$$E_i^n = \left[(a_i W_s^n - b_i)Y^n + c_i W_s^n + d_i \right]^{-3} \qquad \text{for } i=1,2 \tag{69}$$

The stationary optimal gas humidity X_g^n and gas temperature T_g^n are determined from the relations

$$\frac{\partial H^{n-1}}{\partial X_g^n} = B(X_g^n - X_o) + z_1^{n-1}c_w(T_g^n - 273.15) + 2502.7 + z_2^{n-1} = 0 \tag{70}$$

$$\frac{\partial H^{n-1}}{\partial T_g^n} = A(T_g^n - T_o) + z_1^{n-1}(c_p + X_g^n c_w) = 0 \tag{71}$$

The solution of the system eqns (70) and (71), with respect to X_g^n and T_g^n, is

$$X_g^n = \left[BX_o + 273.15 z_1^{n-1}c_w - 2502.7 - z_2^{n-1} - z_1^{n-1}c_w T_o + \right.$$
$$\left. + (z_1^{n-1})^2 c_w c_p/A \right] \left[B - (z_1^{n-1}c_w)^2/A \right]^{-1} \tag{72}$$

$$T_g^n = T_o - z_1^{n-1}(c_p + X_g^n c_w)/A \tag{73}$$

When the stationary value of X_g^n, computed from eqn.(72), is less than the minimum allowable humidity X_*, then the optimal humidity is equal to X_*, i.e.

$$X_g^n = X_* \tag{74}$$

In this case eqn.(73) simplifies to

$$T_g^n = T_o - z_1^{n-1}(c_p + X_* c_w)/A \tag{75}$$

The equations derived above enable the preparation of the computational block scheme, fig.4. The computational procedure in which

224

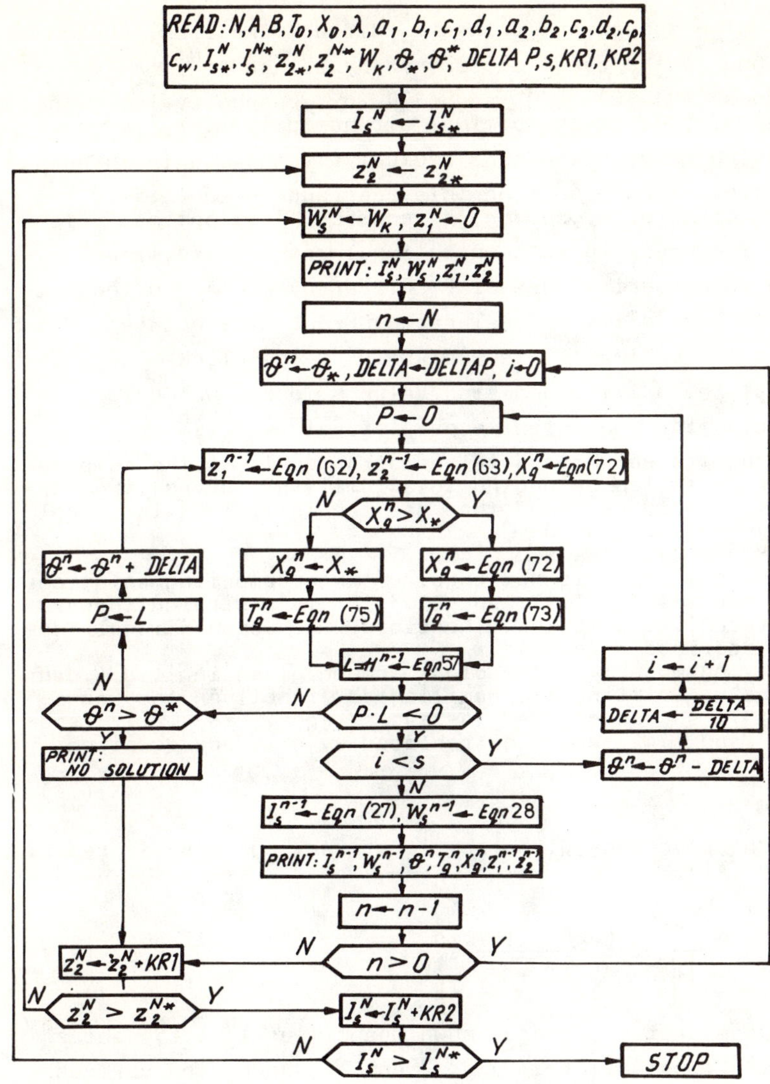

Fig.4. The computational block scheme for the drying mode I.1.

the unknown quantities I_s^N and z_2^N are assumed leads to the family of optimal trajectories having the common final moisture content $W_s^N = W_k$ and various initial states I_s^0, W_s^0. The initial state I_s^0, W_s^0 of every computed trajectory is optimal in relation to the assumed values I_s^N and z_s^N. The subsequent values I_s^N differ one from another by KR2, fig.4, and they belong to the computational range determined by I_{s*}^{N} and I_{s}^{N*}. The subsequent values z_2^N differ one from another by KR1 and belong to the range determined by z_{2*}^{N} and z_2^{N*}.

After use of eqns (62),(63),(72) and (73) (or, respectively, eqns (62),(63),(74) and (75) if the optimal air humidity is X_*) in eqn. (57),the resulting formula contains exclusively the single unknown variable θ^n. However, when solving this rearranged eqn. (57) the use of a trial-and-error procedure is required. The optimal flow θ^n is found in the preassigned computational range determined by θ_* and θ^*. The subsequent values of θ^n differ by DELTA and the initial value of this quantity is designated DELTAP. The accuracy of computations of the optimal θ^n is DELTAP$*10^{-s}$. The block "NO SOLUTION" means that the optimal decision value θ^n does not belong to the range θ_* to θ^* for the definite pair of values I_s^N, z_2^N.

Among the various trajectories obtained as a result of the computations (these trajectories belonging to a family of curves),the specific trajectory should be found that has the initial state equal to the preassigned state I_s^0, W_s^0.

The maximum principle is not required when computing solutions for the drying modes I.2., II.2. and II.3. It is a straightforward problem in differential calculus. Therefore, the computational details are not described here.

For the drying mode I.3, the block scheme given in fig.4 can be used. Obviously, writting the Hamiltonian function the eqns. (34),(36) and (37) should be used instead of eqns. (26) , (27) and (28). Eqns (72) and (74) should be replaced by only one equation

$$X_g^n = X_o \tag{76}$$

Moreover, for the processes with recycle stream, eqn.(60) is replaced by (see eqn.(41))

$$z_1^N = \frac{dM_1(I_s^N)}{dI_s^N} \, z_1^0 = \frac{R}{S + R} \, z_1^0 = \frac{r_1}{1 + r_1} \, z_1^0 \tag{77}$$

Thus, starting computation, the unknown quantities I_s^N, z_1^N and z_2^N should be assumed (instead of I_s^N and z_2^N only). Among the various trajectories obtained as a result of the computations the specific trajectory should be found that has the initial state equal to the preassigned state I_s^0, W_s^0 and condition (77) is satisfied.

For the drying mode II.1, the analogous (to that, which is given in fig.4) block scheme was prepared, but the commonly known discrete algorithm of the maximum principle,given by Katz and Fan (Fan 1966), was applied.

RESULTS

The optimal control and optimal trajectories were obtained for the following numerical data: N=3, W_s^f=0.2 kg/kg, $I_s^f = - 11.2$ kJ/kg

(this enthalpy, for W_s^f given above
corresponds to temperature $T_s^0 =$
$= 293.15$ K), $W_k = 0.1$ kg/kg, $c_p =$
$= 1.00$ kJ/(kgK), $c_w = 1.88$ kJ/(kgK)
$T_0 = 293.15$ K, $X_0 = 0.008$ kg/kg, $X_* =$
$= 0.00001$ kg/kg. Various values of
λ (various unit apparatus prices)
were tested in the computations.
Note: For the crosscurrent processes without product recycle we
have $W_s^0 = W_s^f$ and $I_s^0 = I_s^f$ whereas for
the crosscurrent with product re-
cycle, the values of W_s^0 and I_s^0
result from eqns (42) and (41) -
respectively. For every crosscurrent process $W_s^N = W_k$. For every
countercurrent process $W_s^N = W_s^f$,
$I_s^N = I_s^f$ and $W_s^0 = W_k$.

Drying mode I.1.

The results of the optimization computations are shown in
Figs.5-6 and in Table 2.
The optimal temperatures T_g^n
of drying air that is supplied
to the three-stage cascade of
fluidized dryers are presented
in Fig.5a. The following can be
noted:
(1) Air temperature decreases
monotonically with the stage number n for any value of λ. Values
of T_g^1 and T_g^2 are close to one
another.
(2) All optimal air temperatu-
res (i.e. T_g^1, T_g^2 and T_g^3) increase
with λ. This means that the more
intensive process should be conducted in the more expensive apparatus.
(3) The differences between inlet gas temperatures in the subsequent stages increase with λ.
(4) For $\lambda = 0$ (negligible costs
of equipment and gas pumping)
the optimal air temperature is
the ambient temperature T_0 (see
also farther discussion on the
case when the optimal air humi-

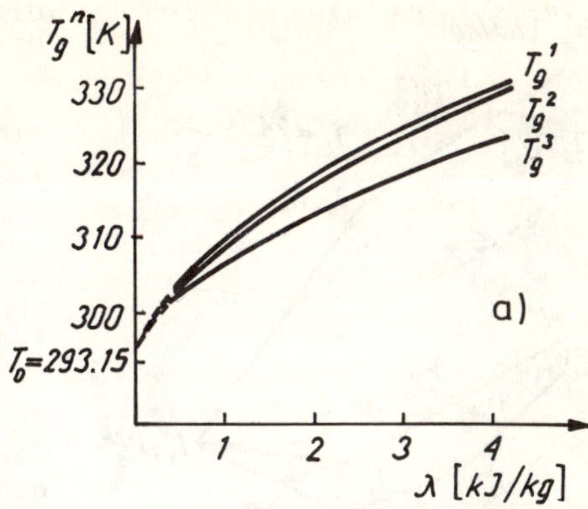

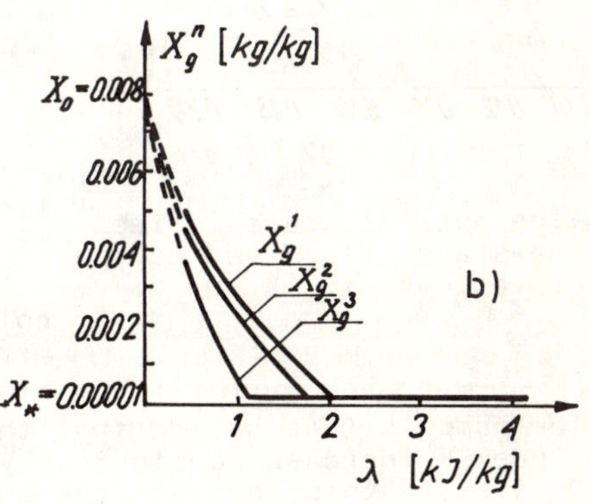

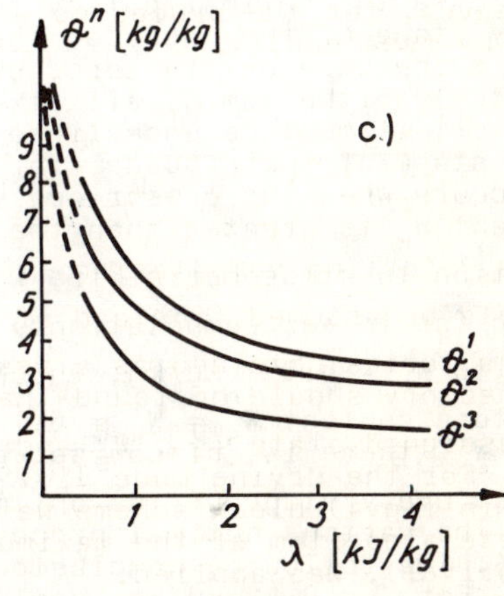

Fig.5. The optimal inlet air temperatures T_g^n, humidities X_g^n and
dimensionless flows θ^n (n=1,2,3)
versus λ.

227

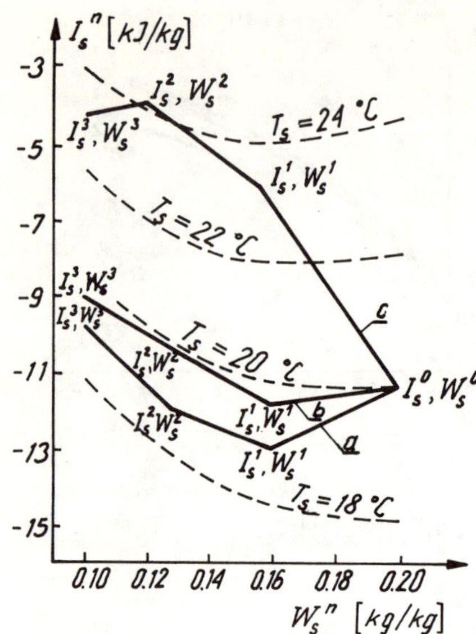

Fig.6. The optimal trajectories of three-stage fluidized drying for various λ.

dity is equal to the ambient humidity X_o).

The optimal inlet air humidities X_g^n are presented in Fig.5b. Their properties are as follows:

(1) The optimal X_g^n assume stationary values only when the unit apparatus price is sufficiently low (i.e. small λ).

(2) The largest stationary values of X_g^n correspond to the first stage whereas the smallest correspond to the last one.

(3) For $\lambda=0$ the optimal air humidity is equal to the ambient humidity X_o.

The optimal dimensionless air flows θ^n are presented in Fig. 5c. The following can be seen:

(1) The optimal dimensionless flows θ^n decrease monotonically with the stage number n.

(2) Increase of λ reduces the optimal θ^n at every process stage.

(3) For $\lambda=0$ the optimal values θ^1, θ^2 and θ^3 become large.

The computations showed that the optimal values of the performance index F increase monotonically with the factor λ (F $\rightarrow$ 0 when $\lambda \rightarrow$ 0, i.e., in this special case only fixed costs are important).

Information obtained for $\lambda \rightarrow$ 0 indicates that in this limiting case the optimal policy relies on exploiting ambient air at every stage of the cascade. The total flow of such air, where the unit price is equal to zero, becomes large. The total process cost is equal to the sum of all fixed costs connected with the process. However, it must be underlined that the case of ambient air for every stage of the cascade (for λ =0) being the optimal solution only occurs when the prescribed value of the final solid moisture content W_k is greater than the equilibrium value of the solid in relation to atmospheric air, W_e. For T_o = 293.15 K and X_o = 0.008 kg/kg (as in our computations), the equilibrium value of W_s is equal to W_e = 0.091 kg/kg, which is less than the prescribed final moisture content W_s^N = W_k = 0.1 kg/kg. In the other case i.e. when $W_e > W_k$, there is, of course, no possibility of using atmospheric air at every stage of the optimal process.

The various optimal process trajectories are presented in Fig.6. The following conclusions can be derived

(1) For λ = 0.42 kJ/kg, cooling of the solid is observed in the first stage of the cascade. The process is practically isothermal in the two next stages.

(2) For λ = 2.10 kJ/kg, the whole process is approximately isothermal.

(3) For λ = 4.20 kJ/kg, the solid is heated in the first two

228

TABLE 2. Optimization Results for the Unconventional and Conventional Crosscurrent Processes without Product Recycle

| | | $\lambda = 0.42$ kJ/kg | | | $\lambda = 4.20$ kJ/kg | | |
| | | Drying mode | | | Drying mode | | |
		I.1.	I.2.1.	I.2.2.	I.1.	I.2.1.	I.2.2.
T_g^1	K	303.31	302.48	304.47	330.83	328.77	337.30
T_g^2	K	302.46	302.48	304.47	330,03	328.77	337.30
T_g^3	K	301.83	302.48	304.47	323.21	328.77	337.30
X_g^1	kg/kg	0.0050	0.0046	0.0080	0.00001	0.00001	0.0080
X_g^2	kg/kg	0.0047	0.0046	0.0080	0.00001	0.00001	0.0080
X_g^3	kg/kg	0.0040	0.0046	0.0080	0.00001	0.00001	0.0080
θ^1	kg/kg	8.14	7.02	8.61	3.39	2.64	2.85
θ^2	kg/kg	7.06	7.02	8.61	2.81	2.64	2.85
θ^3	kg/kg	5.68	7.02	8.61	1.67	2.64	2.85
W_s^0	kg/kg	0.2000	0.2000	0.2000	0.2000	0.2000	0.2000
W_s^1	kg/kg	0.1589	0.1643	0.1654	0.1552	0.1663	0.1678
W_s^2	kg/kg	0.1259	0.1311	0.1309	0.1188	0.1325	0.1331
W_s^3	kg/kg	0.1000	0.1000	0.1000	0.1000	0.1000	0.1000
I_s^0	kJ/kg	-11.20	-11.20	-11.20	-11.20	-11.20	-11.20
I_s^1	kJ/kg	-12.91	-14.05	-8.66	-5.92	-7.08	3.93
I_s^2	kJ/kg	-11.97	-12.63	-6.81	-3.84	-5.47	5.66
I_s^3	kJ/kg	-9.87	-8.78	-3.51	-4.02	-2.18	8.20
F	kJ/kg	13.952	13.977	16.486	54.484	54.593	64.443

stages and is cooled in the last one.

(4) The final optimal temperature or enthalpy of the solid increases with λ.

(5) The largest changes of solid moisture content $W_s^n - W_s^{n-1}$ appear at the first dryer and are lowest at the last dryer. This property becomes more obvious with increasing λ.

The general property of the optimization solutions is that the optimal process intensity and, consequently, the optimal operational costs of drying increase with the measure of the unit apparatus price λ. A large energy consumption is therefore allowed only in relatively costly equipment.

Drying mode I.2.

The results of the optimization computations for this drying mode (for the both variants, i.e.Variant I.2.1. and Variant I.2.2.) are shown in Table 2. By considering this data,the reader can easily compare the properties of the optimal trajectories and the optimal decision values for the conventional and unconventional (control) processes.

For us, the comparison of costs for the conventional and unconventional processes is the most important matter.The optimal data for the performance indices for the various modes and variants of drying (here denoted as $F_{I.1}$, $F_{I.2.1}$ and $F_{I.2.2}$ respectively) are given at the foot of Table.2. It should be remembered that our performance index describes (in exergy units) only the variable part of the total cost which depends on the process decisions.However, when changing from one drying mode to another (for λ =const) the difference between the related performance indices is equal to that for the corresponding total costs. This is,of course,a consequence of the fact that the fixed costs are the same for all modes of drying considered.

To judge the superiority of one mode versus another,knowledge of the relative increments $\Delta F_t/F_t$ is preferable. By introducing the factor μ

$$\mu = F_t/F \qquad (78)$$

i.e.,the ratio of the total exergy cost to its variable part,and taking into account that the fixed costs are the same for each mode of drying, we can evaluate the index ϕ which characterizes the relative change of the process costs when changing from the unconventional process to variant I.2.1. of the conventional process:

$$\phi = \frac{F_{I.2.1} - F_{I.1}}{\mu F_{I.1}} * 100 \% \qquad (79)$$

TABLE 3. The relative differences between the process costs due to various modes of optimization

λ	kJ/kg	0	0.42	4.20	42.00
ϕ	%	0	0.18	0.20	1.07
γ	%	0	17.95	18.04	4.57

The data for ϕ should,in general,be stored for various λ and μ. To avoid voluminous tables we propose to store exclusively "standardized" data for μ=1 (for various λ). The standardized data for ϕ are given in Table 3. The data for an arbitrary μ can simply be found by division of the data in Table 3 by μ.

The values of ϕ increase with λ,but they are generally not too large, even for large λ. This means that replacement of the optimal control process by variant I.2.1 of the optimal conventional process (with optimal chosen levels of T_g and X_g) does not decrease to any great extent the economics of the process considered.

In the same manner,the two conventional processes (respectively variants I.2.1. and I.2.2.) can be compared by defining the index

230

$$\Psi = \frac{F_{I.2.2} - F_{I.2.1}}{\mu F_{I.2.1}} * 100 \ \% \tag{80}$$

The standardized data for Ψ (for $\mu=1$ and various λ) are given in Table 3. The most important conclusion drawn from these data is that the relative increment of the total cost, Ψ, attains large values (the change of Ψ with λ has a maximum, but this effect is not general). This means that the acceptance of an arbitrary (non-optimal) level of the decision variable (the gas humidity in our case) makes the process uneconomical. Therefore, the importance of gas humidity level optimization in drying practice is an essential factor in the light of this work.

<u>Drying mode I.3.</u>

The results of the optimization computations for this drying mode (with product recycle) are presented in Table 4 and in fig.7.

TABLE 4. Data of the Optimal Decisions and the Optimal Performance Index for Various λ and Various Product Recycle Ratio r_1

λ	r_1	T_g^1	T_g^2	T_g^3	θ^1	θ^2	θ^3	F
kJ/kg	kg/kg	K	K	K	kg/kg	kg/kg	kg/kg	kJ/kg
	0	304.31	304.75	304.65	9.82	8.32	7.48	16.53
0.42	0.1	304.37	304.79	304.65	9.77	8.41	7.62	16.69
	0.2	304.43	304.82	304.66	9.66	8.44	7.71	16.73
	0.3	304.46	304.85	304.67	9.56	8.46	7.78	16.75
	0	338.45	336.88	334.52	3.51	2.77	2.21	64.73
4.20	0.1	338.54	338.90	334.64	3.45	2.79	2.25	64.82
	0.2	338.62	338.90	334.74	3.41	2.81	2.26	64.90
	0.3	338.69	338.91	334.83	3.38	2.83	2.29	65.03
	0	452.63	447.45	376.02	1.46	0.96	0.16	217.04
42.0	0.1	452.66	447.95	387.29	1.44	0.98	0.20	217.29
	0.2	452.66	448.37	394.86	1.41	0.99	0.23	217.57
	0.3	452.71	448.90	401.92	1.39	1.00	0.26	217.92

In analysis of the computational data the following conclusions are found:
(1) When unit apparatus price is sufficiently low (small λ), the optimal temperature of gas supplied to subsequent stages of cascade are close one to another and the highest T_g^n is at the first stage and the lowest at the second stage. However, the temperatures T_g^n increase with λ, this effect being most significant for T_g^1 and T_g^2 as well as less significant for T_g^3. As a result for intermedia-

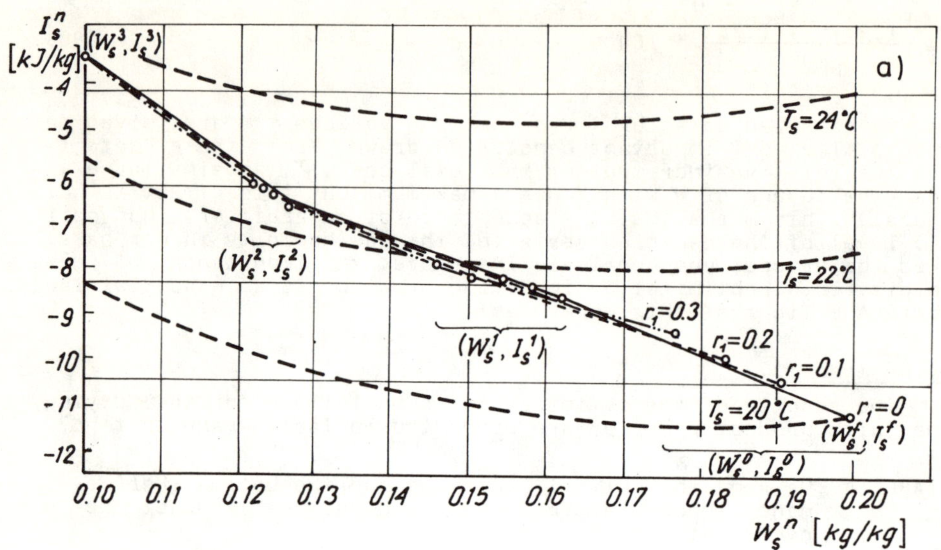

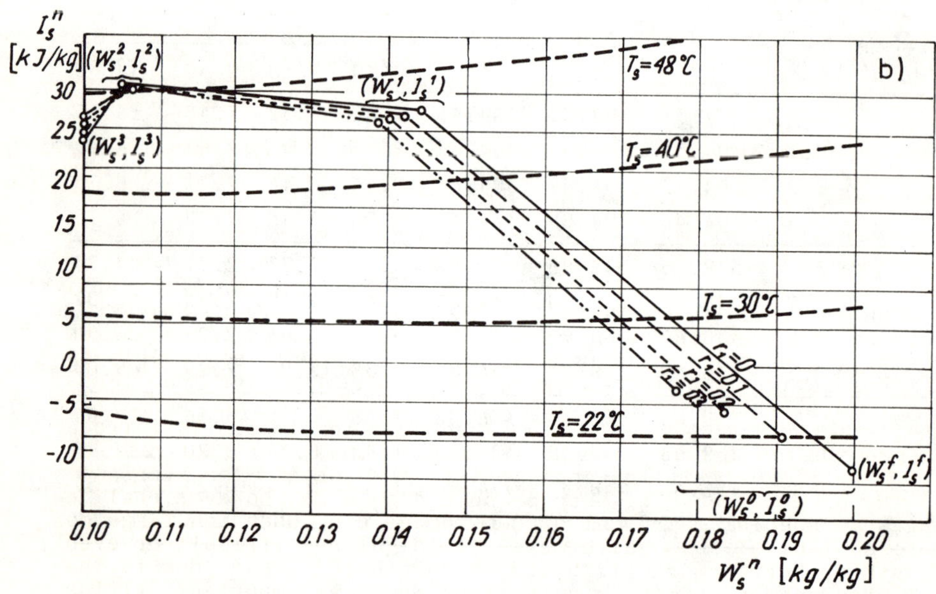

Fig.7.Optimal trajectories of the processes with product recycle
for a) λ=0.42 kJ/kg and b) λ=42.0 kJ/kg.

te and large λ, the temperatures T_g^1 and T_g^2 attain approximately
the same values,whereas the temperature T_g^3 comparable with T_g^1
and T_g^2 for small λ becomes the lowest.When recycle ratio becomes

232

larger an increases of the temperatures T_g^n is observed. This concerns mainly the third stage of cascade, especially for equipments characterized by large unit prices.

(2) Optimal drying gas flows decrease when λ increases. For the apparata characterized by small unit price (small λ) a weak tendency for decreasing of gas flows through the subsequent stages appears - this tendency becoming more explicit when λ increases.This observation concerns mainly the last (third) stage where the lowest gas flows are obtained for sufficiently costly equipment. It is of interest that,when recycle ratio increases, the total dimensionless gas flow, $\sum_{n=1}^{N}\theta^n$, remains practically unchanged and its allocation between the various stages of the cascade becomes slightly more uniform.

(3) It results from the above conclusions that the optimal process intensity increases with the unit apparatus price (as the drying gas temperatures increase and its flows decrease with λ).Therefore the requirement of energetic drying agents is acceptable only for the processes undergoing in a sufficiently costly equipment.

(4) A tendency appears (see fig.7) for the largest changes of the solid moisture content at the first stage and for the lowest ones at the third stage of cascade. This tendency becomes stronger when the unit price of apparatus, λ, increases and when the recycle ratio,r_1, is larger.

(5) At each stage of a sufficiently inexpensive cascade (small λ) the dried solid is slightly heated. For more expensive equipment, the heating (more explicit than in the previous situation) appears only at the two first stages,whereas at the third stage a slight cooling of solid takes place.

(6) The optimal value of the final enthalpy (final temperature) of solid increases with λ. The influence of recycle ratio on the final enthalpy of solid is insignificant. The noticeable differences (an increase of I_s^N with r_1) appear for an expensive equipment, characterized by the large unit price.

(7) The optimal values of the performance index increase with λ and are practically independent of the recycle ratio r_1.However,it should be stressed that when recycle ratio increases, tha fixed costs of the process must increase too as the costs of recycle solid transmission (considered up to now as constans for the constant r_1) will become larger with r_1. When the change of fixed costs is taken into account (i.e. when additional term is introduced into the performance index) it results that, for our process, the economical calculations not justify the use of the variant with the product recycle. Such process should be, however,realized under some special technical circumstances e.g. when too large moisture content of feed solid makes fluidization difficult or even impossible. Furthermore it should be remembered that due to the equilibrium nature of the single stage model an important virtue of recirculation,namely its positive influence on the process kinetics, was lost in our case. Therefore the role of recycle will become more significant when the kinetical terms will predominate in drying.

Drying mode II.1.

The results of the optimization computations for the unconventional countercurrent drying are presented in Table 5 and in fig.8.

TABLE 5. Data of the Optimal Decisions,Trajec-
tories and Performance Index for various λ (X_o = 0.008 kg/kg).

λ	kJ/kg	0.42	4.20	21.0
θ	kg/kg	11.0	3.0	0.9
T_g^1	K	305.65	335.74	365.51
T_g^2	K	306.89	353.08	453.32
T_g^3	K	306.48	358.14	530.93
W_s^3	kg/kg	0.2000	0.2000	0.2000
W_s^2	kg/kg	0.1789	0.1769	0.1626
W_s^1	kg/kg	0.1434	0.1365	0.1223
W_s^0	kg/kg	0.1000	0.1000	0.1000
I_s^3	kJ/kg	-11.20	-11.20	-11.20
I_s^2	kJ/kg	0.75	26.15	54.45
I_s^1	kJ/kg	-1.99	18.70	39.89
I_s^0	kJ/kg	-2.86	8.84	20.11
F	kJ/kg	22.49	77.91	152.70

The data of the optimal trajectories are presented in Table 5
as well in Fig.8, to remember to the reader that the stages of the
cascade are numbered according to the direction of the gas flow.
Thus,the drying gas attains the respective stages of the cascade
in the following sequence: stage n=1, then stage n=2 and then n=N=
=3, whereas the dry solid attains the respective stages of the cascade in the inverted sequence,i.e.: stage n=N=3, then stage n=2,
and then n=1.

The following conclusion can be derived:

(1) Increase of λ reduces the optimal dimensionless air flow θ.
The optimal values of the air flow θ are higher (more explicit for
small λ) than those ones for the respective crosscurrent drying,
and the same value of λ.

(2) The optimal inlet air temperatures to the respective stage
of the cascade increase with λ, this effect being most significant
for T_g^3 as well as least significant for T_g^1. As a result,the temperatures T_g^1, T_g^2 and T_g^3, comparable one to another for small λ,
are considerably different for large λ - the highest temperature
T_g^3 and the lowest one T_g^1. The considerable differences between T_g^1,
T_g^2 and T_g^3 can suggest that the unconventional drying is considerab-

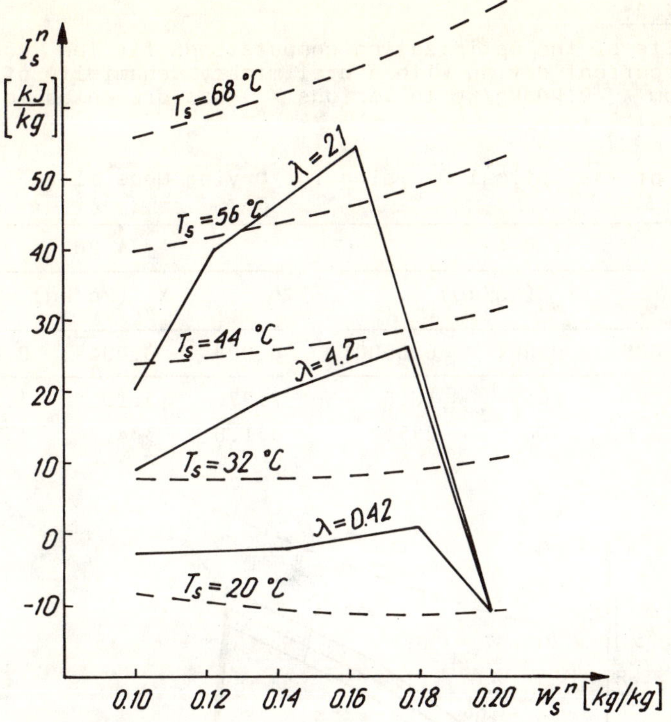

Fig.8.Optimal trajectories of the unconventional
countercurrent drying   (X_o = 0.008 kg/kg)

ly more economic than the conventional one (see comparison of the performance indices in farther text).

(3) For small λ, the lowest changes of the solid moisture content take place at the stage n=N=3 (the first contact of dry solid with drying air) and the largest ones at the stage n=1. For large λ reverse situation is observed - the largest changes of the solid moisture content take place at the stage n=N=3 and the lowest ones at the stage n=1. The changes of the solid moisture content at the stage n=2 is practically value of λ independent. Thus,one can confirm that the relative process intensity increases at the stage n=3 and decreases at the stage n=1, when λ increases. Note that the above statement is connected with the relative process intensity. The absolute process intensity increases at every stage when λ increases.

(4) For every value of λ, the dried solid is heated at the stage n=3 and cooled at the stages n=2 and n=1. These effects increase with λ.

(5) The optimal value of performance index increases with λ (see comparison of the performance indices for this drying mode and conventional one for X_o=0.008 kg/kg, farther text).

<u>Drying mode II.2.</u>

The results of the optimization computations for the conventional countercurrent drying with a preliminary dehumidity of the inlet air (from X_o=0.008kg/kg to various X_w) are presented in Table 6 and in Fig.9.

TABLE 6. Data of the Optimal Decision for Drying Mode II.2.

λ kJ/kg	0.42			4.20		
	X_w (kg/kg)			X_w (kg/kg)		
	0.008	0.004	0.00001	0.008	0.004	0.00001
θ kg/kg	11.2	10.5	10.0	2.97	3.10	3.25
T_g K	305.8	300.9	295.2	351.8	344.7	336.9

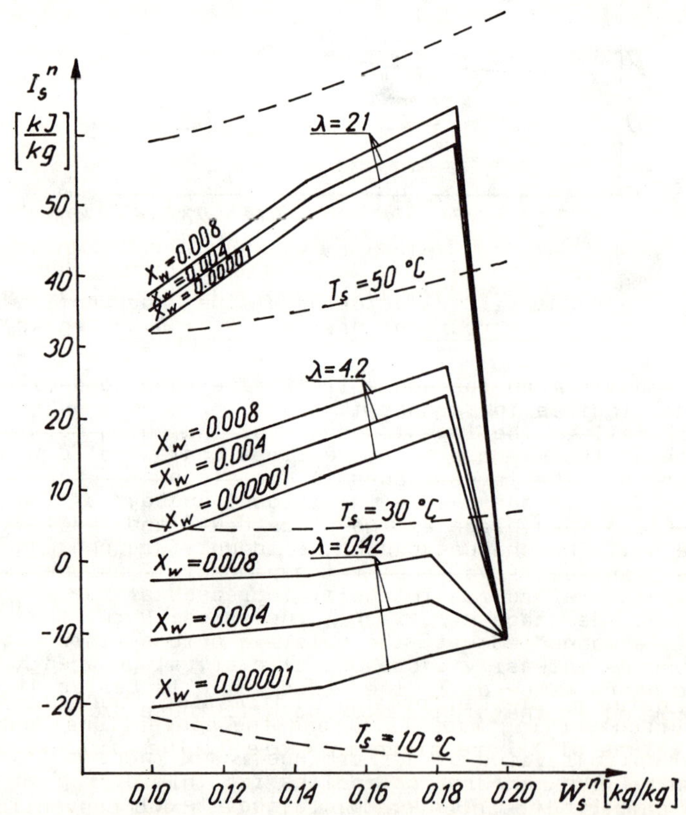

Fig.9.Optimal trajectories of the conventional countercurrent drying with preliminary dehumidity of the inlet air from X_o to various X_w

The following conclusions can be derived:

(1) Increase of λ reduces the optimal dimensionless air flow θ for every X_w=constant, whereas for λ = constant the following situation is observed: for small λ, θ decreases when X_w decreases and for larger λ, θ increases when X_w decreases.

(2) For every X_w=constant, the optimal inlet air temperature (the same for every stage) increases with λ, whereas for every λ =constant, the optimal inlet air temperature decrease when X_w decrease (process intensity reduction, because of the inlet air temperature lowering, is compensated by the inlet air humidity X_w lowering).

(3) For every λ and every X_w, the largest changes of the solid moisture content take place at the stage n=1 and the lowest ones at the stage n=N=3 (here, in the conventional drying mode, the inlet air temperature is the same for every stage, whereas the inlet air humidity increases with stage number. It is the reason, the process intensity decreases when the stage number increases). The changes of the solid moisture content at the stage n=1 are value of λ and X_w independent. The changes of the solid moisture content at the stages n=2 and n=N=3 are value of X_w independent if λ =constant, whereas they increase at the stage n=2 and decrease at the stage n=N= =3, when λ increases.

(4) For every value of λ and X_w, the dried solid is heated at the stage n=N=3 and cooled at the stages n=2 and n=1. The effect of heating increases with λ and X_w.

TABLE 7. Data of the Performance Index for the Unconventional Drying Mode II.1.and the Conventional Drying Mode II.2. for Various X_w (Performance Index F expressed in kJ/kg).

λ	0.42 kJ/kg				4.20 kJ/kg			
	Drying Mode				Drying Mode			
	II.1.	II.2.			II.1.	II.2.		
	X_w	X_w			X_w	X_w		
	0.008	0.008	0.004	0.00001	0.008	0.008	0.004	0.00001
F	22.49	22.57	16.68	13.13	77.91	78.94	74.05	69.36

The datas of the performance indices for the unconventional countercurrent drying mode II.1. and the conventional one II.2.for various values of X_w are given in Table 7. The following conclusions can be derived:

(1) To compare the unconventional process and conventional one, we must take into account the drying modes II.1. and II.2. with X_w=0.008. We can confirm that there is an economic superiority of the unconventional drying mode over conventional one, this superio-

rity being more important for large λ, because

$$\frac{F_{II.2} - F_{II.1}}{F_{II.1}} * 100\% = \frac{22.57 - 22.49}{22.49} * 100\% = 0.35\%$$

for λ = 0.42 kJ/kg, and

$$\frac{F_{II.2} - F_{II.1}}{F_{II.1}} * 100\% = \frac{78.94 - 77.91}{77.91} * 100\% = 1.32\%$$

for λ = 4.20 kJ/kg. This superiority (1.32%) can be important in industry practice.

(2) For every λ , the performance index (for drying mode II.2) decreases when X_w decreases, thus the preliminary dehumidity of the inlet air is recomended.

To compare the crosscurrent drying processes and countercurrent ones, we can take into account the conventional processes, i.e. drying mode I.2. variant I.2.2. ($X_g^n = X_o = 0.008$ kg/kg for n=1,..N) and drying mode II.2. with $X_w = X_o = 0.008$ kg/kg. The datas are presented in Table 8 (look for these datas in Table 2 and in Table 7).

TABLE 8. The comparison of performance indices for the conventional co-untercurrent and crosscurrent processes

λ	0.42 kJ/kg		4.20 kJ/kg	
	Drying mode		Drying mode	
	I.2.2.	II.2.	I.2.2.	II.2.
F	16.486	22.57	64.443	78.94

The datas, stored in Table 8, suggest that for the silica gel-air--water system, in the range of the solid moisture content and temperature investigated here, the crosscurrent drying processes are recommended.

Drying mode II.3.

The results of the optimization computations for the conventional countercurrent drying with the drying air recycle are presented in Table 9 and in Fig.10 - for λ =0.42 kJ/kg, as well in Table 10 and in Fig.11 - for λ =4.20 kJ/kg, respectively.

The following conclusions can be derived:

(1) For every λ, the optimal inlet air temperature increases with drying air recycle ratio r_2. For every r_2, the optimal inlet air temperature (the same for every stage) increases with λ.

(2) For every drying recycle ratio r_2, the dimensionless air flow

TABLE 9. Datas of the Optimal Decisions and Performance
         Indices for λ=0.42 kJ/kg and Various Drying Air
         Recycle Ratio r_2.

r_2	kg/kg	0	0.5	1.0	1.5	2.0
T_g	K	305.8	307.8	309.5	311.6	313.9
g_2	kg/kg	11.2	13.8	16.0	17.5	18.5
θ	kg/kg	11.2	9.2	8.0	7.0	6.2
F	kJ/kg	22.6	29.3	34.9	39.8	44.1

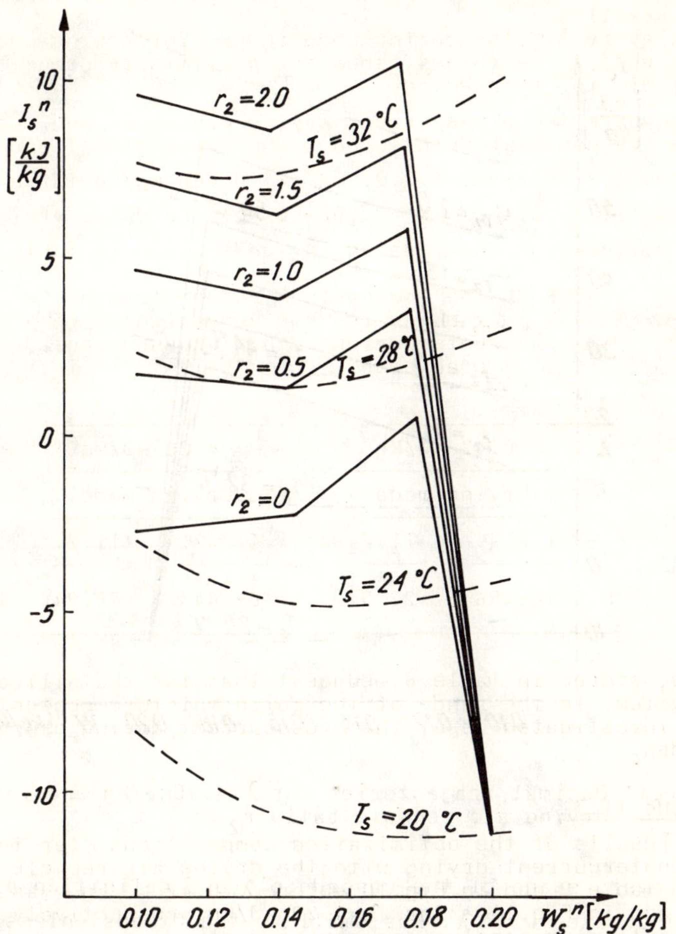

Fig.10. Optimal trajectories for λ=0.42kJ/kg and various
        drying air recycle ratio r_2

TABLE 10. Datas of the Optimal Decisions and Performance
Indices for λ=4.20kJ/kg and Various Drying Air
Recycle Ratio r_2.

r_2	kg/kg	0	0.5	1.0	1.5	2.0
T_g	K	351.8	360.1	369.5	382.1	402.0
g_2	kg/kg	2.97	3.35	3.48	3.30	2.70
θ	kg/kg	2.97	2.23	1.74	1.32	0.90
F	kJ/kg	78.9	96.1	108.5	117.6	124.0

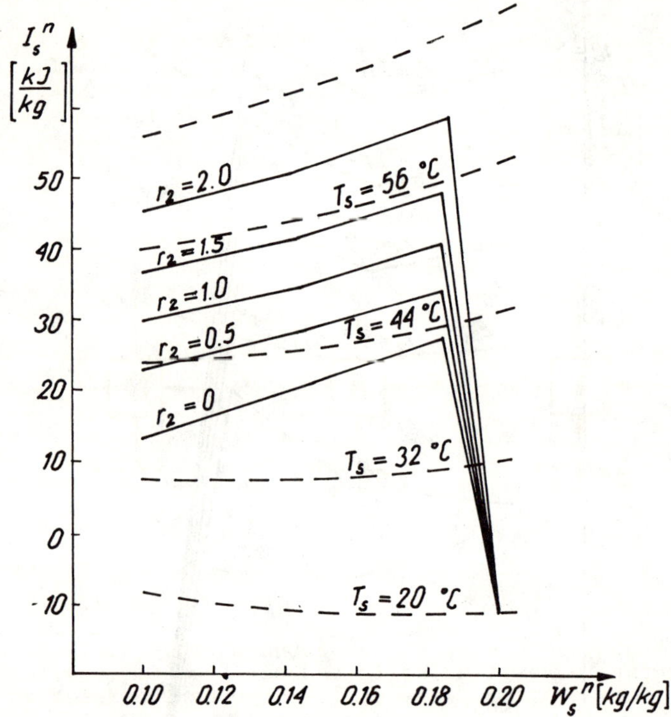

Fig.11.Optimal trajectories for λ=4.20kJ/kg and various
drying air recycle ratio r_2

through every stage of the cascade,g_2 (see definition of g_2 , eqn.
(17)), increases with λ. For small λ, the dimensionless flow g_2
inreases with r_2, whereas for large λ, the maximum of g_2 appears
when r_2 increases. More important than g_2, the dimensionless air
flow θ (describing the fresh air consumption per unit dried solid
stream) decrease with λ as well with r_2.

240

(3) The general shape of the optimal trajectories is practically
value of λ and r_2. However, at the stage n=N=3, the dried solid is
heated to higher and higher temperature when the air recycle ratio
r_2 increases.

(4) For every r_2=constant, the performance index increases with
λ. For every λ=constant, the performance index increases with r_2;
it means that the countercurrent drying with the drying air recyc-
le is not recommended (if performance criterion considered here is
taken into account).

SOLID PARTICLES HEATING

Now, we will consider the solid solid particles heating optimization problem. It is a simplification of the heat and mass transfer problems considered above.

A solid particles (silica gel) stream has to be heated by a
gas (air) in the Nth stage of a crosscurrent cascade of fluidized
dryers. Thermodynamic equilibrium between the gaseous and solid
phases leaving the fluidized stage is assumed. It is desired to define the optimal air temperature and flow to each stage in order
to minimize the process cost. The air temperature is not limited
and the air flow to each stage and the total air flow are unconstrained (Sieniutycz 1983).

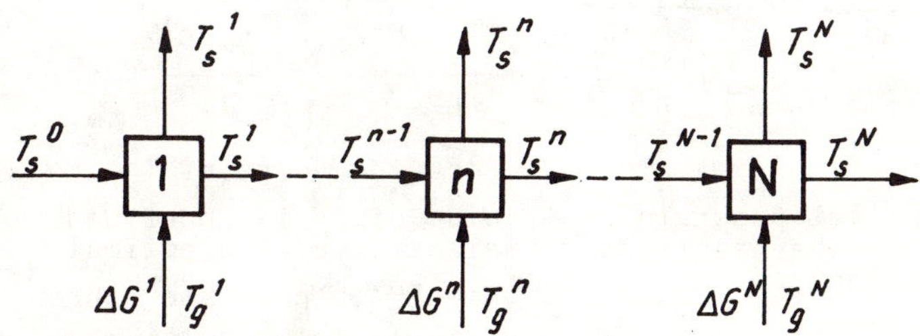

Fig.12. Solid particles heating in the N-stage crosscurrent
        fluidized cascade

The thermodynamic equilibrium assumed above implies an equal
temperature for the exiting solid and air (see Fig.12). The heat balance across the n-stage (for n=1,..N) can be expressed as

$$S (T_s^n - T_s^{n-1}) c_s = \Delta G^n (T_g^n - T_s^n) c_p \tag{81}$$

Eqn.(81) can be written as the state equation for n=1,..N

$$T_s^n - T_s^{n-1} = K (T_g^n - T_s^n) \theta^n \tag{82}$$

This is only equation of state. The boundary condition are defined,
the values of T_s^0 and T_s^N are known.

The dimensionless air flow $\theta^n = \Delta G^n / S$ and coefficient $K = c_p / c_s$.

Two variants of this process will be considered:
(1) The air from every stage of the cascade is released into the atmosphere.
(2) The heat content of the outlet air is recovered.

Variant 1.

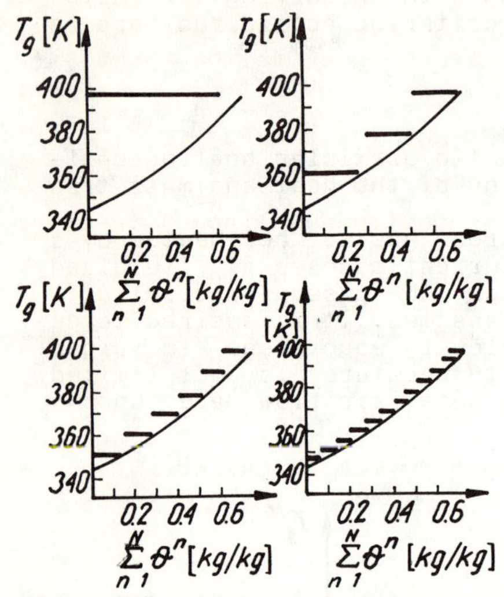

Fig.13.The optimal discrete pro-
        files of the heating air
        for variant 1

The performance index can be written in the following form:

$$F = \sum_{n=1}^{N} \left[\frac{1}{2} A (T_g^n - T_o)^2 + \lambda \right] \theta \quad (83)$$

The Hamiltonian function (n=1,..N)

$$H^{n-1}(T_s^n, T_g^n, z^{n-1}) = \frac{1}{2} A (T_g^n - T_o)^2 +$$

$$+ \lambda + z^{n-1} K (T_g^n - T_s^n) = 0 \quad (84)$$

In view of the unconstrained total flow of air, the value of the Hamiltonian (for autonomous process) equals zero.

The adjoint equation for n=1,..N

$$\frac{z^n - z^{n-1}}{\theta^n} = - \frac{\partial H^{n-1}}{\partial T_s^n} = z^{n-1} K \quad (85)$$

Since the value of T_s^N is given, the value of z^N is undefined.
Selection of the optimal temperature T_g^n is possible through the following expression for n=1,..N:

$$\frac{\partial H^{n-1}}{\partial T_g^n} = A (T_g^n - T_o)^2 + z^{n-1} K = 0 \quad (86)$$

After applying eqn.(86) in formula (84) we have the following additional equation:

$$T_g^n = T_s^n \pm \sqrt{(T_s^n - T_o)^2 + \frac{2\lambda}{A}} \quad (87)$$

The eqns (82) to (86) enable the preparation of the computational block scheme. The computational procedure in which the unknown quantity z^N is assumed leads to the family of optimal trajectories having the common final temperature T_s^N and various initial temperature T_s^0. The initial temperature T_s^0 of every computed trajectory is optimal in relation to the assumed value z^N.
For n=N, the computational procedure starts with given value of T_s^N

242

and assumed value of z^N, thus value of T_g^N can be computed from eqn.(87), then z^{N-1} from (84), θ^n from (85) and T_s^{N-1} from (82). Then the computational procedure can be repeated for n=N-1 because values of T_s^{N-1} and z^{N-1} are known (computed above).Then the computational procedure is repeated for n=N-2, n=N-3,...n=1 computing the value of T_s^0 at the end. Among the various trajectories obtained as a result of the computations, the specific trajectory should be found that has the initial temperature equel to the preassigned temperature T_s^0.

The results of the optimization strategy for T_s^0=293.15K, T_s^N==333.15K, T_o=293.15K, λ=4.20kJ/kg, $c_p=c_s$=1kJ/(kgK) and a various number of stages are given in Fig.13. The continuous line on these graphs represents an optimal strategy for a limiting multistage configuration,namely, a horizontal heat exchanger. In this case the calculations were made using an algorithm for the continuous version of the maximum principle. It follows from Fig.13.that the optimal discrete profile is an increasing one.For consecutive stages of the process,the air flows decrease. For an increasing number of stages, the total air flow increases,thus approaching in the limit a continuous heat exchanger. It appears that a 12-stage approximation approaches the optimum achieved for a continuous process.

Variant 2.

The performance index can be written in the following form:

$$F = \sum_{n=1}^{N} \left[\frac{1}{2} A (T_g^n - T_o)^2 - \frac{1}{2} A (T_s^n - T_o)^2 + \lambda \right] \theta^n \tag{88}$$

The Hamiltonian function for n=1,..N

$$H^{n-1}(T_s^n,T_s^n,z^{n-1}) = \frac{1}{2} A (T_g^n - T_o)^2 - \frac{1}{2} A (T_s^n - T_o)^2 + \lambda +$$
$$+ z^{n-1} K (T_g^n - T_s^n) = 0 \tag{89}$$

The adjoint equation for n=1,..N

$$\frac{z^n - z^{n-1}}{\theta^n} = - \frac{\partial H^{n-1}}{\partial T_s^n} = A (T_s^n - T_o) + z^{n-1} K \tag{90}$$

As in variant 1, the Hamiltonian equals zero,and the value of Z^N is undefined.
Selection of the optimal temperature T_g^n is possible through the following expression for n=1,..N:

$$\frac{\partial H^{n-1}}{\partial T_g^n} = A (T_g^n - T_o) + z^{n-1} K = 0 \tag{91}$$

After applying eqn.(91) in formula (89) we have the following equation:

$$T_g^n = T_s^n \pm \sqrt{\frac{2\lambda}{A}} \qquad (92)$$

The sign "+" is connected with solid particles heating, and "-" with solid particles cooling.
From eqn.(89), for n=1,..N, it follows that

$$z^{n-1} = -\frac{\frac{1}{2} A (T_g^n - T_o)^2 - \frac{1}{2} A (T_s^n - T_o)^2 + \lambda}{K (T_g^n - T_s^n)} \qquad (93)$$

and after exchange of subscript n for n-1 (thus, n=0,1,..N-1)

$$z^n = -\frac{\frac{1}{2} A (T_g^{n+1} - T_o)^2 - \frac{1}{2} A (T_s^{n+1} - T_o)^2 + \lambda}{K (T_g^{n+1} - T_s^{n+1})} \qquad (94)$$

Utilizing eqns.((82),(92),(93) and (94) in eqn.(90) results in, for n=1,..N-1

$$T_s^{n+1} - T_s^n = T_s^n - T_s^{n-1} \qquad (95)$$

Eqn.(95) shows the the increase in the temperature of the solid is the same for all stages. Thus for n=0,1,..N

$$T^n = T^0 + \frac{n (T_s^N - T_s^0)}{N} \qquad (96)$$

When the temperature T_s^n of the solid is known, the optimal inlet air temperature T_g^n can be calculated from eqn.(92), and the optimal air flow from eqn.(82). It turns out that

$$\theta^1 = \theta^2 = \ldots = \theta^N = \frac{T_s^N - T_s^0}{N K \sqrt{\frac{2\lambda}{A}}} \qquad (97)$$

i.e. the air flow is the same for all stages.

SUMMARY

For unconventional and conventional steady-state multistage crosscurrent and countercurrent fluidized drying, as well as crosscurrent solid particles heating, the optimization problem has been formulated and solved.
The economic effectiveness index, corresponding to the sum of

the investment and operational costs, was accepted as a performance criterion. To express the cost of the drying (heating) gas in terms of the drying parameters (gas temperature and gas humidity) an economic balance of the dryers investigated was made. In this balance the so-called exergy tariff of prices was applied for the purpose of quantitative evaluations. Then, the expression describing the performance criterion was transformed into some special form in which, together with physiochemical variables,only one parameter λ, connected with the sum of the investment and gas pumping costs, appeared. The values of this parameter for a concrete drying process can be computed on the basis of formula (7) given in this paper ($\lambda = \lambda_e/e$).

The original discrete algorithm with the constant Hamiltonian - for crosscurrent processes, and commonly known discrete algorithm of maximum principle (given by Katz and Fan)- for countercurrent processes, were used in the optimization computations.On this basis a computer program was developed and the optimization results were obtained for the family of drying processes corresponding to various values of the parameter λ. These results are quite general, since amongst the whole family of optimal solutions a special solution can always be found which pertains to the concrete value of λ calculated for the specified drying (heating) process.

The computations were performed for the case of drying (heating) of silica gel by air in a three-stage cascade. From the computer outputs the following conclusions were reached:

(1) The large energy consumption is reasonable only in sufficiently costly apparatus.

(2) The superiority of the optimal unconventional (control) processes (optimal inlet air temperature is varied with stage number) over optimal conventional processes (optimal inlet air temperature is constrained to be the same for every stage) is negligible one when the unit apparatus price is sufficiently low (i.e.small λ). This superiority increases with λ. Thus,the processes with the inlet air temperature controlled can be recommended for the cascade of expensive apparata only.

(3) For the crosscurrent processes, the optimal inlet air humidities X_g^n assume stationary values only when the unit apparatus price is sufficiently low, these humidities being equal to the minimal air humidity for more expensive apparata.

(4) The optimal inlet air humidity level for crosscurrent processes as well as inlet air humidity for countercurrent processes are dominant factors which make it possible to improve the economics of the process.

(5) The process intensity increases with λ,thus the optimal air flows decrease with λ (for cross- and countercurrent processes).

(6) For the silica gel-air-water system, for the solid moisture content and temperature investigated here, the crosscurrent processes are cheaper than respective countercurrent ones.

(7) For the crosscurrent processes with product recycle.the performance index increase with the product recycle ratio.

(8) For the countercurrent processes with drying air recycle,the performance index increase with the air recycle ratio.

NOMENCLATURE

A_a - apparatus cross-sectional area,m^2

A,B - expressions appearing in eqn.(21), $kJ/(kgK^2)$ and kJ/kg
b - exergy of drying gas, kJ/kg
c_g - unit price of drying gas, $/kg
c^g - heat capacity, $kJ/(kgK)$
e - economic value of unit exergy, $/kJ
e_{el} - unit price of electrical energy, $/J
F_{el} - exergy performance index, kJ/kg
F',F'' - economic performance index, $/kg
F_t - total cost expressed in terms of exergy, kJ/kg
G^t - fresh gas flow rate for countercurrent drying, kg/h
G_r - recycle gas flow rate, kg/h

G^N - total gas flow rate for crosscurrent drying, kg/h
ΔG^n - drying gas flow rate through stage n, kg/h
g_1,g_2 - dimensionless gas flow rate, see eqns.(12) and (17) resp.
H - Hamiltonian
I_s - solid enthalpy per unit mass of dry material, kJ/kg
i_g - gas enthalpy per unit mass of dry gas, kJ/kg
i_s - enthalpy of gas in equilibrium with solid, kJ/kg
i_{sg} - enthalpy of gas in equilibrium with solid for countercur-
             rent drying, kJ/kg
I^f - enthalpy of feed solid, kJ/kg
J,j_o - total cost of cascade and fixed limiting value of J, $
K $= c_p/c_s$
M - molar mass, $kg/kmol$
M_1,M_2 - feed and recycled solid mixing function, see eqns.(41-42)
N,n - total number of stages and current stage number
ΔP - pressure drop in fluidized layer, Pa
p - fluidized drying apparatus price per unit area sieve bot-
             tom, $/m^2$
R - dry solid recycle flow rate, kg/h
R - gas constant, $kJ/(kmolK)$
r_1 $= R/S$ - solid recycle ratio, kg/kg
r_2 $= G_r/G$ - gas recycle ratio, kg/kg
T_g,T_s - inlet gas temperature and solid temperature, respectively, K
T_o - ambient temperature, K
T_{gr} - maximum acceptable payout time, year
T_u - utilization time of dryer during year, $h/year$
T_{sg} - temperature of gas in equilibrium with solid for counter-
             current drying, K
v - superficial gas velocity, m/h
W_s - absolute moisture content of solid (mass of moisture per
             unit mass of dry solid), kg/kg
W^f,W_k - initial and final value of W_s, respectively, kg/kg
W_e - value of W_s in equilibrium with ambient air, kg/kg
X_g,X_s - absolute humidity of inlet gas stream and humidity of gas
             in equilibrium with solid, respectively, kg/kg
X_{sg} - humidity of gas in equilibrium with solid for countercur-
             rent drying, kg/kg
X_o - ambient humidity, kg/kg
$\tilde{z}$ - factor describing "freezing" of capital costs
z^n - adjoint variable at stage n

Greek symbols

β - coefficient describing renovations, year^{-1}

η - pumping efficiency

θ^n = $\Delta G^n/S$, dimensionless gas flow at stage n, kg/kg

λ = λ_e/e, exergy coefficient of investment and gas pumping
     costs, kJ/kg

λ_e - economic coefficient of investment and gas pumping costs,\$/kg

μ - ratio of total cost to variable part of total cost

ς - gas density, kg/m^3

Auxiliary variables appearing in computer program
DELTA, DELTAP, KR1, KR2, i, s, L, P

Superscripts

n,N - stages n and N, respectively
*   - highest allowable value

Subscripts

g   - gas
p   - dry air
s   - solid, gas in equilibrium with solid
w   - moisture
*   - minimum allowable value

REFERENCES

Fan,L.T. 1966. The Discrete Maximum Principle, New York, Wiley,1966.
Reda,S. 1983. M.Sc.Thesis, Warsaw Univ. of Technology, 1983.
Sieniutycz,S. 1973. AIChE J., 19 (1973) 277.
Sieniutycz,S. 1973a. Rep. Inst. Chem. Eng., Warsaw Univ. of Tech-
nology, 2 (3) (1973) 17.
Sieniutycz,S. 1978. Optimization in Process Engineering, WNT, War-
saw, 1978.
Sieniutycz,S.,Szwast,Z. 1982. Chem. Eng. J., 25 (1982) 63-75.
Sieniutycz,S.,Szwast,Z. 1983. AIChE J., 23 (1983) 155.
Sieniutycz,S.,Szwast,Z. 1984. Drying'84 (1984) 49-61, Edited by
Mujumdar,A.S.
Szargut.J.,Petela,R. 1965. Exergy, WNT,Warsaw, 1965.
Szwast,Z. 1978. Ph.D.Thesis, Warsaw Univ. of Technology, 1978.
Szwast,Z. 1984. V Symposium of Drying, 1 (1984) 26-34,Wrocław.

APPENDIX

    The discrete optimization algorithm of the maximum principle
with a constant discrete Hamiltonian can be applied for optimization of a large class of multistage processes,existing in chemical
engineering, which exhibits a linear property in relation to some

important decision variable θ^n describing the increment of the independent variable t at stage n for n=1,..N:

$$\theta^n = t^n - t^{n-1} \tag{A-1}$$

The general performance index and discrete state equations for n=
=1,..N and i=1,..s are,respectively, of the form

$$F = \sum_{n=1}^{N} \theta^n f_o^n(x^n, u^n, t^n) \qquad\qquad (A-2)$$

$$x_i^n - x_i^{n-1} = \theta^n f_i^n(x^n, u^n, t^n) \qquad\qquad (A-3)$$

where x^n is the s-dimensional state vector at stage n, x_i^n is the i-th coordinate of x^n, u^n is the r-dimensional decision vector at stage n, and t^n is independent variable at stage n.
The mathematical model, eqns.(A-1) - (A-3), is linear with regard to the decision θ^n and can be arbitrary (e.g. nonlinear) in relation to the decision vector u^n - hence, the special character of the decision variable θ^n.

After introducing the adjoint vector z^n and Hamiltonian H^{n-1}

$$H^{n-1}(x^n, z^{n-1}, u^n, t^n) = f_o^n(x^n, u^n, t^n) + \sum_{i=1}^{s} z_i^{n-1} f_i^n(x^n, u^n, t^n) \qquad (A-4)$$

for n=1,..N, the necessary optimality conditions take the form

$$\frac{x_i^n - x_i^{n-1}}{\theta^n} = \frac{\partial H^{n-1}}{\partial z_i^{n-1}} \qquad n=1,..N \quad ; \quad i=1,..s \qquad\qquad (A-5)$$

$$\frac{z_i^n - z_i^{n-1}}{\theta^n} = -\frac{\partial H^{n-1}}{\partial x_i^n} \qquad n=1,..N \quad ; \quad i=1,..s \qquad\qquad (A-6)$$

$$\frac{\partial H^{n-1}}{\partial u_1^n} = 0 \qquad n=1,..N \quad ; \quad 1=1,..r \qquad\qquad (A-7)$$

$$\frac{H^n - H^{n-1}}{\theta^n} = \frac{\partial H^{n-1}}{\partial t^n} \qquad n=1,..N \qquad\qquad (A-8)$$

Equation (A-5) is the state equation (A-3) in canonical representation. Equation (A-6) is the adjoint equation. The undetermined coordinates of the vector x^N correspond to the vanishing coordinates of the vector z^N - if process without product recycle is considered, or the following equation has to be satisfied

$$z_i^N = \frac{\partial M_i(x^N)}{\partial x_i^N} z_i^0 \qquad\qquad (A-9)$$

when process with product recycle is considered. The prescribed coordinates of the vector x^N always correspond to the undetermined coordinates of the vector z^N.
For t^N undetermined, $H^N=0$. Moreover, for autonomous processes (explicitly independent of t^n) Hamiltonian is constant along the optimal trajectory (see eqn.(A-8)).

Energetic and Economic Optimization of Industrial Systems Compared

P. Le Goff, R. Rivero, S. de Oliveira *and* B. Schwarzer
Laboratoire des Sciences du Génie Chimique
CNRS-ENSIC-INPL, Nancy, France

ABSTRACT
Industrial processes can be represented as open systems where available energy (exergy) enters, utilized energy (exergy content of goods and services) exits and degraded energy (destroyed exergy) is rejected into the environment.

The overall evaluation of each process can be made according to an energy accounting (in terms of exergy or of primary fossil energy) or a financial accounting (in terms of national or foreign currency). In any case, the process can be optimized by determining the best compromise between the "costs" of the various input fluxes.

According to the needs of the decision maker one can choose the most convenient value criterion which would be to minimize :
- the total cost in national currency, for a project engineer working for a private company,
- the payments in foreign currency, for a public organization,
- the total energy consumption, for a hypothetical decision maker responsible for the world's non renewable fossil energy,
- the entropy production in order to increase the thermodynamic efficiency of the process.

We present and compare some examples of energetic and economic optimizations, according to various criteria. The examples given concern a heat exchanger, the drying of beet pulp and some heat transformers for upgrading industrial waste heat.

Furthermore, we generalize the concept of "pay-back time" through the ratio of the invested capital cost and of the operating cost given in terms of exergy or of primary energy. Numerical values are given for an oil-fired boiler.

1. INTRODUCTION

Over the last two centuries, the work of the machine has progressively replaced the work previously supplied by human being's muscles and by those of the animals at man's service.

The "(steam)horse-power" has replaced the horse-drawn carriage ! So man needed to know how to measure and compare the "value" of different energy sources used for those engines. In certain cases, this determination can be made readily :
- for example, when water flowing down a mountain towards the ocean is used to drive a turbine, its energy value is obviously higher the higher the altitude of the water source ;
- when wind is used to drive a windmill, its energy value is an increasing function of wind's velocity and density.

For thermal energy sources, the evaluation was a little more delicate. It is the development of thermo-mechanical engines in the last century, which led scientists to state that any amount of heat has a higher value the higher the temperature of its source.

The concepts of "Energie Utilisable" (Gouy, 1889), then "Available Energy" (Keenan, 1932) and finally "Exergie" (Rant, 1956) are a result of this aim to measure the "fire power" according to its capacity to drive an engine. This commodity should then be called the "Mechanisable Heat".

But heat is able to perform not only mechanical work ! Generally speaking, the different sources of energy available on earth are not only useful to replace muscular work. Quite the opposite, those sources are mainly used to produce material goods and services in order to satisfy the physiological needs and the well-beings of humankind.

Every energy source is then an "economic good" whose value must be compared, according to a proper "value scale", to that of other goods useful to man (raw-materials, available space, etc.). Unfortunately, such a value scale has no objective definition. It is, in each case, a subjective question, defined by each user. For instance, for a given industrial operation, different value scales will be proposed and used :
- by an engineer, responsible of the project in a private company,
- by a government officer, responsible for the payment balance in foreign currency of his country,
- by a hypothetical world decision-maker, responsible for the optimization of the non-renewable energy capital available in the planet.

The first two options deal with financial accounting balances (in national or foreign currency). The third one deals with an energy accounting balance (in exergy or fossil primary energy). These two types of accounting have many analogies but also a fundamental difference.

Actually, in the first case, there is always or usually an **increase in the overall economic value** : the operation produces added-value goods, from low-value raw materials. On the other hand, in the second case, there is never an increase in value, but always an overall decrease : every operation, intrinsically irreversible, can only decrease the exergy stock (or non-renewable energy stock) of the system formed by the earth and the sun.

In the following these different accounting procedures will be presented in such a way that they can be compared, from the point of view of their base concepts as well as their application to real problems.

2. THE ENERGY VALUE CONCEPT

The value of economic goods can be estimated by reference to various value scales (Le Goff *et al.*, 1990). For energy systems we can state several scales of value based on concepts other than purely economic ones such as thermodynamics, ecology, physiology, etc.

Before discussing the energy value scales we must establish the balance equations of energy, exergy and value.

2.1 Energy Balance

We will restrict ourselves to the study of the case where the system is a "technical structure" (a factory, a machine, some equipment...) which is bought, installed and started up at time zero and which afterwards continues to operate for N years, in steady state with constant annual energy fluxes of E_s, E_u, E_r (in joules per year) (Figure 1).

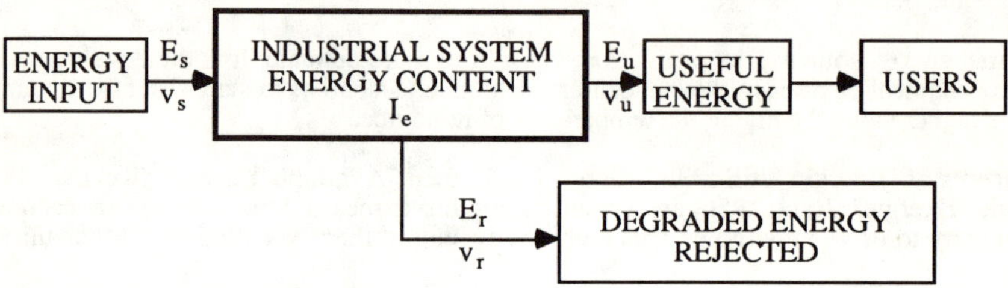

Figure 1. Energy balance.

E_s is the flux of energy supplied to the system, E_u is the useful energy leaving the system : it is the energy content of the functional products. E_r is the flux of energy degraded and rejected into the environment (into the air, the sea, waste deposits...).

The energy balance can be written :

$$E_s = E_u + E_r \tag{2.1}$$

This is the first law of thermodynamics.

2.2 Exergy Balance

The second law of thermodynamics may be expressed in terms of a balance of exergy (Ex), that is the portion of a given quantity of energy that may be converted into mechanical work. With the same hypotheses of the energy balance, we obtain :

$$Ex_s = Ex_u + Ex_r + Ex_d \tag{2.2}$$

where Ex_d is the exergy destroyed during the process due to irreversibilities (entropy production).

2.3 Value Balance

An economist sees the system differently : the industrial system transforms economic goods of low commercial value into more useful goods, value-added goods. In addition, goods rejected are goods of reduced value - or even of negative value in the case of nuisances.

Lastly the operation requires the supply of expensive quantities (material, energy, information (see Figure 2).

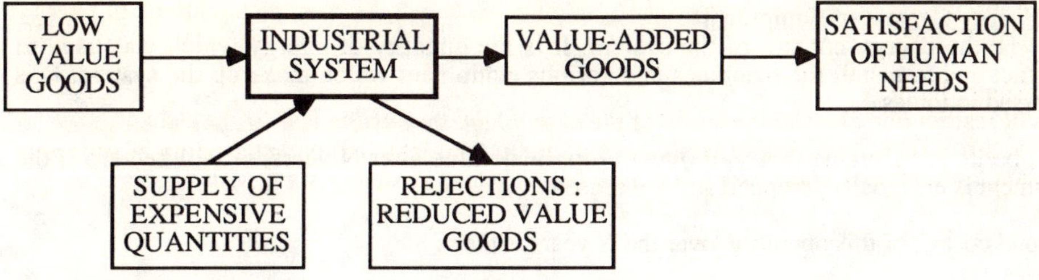

Figure 2. Balance of value.

Let us write the "profit" P of the operation as the difference between the sum of the values of the output quantities and that of input quantities :

$$(\text{PROFIT}) = \begin{pmatrix} \text{SUM OF VALUES} \\ \text{AT OUTPUT} \end{pmatrix} - \begin{pmatrix} \text{SUM OF VALUES} \\ \text{AT INPUT} \end{pmatrix}$$

thus

$$P = (E_u v_u + E_r v_r) - (E_s v_s - F) \tag{2.3}$$

where "v" is the value of one energy unit (one joule) with respect to a given scale of values, defined a priori and chosen as a reference scale. v_s, v_u, v_r are the values of one Joule in the supplied energy, the useful energy and the rejected energy, respectively. F is the value of expensive supplies necessary to carry out the operation.

For the operator, the net annual operating cost C_o is the difference between the value of expensive quantities supplied to the system and the profit he gets back :

$$C_o \equiv F - P \tag{2.4}$$

Combining the value balance (2.3) and the energy balance (2.1), we get :

$$C_o = E_r (v_s - v_r) - E_u (v_u - v_s) \tag{2.5}$$

$(v_s - v_r)$ is the value lost by each joule rejected into the environment ; $(v_u - v_s)$ is the value added to each useful joule, via the operation.

Introducing the specific energy consumption $r \equiv E_r/E_u$, the balance (2.5) becomes :

$$\frac{C_o}{E_u} = r(v_s - v_r) - (v_u - v_s) \tag{2.6}$$

C_o/E_u is the net operating cost per joule of useful energy.

Equations (2.5) and (2.6) can be expressed as follows :

$$\begin{pmatrix} \text{NET} \\ \text{OPERATING} \\ \text{COST} \end{pmatrix} = \begin{pmatrix} \text{LOSS OF} \\ \text{VALUE IN} \\ \text{REJECTS} \end{pmatrix} - \begin{pmatrix} \text{GAIN OF VALUE} \\ \text{IN FUNCTIONAL} \\ \text{PRODUCTS} \end{pmatrix}$$

Note : the balance of value is presented here in the opposite way to that used by financial management. In fact in a usual commercial operation, the gross profit P is greater than the cost of supplies F. The difference P-F, the net profit, is positive. The "net operating cost" as defined here, is the opposite quantity, therefore negative. We will see later the reason for this unusual presentation.

2.4 <u>Energy Content of Equipment</u>
Let I_e be the energy content of the system. It is the quantity of energy which was used to construct and to install the machines and various equipment which make up the system. I_e is expressed in joules.

Let v_i be its initial value (per joule) and v_f its final value after N years, v_f will be zero if the equipment is eventually scrapped and not reused.

The total cost C_t of this operation over the N years is then :

$$C_t = NC_o + I_e \, (v_i\text{-}v_f) \tag{2.7}$$

or, by taking into account (2.6) :

$$\frac{C_t}{NE_u} = r(v_s\text{-}v_r) - (v_u\text{-}v_s) + \frac{I_e}{NE_u}\,(v_i\text{-}v_f) \tag{2.8}$$

which can be expressed as :

$$\begin{pmatrix}\text{TOTAL}\\ \text{COST PER UNIT}\\ \text{OF USEFUL}\\ \text{ENERGY}\end{pmatrix} = \begin{pmatrix}\text{LOSS OF}\\ \text{VALUE}\\ \text{IN}\\ \text{REJECTS}\end{pmatrix} - \begin{pmatrix}\text{GAIN OF}\\ \text{VALUE IN}\\ \text{FUNCTIONAL}\\ \text{PRODUCTS}\end{pmatrix} + \begin{pmatrix}\text{UNIT OF}\\ \text{AMORTIZEMENT}\\ \text{OF NON RE-USED}\\ \text{EQUIPMENT}\end{pmatrix}$$

We will call i_e the "specific energy content of equipment". It is :

$$i_e \equiv \frac{I_e}{NE_u} = \frac{(\text{energy content of equipment})}{\left(\begin{array}{c}\text{total quantity of useful}\\ \text{energy produced in N years}\end{array}\right)} \tag{2.9}$$

An objective, undoubtedly universal among management, is to minimize this total cost per unit of useful energy with respect to whatever will be the reference scale for the value of energy.

2.5 <u>A General Efficiency Definition</u>
We can obtain a general efficiency expression of a process/machine from the following equation :

$$\text{efficiency} = \frac{\text{sum of values at output}}{\text{sum of values at input}}$$

$$\eta = \frac{E_u \cdot v_u}{E_s \cdot v_s} \tag{2.10}$$

In this way, the energetic performance of a process/machine, operating at steady-state, can be evaluated according to different value scales.

3. ENERGY VALUE SCALES
As we have stated in the previous section it is possible to assign to one energy unit different value scales. In this section, we will present four groups of energy value scales : enthalpy, entropy, ecology and economy.

3.1 <u>Enthalpy Based Scales</u>
A First Scale, the simplest, consists of postulating that all forms of energy have the same value : thus a calorie is always "worth" a calorie, no matter what its temperature. This is the first law of thermodynamics.

Then we have : $v_s = v_u = v_r = v_i = v_f = 1$.

In this case, the value balance becomes indistinguishable from the energy balance. The costs, given by (2.6) and (2.8) are zero. The operation "costs" nothing in energy... since this is a conservative quantity !

The expression of η given by (2.10) is the first law efficiency.

A Second Scale, a little more complicated, consists of postulating that the residual heat, rejected into the environment, has no value : $v_r = 0$. Also, the equipment will be scrapped at the end, with no further use ; its final value will be zero : $v_f = 0$. Expressions (2.6) and (2.8) for the costs then reduce to :

$$\frac{C_o}{E_u} = r \quad \text{and} \quad \frac{C_t}{NE_u} = r + i_e \tag{3.1}$$

The ratio C_t/NE_u is the "specific total consumption of energy", that is the number of joules which are lost, per useful joule. Equation (3.1) recalls that in order to calculate the total consumption, not only the residual heat rejected into the air or rivers has to be counted, but also the energy content of the equipment scrapped at the end.

3.2 Entropy Based Scales
The Third Scale consists of assigning the value $v = 1$ to each joule of "noble" energy (mechanical, electrical) and a lower value v_t to each joule of thermal energy at whatever temperature ; this value v_t will correspond to the efficiency of industrial production of noble energy (and its delivery to the consumer) from thermal energy.

For example, it is usually accepted that 3 thermal joules give one electrical joule.

Thus we write : $v_t = 0.33$.

This scale is commonly used in the comparison of various systems for heating houses with electricity or with fuels.

The Fourth Scale takes into account the fact that heat has a higher value, the higher its temperature. The value of a quantity of heat Q is then given by the CARNOT coefficient $v_t = 1-(T_o/T)$ where T_o is a reference temperature. Remember that thermal EXERGY ($\equiv$ available energy) is defined by :

$$Ex \equiv Q\left(1 - \frac{T_o}{T}\right) \tag{3.2}$$

The result is that the balance of value becomes here identical to the balance of exergy. Noble energies will be considered as pure exergy, with value $v = 1$.
In this case, η (see (2.10)) is the exergy efficiency of the process/machine.

3.3 Ecology Based Scales
The earth + sun is a semi-closed system, in the sense that it receives practically no energy from the surroundings, whereas it is continually losing energy by radiation to interstellar space. On another hand, energy exchanges between the different sources and sinks of the system by irreversible processes, are described as exergy losses (Figure 3).

An hypothetical world government which would have, as its only objective, to maximize the well-being of the maximum number of humans, could decide to substitute certain rare energy sources by other less rare ones, in accordance with an overall scale of values of energy sources in the semi-closed ecological system which is our planet.

A Scale with 2 values : 0 = Renewable 1 = Non-renewable. A very simple (without doubt too simple) mathematical model consists of giving the value of zero to renewable energies and also to fusible isotopes (deuterium) which exist in quasi-infinite quantities, and the value of one to all other, non-renewable energies.

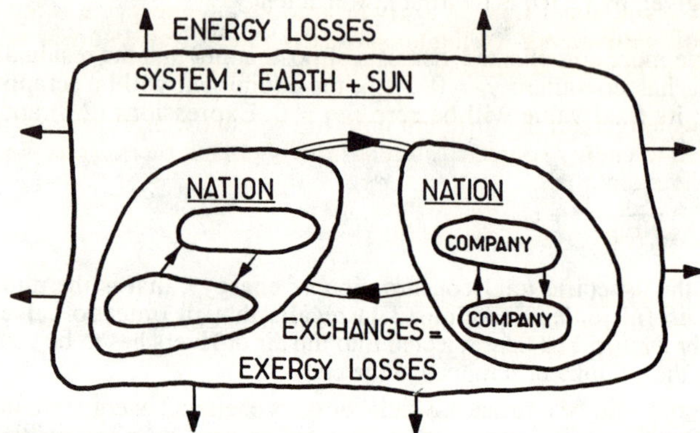

Figure 3. Exergy losses.

This simplistic scale is often more or less unconciously adopted by the man-in-the-street when he claims that solar energy "costs nothing" and "we only have to use it" instead of fuel or electricity... The best rejoinder to make to him is to remark that rainwater collected in the mountain behind a dam is also solar energy and that it shouldn't cost anything either ! Depreciation of invested capital, which is the principal cost in this type of system, is curiously unknown to a lot of consumers.

The Ecological Value Scale. As discussed in previous publications (Le Goff, 1979a, b), any energy source may be characterized by the following factors :

* the volume of reserve : R (small for petroleum, quasi-infinite for sun radiation) ;
* the maximum available power : P_{max}, per unit of land area ;
* the usefulness : U, at a specific place and at each precise day and hour ;
* the energy content of energy converters : i_e.

The value of an energy source, should, in general be an increasing function of P_{max} and U and a decreasing function of R and i_e.

The simplest function (Le Goff, 1979b) would be :

$$v \equiv \frac{U(1-i_e).P_{max}}{R} \tag{3.3}$$

This is a very crude approach to the problem, given here only to draw attention to the fact that it is possible to define an ecological value of energy, even before thinking of technical feasibility or of economic profitability.

3.4 Economy Based Scales

The fourth group of value scales is based on economics. This is the exchange value between open systems, and consequently the classical definition of an economic good.

We will not deal in detail with this subject since it is already very well-known. Let us only recall that the main value scales for an economical good are based on :

* its production cost ;
* its price in francs on the french market ;
* its price in foreign currency on the international market.

At the international level, the value of a given amount of energy is often expressed in U.S. Dollars, or in "oil-equivalents" taking an specified Saudi Arabian crude oil as reference. This value can also be expressed in terms of "tons of coal equivalents".

Anyway the equations to convert the different units are somehow arbitrary, mainly in the case where they are applied to energy sources different from chemical fuels. Thus the derived value scales may be quite divergent.

4. ECONOMIC AND ENERGY ACCOUNTING

4.1 Energy equivalent period of a heat converter

Let us consider the special case where an industrial system as shown in Figure 2 is presented as an energy converter (Le Goff, 1980) (Figure 4).

For example the system could be a furnace which converts chemical energy into thermal energy, a motor which converts electrical energy into mechanical energy, or an electric light bulb which converts electrical energy into light energy, etc.

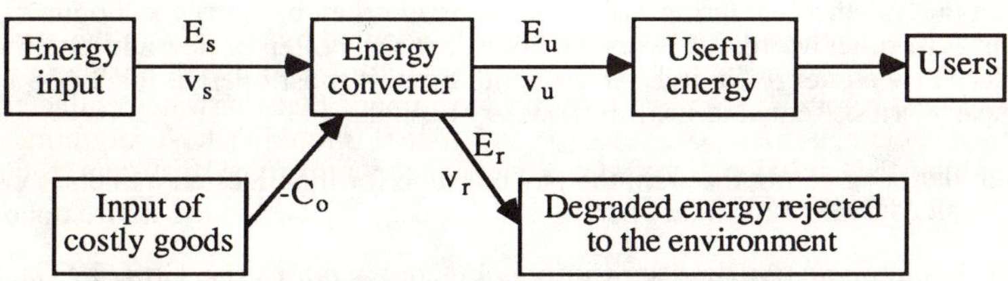

Figure 4. Energy converter.

The values v_i and v_f of Eq. (2.7) are substituted here by the energetic content per kg of the energy converter, CE_k. Let K_m be the mass of the technical structure. Then the energetic content of the energy converter is :

$$I_e \equiv K_m (CE_{ki} - CE_{kf}) \tag{4.1}$$

Let v_a, v_u, v_r and v_k be the value of a joule respectively in the input, the useful output, the rejects, and in the invested capital (energy converter).

Generalizing the concepts introduced in the previous chapter leads us to define one, or several, energy equivalent periods for the energy converter.

That is, the value of the energy capital can be related to any of the three fluxes of energy.

$1 - \tau_{es} \equiv \dfrac{I_e}{\dot{E}_s}$ is the periodic equivalent to the energy consumed : that is the time the converter takes to consume an amount of energy equal to its own energy content.

$2 - \tau_{eu} \equiv \dfrac{I_e}{\dot{E}_u}$ is the periodic equivalent to the useful energy produced : that is the time the converter takes to produce an amount of useful energy equal to its own energy content.

$$3 - \tau_{er} \equiv \frac{I_e}{\dot{E}_s - \dot{E}_u}$$ is the periodic equivalent to the energy rejected to the environment.

We have applied these definitions to the case of a boiler producing 150 ton/hour of steam at 30 bar and having the characteristics given in Table 1. It can be seen that the three energy equivalent periods are respectively 15, 21 and 61 hours. It is interesting to note that these periods are remarkably short. For example, this means that it only takes 21 hours for the boiler to produce an amount of energy, in the form of 30 bar steam, equal to that required for its own construction. It is tempting to call this the "energy pay out time" but this is something of an illusion since at the same time as steam is produced energy is also rejected to the environment and there is an overall consumption and not a production of primary energy.

4.2   Cost equivalent periods

To make a fruitful comparison between energy and monetary accounting, we have calculated the 3 monetary equivalent periods for the boiler presented in Table 1. It can be seen that these 3 equivalent periods are respectively 1 121 h, 857 h and 3 636 hours (Le Goff, 1980).

1. It seems that the last period (τ_{mr}) is negative ! This is due to the fact that the monetary cost ($E_s v_s - E_u v_u$) of the operation is negative which is perfectly normal since its opposite, the profit, must obviously be positive. The value of the steam produced ($E_u v_u$) is greater than that of the fuel consumed ($E_s v_s$).

2. It can be seen that, in absolute terms, τ_{mr} is much greater than its energy analogue τ_{er} which is only 61 hours. In other words, it takes 61 hours for the boiler to degrade an amount of energy equal to its own energy capital. At the same time, it must operate for 3 636 hours to produce a monetary "value" equal to its own monetary capital.

3. It can be seen that, neglecting the sign, the pay out time for the invested monetary capital is the same as τ_{mr}, the equivalent monetary period of the energy rejected to the environment.

4. The value of τ_{mr} obviously depends to a large extent on the selling price of steam with respect to the cost price of fuel. If we assume that the steam is sold for 70 FF/ton[1], and the fuel is bought for 750 FF/ton, we obtain : $\tau_{mr} = -$ 3 636 h.

If the steam is sold for 60 FF/ton, the τ_{mr} goes to 9 230 hours. The minimum selling price for the steam is 53.5 Francs/tonne which would give zero profit and therefore an infinite value for τ_{mr}.

5. To eliminate this sensitivity to variations in market price it is preferable to use the two other monetary equivalent periods τ_{ms} and/or τ_{mu} which are independent of the selling price of the product. These two quantities only differ from each other by the conversion yield of the machine.

Table 2 gives some typical examples of these quantities. It should be remembered that these are the periods over which the converter consumes (or produces) an amount of energy which costs as much as its own construction cost.

---

[1] The prices given in the text are in French francs and date from 1979-1980. In order to convert them into prices of 1990 the inflation rate for the period of 1980 until 1989 has to be considered, which amounts to 94.5 %. The currency exchange rate for February 1990 is : 1 US $ \equiv 5.62$ FF.

Consider a boiler complete with its accessories,

> * weight 140 ton,
> * which produces 150 tons/hour of saturated steam at 30 bar,
> * which consumes 10.7 ton/hour of n° 2 fuel,
> * cost price fully installed 9x10⁶ Francs (1979 value before tax)[2].
> * which has a residual value of zero.

In addition, we also have the following data :

> * the enthalpy content of steam at 30 bar   : 530×10^6 cal/ton
> * the lower calorific value of the fuel   : 10 000 kcal/kg
> * the energy content of the boiler   : 50 MJ/kg
> * the selling price of the steam   : 70 FF/ton
> * the cost price of the fuel   : 750 FF/ton

ENERGY EQUIVALENT PERIODS based on the amount of energy :

> * Primary energy consumed $\tau_{es} \equiv \dfrac{I_e}{E_s} = 15.6$ hours
>
> * Useful energy produced $\tau_{eu} \equiv \dfrac{I_e}{E_u} = 21.0$ hours
>
> * Energy rejected to the environment $\tau_{cr} \equiv \dfrac{I_e}{E_s-E_u} = 61$ hours

MONETARY EQUIVALENT PERIODS based on the price of energy :

> * Primary energy consumed $\tau_{ms} \equiv \dfrac{K_m v_{ki}}{E_s v_s} = 1\ 121$ hours
>
> * Useful energy produced $\tau_{mu} \equiv \dfrac{K_m v_{ki}}{E_u v_u} = 857$ hours
>
> * Energy rejected to the environment $\tau_{mr} \equiv \dfrac{K_m v_{ki}}{E_s v_s-E_u v_u} = -\,3\ 636$ hours

Table 1. Equivalent Periods for a Boiler Producing Steam.

$$\tau_{ms} = \frac{\text{investment cost}}{\text{cost of energy consumed per year}} = \frac{K_m v_{ki}}{E_s v_s}$$

ENERGY CONVERTERS
* Electric light bulb   : 50-500 watts   : 300 ± 100 hours
* Steam generator   : 4 t/h - 150 t/h   : $1\ 700 \pm 700$ hours
* Electric motor   : 10 kW - 1000 kW   : $2\ 000 \pm 1\ 000$ hours
* Turbo Alternator   : 500 kW - 20 000 kW : $35\ 000 \pm 2\ 000$ hours

INFORMATION CONVERTERS
* Television receiver   = 200 000 hours
* Computer (IBM - CII) 5 kW   = 600 000 hours
* Office calculating machine (HP 25) 0.5 W = 20 000 000 hours

Table 2. Equivalent Periods for Equipment consuming Energy.

---

[2] See footnote 1 page 8.

Table 2 shows that the equivalent periods are roughly constant for a given type of converter and for a given state of technology, that is they are independent of the power of the machine.

Table 2 also shows that for all energy converters the equivalent period lies between several hundred hours and several tens of thousands of hours.

However the equivalent period is much greater for "information converters" and can be up to several tens of millions of hours.

In fact, the equivalent period is a sort of measure of the technological complexity of a system. τ_e is all the greater the more the systems contain something other than energy, that is information ; the knowledge and know-how of those who designed and built the system in question.

5. OPERATING COST OPTIMIZATIONS

In order to present some typical cases of operating cost optimizations, we will analyze three examples according to energy, exergy and monetary scales. The first one is a double-pipe heat exchanger (Le Goff and Giulietti, 1982), the second one is a process of drying beet pulps (Le Goff et al., 1980) and the last one is a comparison of four types of heat transformers to upgrade waste heat of a industrial process (Le Goff et al., 1990).

5.1 Optimization of a Heat Exchanger

Let us consider a double-pipe heat exchanger as shown in Figure 5 through which flow two aqueous solutions whose physical properties are identical to that of pure water. In the annular space, the flowrate is $\dot{M}' = 1$ kg.s^{-1} with an inlet temperature $T_1' = 20$ °C. In the cylindrical space, the fluid enters at $T_1 = 120$ °C and its flowrate can be varied in a very large range. The dimensions of the heat exchanger are :

```
* outside tube diameter :      5 cm
* inside tube diameter :       2.5 cm
* total length :               10 m
```

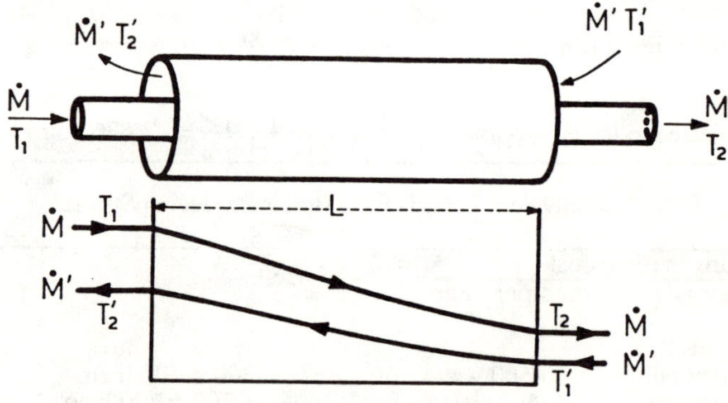

Figure 5. Double pipe heat exchanger.

Let Q be the heat flux transferred from one fluid to the other, with :

$$Q = Mc_p(T_1-T_2) = M'c_p'(T_2'-T_1') \qquad (5.1)$$

The objective is to determine the flowrate of the hot fluid, that minimizes either the consumption of exergy Ex or the monetary operating cost C for a given value of the heat flux Q.

The Specific Operating Consumption of Exergy $\left(\text{SOCEx} \equiv \dfrac{\text{Ex}}{Q}\right)$ results from the addition of three terms :

* the specific operating consumption of *mechanical* exergy due to pressure drops in the hot fluid (SOCMEx) and in the cold fluid (SOCMEx') ;
* the specific operating consumption of *thermal* exergy (SOCTEx) due to the temperature gradients in the boundary layers.

The mathematical expression of the SOCEx has been established elsewhere (Le Goff, 1982). We obtained :

$$\text{SOCEx} \equiv \frac{\text{Ex}}{Q} = \text{SOCMEx} + \text{SOCMEx}' + \text{SOCTEx} \tag{5.2}$$

$$\text{with SOCMEx} = K_u \frac{\text{NTU.M}}{Lf.M_{min}} \qquad \text{with } K_u \equiv \frac{u_m^2 \, Pr^{2/3}}{c_p(T_1 - T_1')}$$

$$\text{SOCMEx}' = K_u' \frac{\text{NTU}' \, M'}{Lf'\eta M_{min}} \qquad \text{with } K_u' \equiv \frac{u_m'^2 \, Pr'^{2/3}}{C_p'(T_1 - T_1')}$$

$$\text{SOCTEx} = \frac{M}{M_{min}} \frac{T_1'}{\eta(T_1 - T_1')} \, Ln \left\{ \left(1 - \eta \, \frac{M_{min}}{M} \left(1 - \frac{T_1}{T_1'}\right)\right) \left(1 - \eta \, \frac{M_{min}}{M} \left(1 - \frac{T_1'}{T_1}\right)\right) \right\} \tag{5.3}$$

On Figure 6, these quantities are plotted as functions of the ratio $\alpha \equiv Mc_p/M'c_p'$.

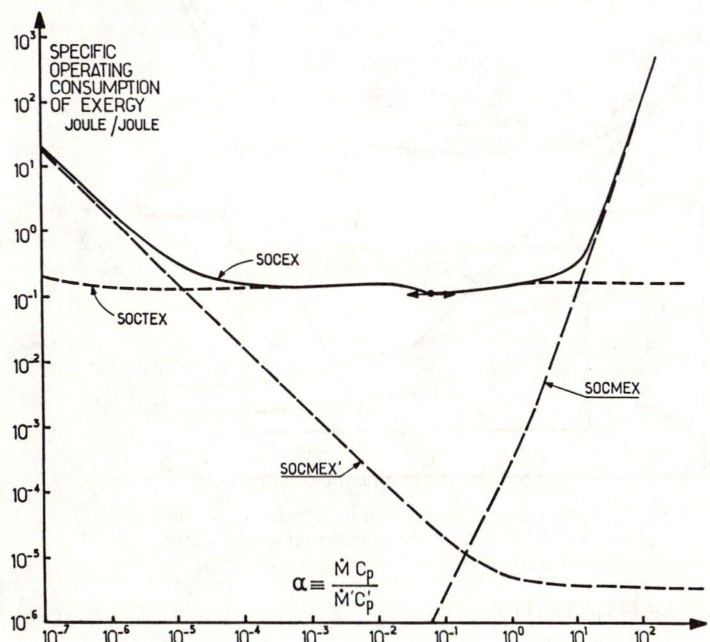

Figure 6. The specific operating consumption of exergy and its thermal and mechanical components as a function of the thermal flux ratio α.

The specific monetary operating cost (SMOC) results from the combination of three terms :

* the buying cost of the electrical energy supplied to the motor of the pump ;
* the buying cost of the thermal energy (hot fluid at 120 °C) ;
* the selling cost of the same fluid after becoming warm (at T_2).

The general expression of the specific monetary operating cost is (Le Goff, 1982) :

$$SMOC = (SOCMEx + SOCMEx')v_w + \frac{\beta}{\eta}(v_{q1} - v_{q2}) + v_{q2} \qquad (5.4)$$

where v_w and v_{q1} are the buying prices of energy :
$\qquad v_w = 0.0555$ FF/MJ for electricity[3]
and $\qquad v_{q1} = 0.0239$ FF/MJ for the hot fluid at 120 °C[4].

For the selling value of the residual warm water, we have considered three possibilities :

$v_{q2} = 0$ (fluid rejected to the environment : no value)
$v_{q2} = v_{q1}$ (value of thermal energy independent of its temperature)
$v_{q2} = v_{q1} \dfrac{T_2 - T_0}{T_1 - T_0}$ $\left(\begin{array}{c}\text{value proportional to the enthalpy content :}\\ \text{FOURIER scale of value}\end{array}\right)$

On Figure 7, the SMOC is plotted as a function of α, for the three cases.

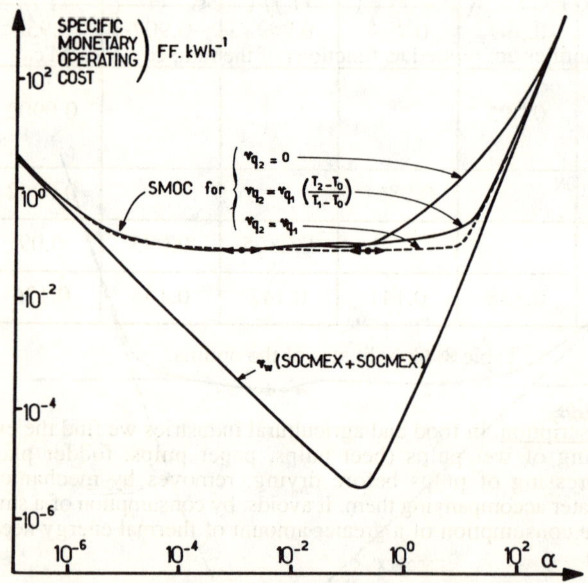

Figure 7. Specific monetary operating cost as a function of the thermal flux
ratio α for three values of the residual warm water.

<hr>

[3] See footnote 1 page 8.
[4] See footnote 1 page 8.

262

Figures 6 and 7 show that for very large and very small values of α (i.e. of M), the mechanical contribution (monetary or exergetic) is much more important than the thermal contribution. In a wide intermediate domain of α (which includes all practical situations) the mechanical contribution is negligible as compared to the thermal one.

The coordinates of the various optima are summarized in Table 3. Note that the monetary and exergetic optima are theoretically different but as they are located outside of the practical domain of α ($0.1 \leq \alpha \leq 20$) imposed by technical constraints, the practical optima are all located at the border line of the range, that is $\alpha \approx 0.1$. In any way, the minima are very flat, so that such an optimization is *not* a decisive procedure for the design of the heat exchanger.

MONETARY MINIMUMS IN FRENCH FRANCS						
1 st case $v_{q2} = 0$		2nd case $v_{q1} = v_{q1}$	3rd case $v_{q2} = v_{q1} \dfrac{T_2 - T_1'}{T_1 - T_1'}$			
Minimum		Minimum	Minimum		EXERGY MINIMUM	
theoretical	practical	theoretical	theoretical	practical	theoretical	practical
α 12×10^{-2}	0.1	0.2	12×10^{-2}	0.1	6×10^{-2}	0.1
η 0.9994	0.907	0.794	0.9994	0.907	0.955	0.907
SMOC 1st case 0.0865	0.095				0.0902	
2nd case		0.0861			0.0862	
3rd case			0.0865	0.095	0.09	
SOCEx 0.141	0.138	0.141	0.141	0.138	0.136	0.138

Table 3. Coordinates of the optima.

<u>5.2 Drying of Beet Pulps</u>
<u>5.2.1 Process description</u>. In food and agricultural industries we find the example of the pressing and the drying of wet pulps (beet pulps, paper pulps, fodder plants, brewing draffes...). The overpressing of pulps before drying, removes by mechanical means an important part of the water accompanying them. It avoids, by consumption of a small amount of mechanical energy, the consumption of a greater amount of thermal energy needed by water vaporization (Figure 8).

If x is the pulp drying rate between pressing and drying (x = percent of dry matter in the pulp), the mechanical energy consumption W of the press is a growing function of x. By contrast, the thermal energy Q needed by the dryer is a decreasing function of x. So there is an optimal value of x, minimizing the primary energy consumption (Figure 9).

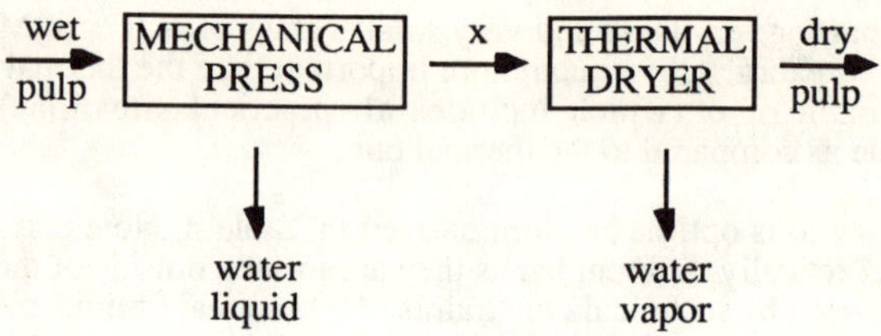

Figure 8. Drying of beet pulp.

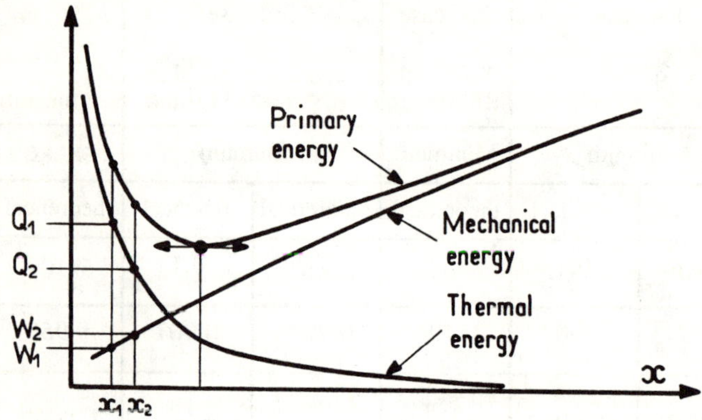

Figure 9. Primary energy and its mechanical and thermal components as a function
of the pulp drying rate x.

Let us assume :

$$\left|\begin{array}{c}\text{Primary}\\ \text{Energy}\\ \text{Consumption}\end{array}\right| = E_p = A_wW + A_qQ \qquad (5.5)$$

Coefficients A_w and A_q depend only on conversion efficiencies of primary energy into
mechanical energy used by the press, and thermal energy used by the dryer. A_w and A_q are
independent of economical factors.

Let us assume :

$$\left|\begin{array}{c}\text{Energetic}\\ \text{operating}\\ \text{cost}\end{array}\right| = C_E = B_wW + B_qQ \qquad (5.6)$$

This is the economical function to minimize. B_w et B_q depend only on fuel, electrical energy...
prices. Besides we shall specify which fraction of each price corresponds to an importation
from foreign countries, payed in foreign currencies. Then we can write a new form of equation
(5.6) in terms of foreign exchange and not in national currency.

Finally, there are three different optimal values of x, corresponding to three minima :

* of primary energy consumption,
* of operating cost in national currency,
* of cost in terms of foreign exchange.

5.2.2 The mathematical model.

a) The total primary energy consumption. Let us consider the system in the operating state number one, taken to be reference state. So the pulp drying leaving the press is x_1, the thermal energy is Q_1, the mechanical energy is W_1.

Let us consider a second operating state slightly different from the first one, and for which :

$$x_2 = x_1 + \Delta x$$

$$W_2 = W_1 + \Delta W \tag{5.7}$$

$$Q_2 = Q_1 - \Delta Q$$

Let us consider only the relative differences of dryness x, Q and W. One introduces the coefficients a and b as follows :

$$\frac{\Delta W}{W_1} = a \frac{\Delta x}{x_1} \qquad \text{and} \qquad \frac{\Delta Q}{Q_1} = - b \frac{\Delta x}{x_1} \tag{5.8}$$

The economists call a and b the "elasticities".

The integration of equations (5.8) leads to :

$$\frac{W}{W_1} = \left(\frac{x}{x_1}\right)^a \qquad \text{and} \qquad \frac{Q}{Q_1} = \left(\frac{x}{x_1}\right)^{-b} \tag{5.9}$$

So equation (5.5) becomes :

$$E_p = A_w W_1 \left(\frac{x}{x_1}\right)^a + A_q Q_1 \left(\frac{x}{x_1}\right)^{-b} \tag{5.10}$$

b) Optimum coordinates. We cancel the derived function of equation (5.10) and we obtain the optimum coordinates :

$$\left|\begin{matrix} \text{optimal} \\ \text{dryness} \end{matrix}\right| \equiv x_{oe} = x_1 \left|\frac{b}{a} \frac{A_q Q_1}{A_w W_1}\right|^{\frac{1}{a+b}} \tag{5.11}$$

$$\left|\begin{matrix} \text{primary energy} \\ \text{minimal} \\ \text{consumption} \end{matrix}\right| = (E_p)_{min} = k |A_q Q_A|^{a/a+b} |A_w W_1|^{b/a+b} \tag{5.12}$$

where : $k \equiv \left(\frac{b}{a}\right)^{-\frac{b}{a+b}} + \left(\frac{b}{a}\right)^{\frac{a}{a+b}}$

By the same way from equation (5.6) the energetic operating cost becomes :

$$C_E = B_w W_1 \left(\frac{x}{x_1}\right)^a + B_q Q_1 \left(\frac{x}{x_1}\right)^{-b} \tag{5.13}$$

This operating cost is at a minimum for :

$$x_{of} = x_1 \left| \frac{b}{a} \; \frac{B_q Q_1}{B_w W_1} \right|^{\frac{1}{a+b}} \tag{5.14}$$

and it is equal to :

$$(C_E)_{min} = k |B_q Q_1|^{\frac{a}{(a+b)}} \; |B_w W_1|^{\frac{b}{a+b}} \tag{5.15}$$

We can derive the same equations (5.13, 5.14, 5.15) with the operating cost expressed in terms of foreign exchange. Let us call y_w the part of the electrical energy price paid in foreign currencies, and y_q the same part for fuel, coal or gas.

The optimal dryness x_{od} corresponding to the minimal cost in foreign currencies is :

$$x_{od} = x_1 \left| \frac{b}{a} \; \frac{y_q}{y_w} \; \frac{B_q Q_1}{B_w W_1} \right|^{\frac{1}{a+b}} \tag{5.16}$$

c) Comparison between the three optima. The three optimal values of x (5.11), (5.14), (5.16) may be sometimes very different. Let us compare them. We have :

$$\frac{x_{of}}{x_{oe}} = \left| \frac{B_q}{B_w} \; \frac{A_w}{A_q} \right|^{\frac{1}{a+b}} \tag{5.17}$$

$$\frac{x_{od}}{x_{of}} = \left| \frac{y_q}{y_w} \right|^{\frac{1}{a+b}} \tag{5.18}$$

<u>5.2.3. The drying process</u>. After sugar extraction by water washing, the beet contains still some nutritive elements, useful for animal feed.

The pulp has the initial dryness x_i equal to 6 per cent. So it will be interesting to eliminate all the water contained in the beet. In fact the equilibrium humidity of the pulp in free atmosphere is equal to 11 or 12 per cent. Therefore the pulp is dried until x_f equal to 88 or 89 per cent.

Water elimination helps to promote preservation and stability during transport. Since 1975, France is the first European producer (834.000 tons). The dehydration units are connected to sugar refineries. Their capacities vary from 2.000 to 6.000 tBj (Beet tons a day). The unit we have studied has the following characteristics : a capacity of 4.000 tBj, it operates only 80 days each year, for 10 years.

Mechanical energy of pressing. From the experimental data supplied by the sugar refineries, for x varying from 10 to 20 per cent, we have established the empirical following relationship :

$$W = (1.95.10^4) x^{4.36} \text{ kwh/ton of pellets} \tag{5.19}$$

266

To convert this energy consumption in terms of primary energy (toe) we use the coefficient proposed by the France National Grid :

$$0.27 \text{ toe} \Rightarrow 1000 \text{ kwh}$$

So the pressing energy required is given by

$$W = 5.25x^{4.36} \text{ toe/ton of pellets} \tag{5.20}$$

Drying energy. One ton of pellets at $x_f = 88$ per cent, comes from a pulp at the dryness x, with mass 0.88/x ton. One ton of pellets requires the vaporization of (0.88/x)-1 ton of water.

The thermal energy consumption of the dryer is proportional to this water quantity. From the experimental data of the unit we have established :

$$Q = \left| \frac{5.10^{-2}}{x} - 0.057 \right| \text{ toe/ton of pellets} \tag{5.21}$$

Marginal energy consumption of the dryers. The operation of the dryers requires some electrical energy E_0 (drum rotation) which is independent of x. From experimental data we measured :

$$E_w = 0.0055 \text{ toe/ton of pellets}$$

Total working energy consumption. This consumption is the sum of the three previous terms :

$$E = W + Q + E_w$$

$$E = (5.25)x^{4.36} + (5 \times 10^{-2})x^{-1} - 0.051 \tag{5.22}$$

Electrical energy boughet from the national grid. Using equations (5.11) and (5.12), we calculate the optimum coordinates (Figure 10) :

$$x_{oe} = 0.32$$

$$(E_p)_{min} = 0.142 \text{ toe/ton} \tag{5.23}$$

Today the pressing rate leads to x = 20 %, with an energy consumption of :

$$E_{p1} = 0.204 \text{ toe/ton}$$

If we pass from this situation to the optimal state, the energy savings reach 0.0055 toe/ton e.g. 27 per cent.

Electrical energy produced in the factory. The France National Grid propose the equivalence coefficient : $0.27 \text{ toe} \Rightarrow 1.000 \text{ kwh}$. The sugar refineries produce all or the most part of their electrical energy, and their equivalence coefficient is :

$$0.124 \text{ toe} \Rightarrow 1.000 \text{ kWh}$$

Let us call u the rate of electrical energy produced in the unit. We can show that the new optimal value x_{oe}' is given by the following expression :

$$x_{oe}' = x_{oe} \left| 1 + u \left(\frac{0.124}{0.27} - 1 \right) \right|^{- \frac{1}{5.36}} \tag{5.24}$$

The new optimum coordinates are now (Figure 10) :

$$x_{oe}' = 0.36 \ \%$$

$$(E_p')_{min} = 0.115 \ \text{toe/t} \tag{5.25}$$

The energy saving is 20 % greater than the previous one.

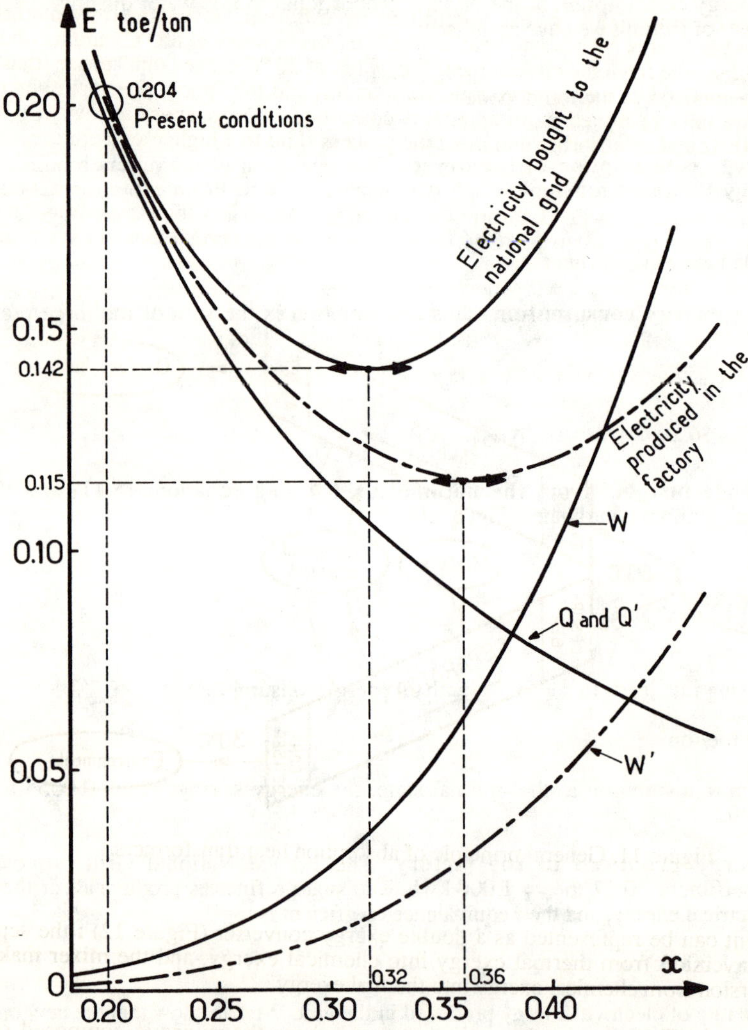

Figure 10. Optimization of the drying of beet pulp.

Note : all these expressions have been extrapolated from experimental values for x inferior to 20 per cent. The presses used today in deshydratation units do not reach the values of 30 or 36 per cent obtained by calculations. Only prototype presses are able to attain these high values of x. So our quantitative conclusions must be taken with caution.

5.3   Comparison of the Performance of Heat Transformers

   5.3.1   Systems description. We will present here a comparative analysis of the performance of four types of heat transformers utilized in an industrial process to upgrade a waste heat source (waste water at 80 °C) to useful heat at 180 °C (9 bar steam).

The general principle of the three first heat transformers (HT) studied in this paper is shown in Figure 11.

A mixture of two components, whose volatilities are different, is separated by distillation, between a desorber heated by a waste heat source (i.e. waste water at 80 °C) and a condenser cooled by a cold source from the environment (i.e. water at 30 °C). The "rich" phase (that is the phase mainly composed of the more volatile component) and the "poor" phase coming out of the separator are later re-mixed in an apparatus composed of an evaporator heated with the waste heat at 80 °C and an absorber that heats the process fluid to a higher temperature, i.e. 180 °C.

Such a heat transformer is a type-II absorption heat pump (AHP) in which some fraction of the waste heat is degraded from 80 to 30 °C, in order to upgrade the complementary fraction from 80 to 180 °C.

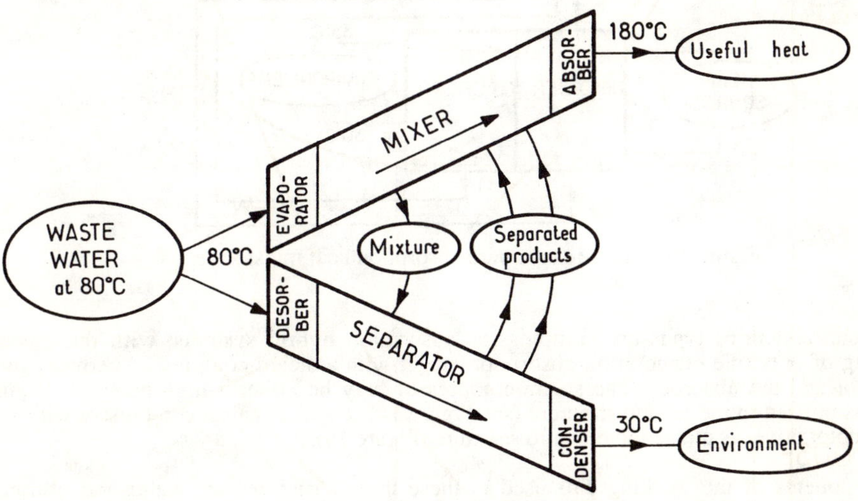

Figure 11. General principle of absorption heat transformers.

This equipment can be represented as a double exergy converter (Figure 12) : the separator makes the conversion from thermal exergy into chemical exergy, and the mixer makes the reverse-conversion from chemical exergy into thermal exergy.

The first type of HT is a pure absorption system, where the mixer is composed of four evaporation-absorption stages that are in thermal series, the mass flowrates being in parallel (Figure 13).

Figure 12. A heat transformer is a double exergy converter.

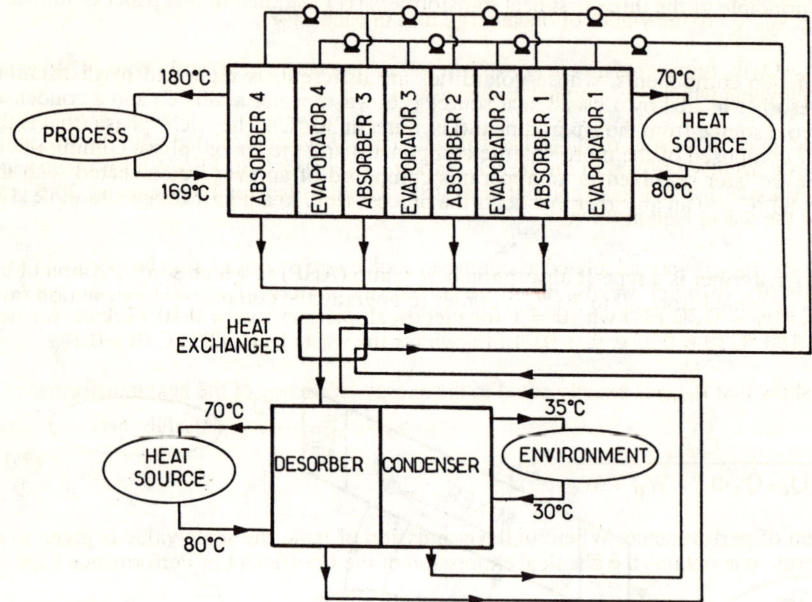

Figure 13. A multi-stage pure absorption heat transformer.

The second and third types are compression-absorption hybrid systems, with the mixer consisting of only one evaporation-absorption stage, with a steam-compressor between the evaporator and the absorber. The steam-compressor may be either a high-pressure steam ejector, as utilized in the second structure (see Figure 14) or a mechanical compressor with an electric motor drive, as utilized in the third structure (Figure 15).

The components of the working pair used in these three structures are water and lithium bromide. The "rich" phase is pure water. The "poor" phase (i.e. : water poor) is a concentrated salt solution (55 % salt mass fraction). In these three structures, the absorption stages that operate above 150 °C, are made of graphite, in order to withstand the corrosion effects of lithium bromide up to 220 °C.

The fourth structure is an open cycle compression heat pump also called mechanical vapor recompression system (MVR), as showed in Figure 16. In this machine, mechanical exergy is employed to upgrade a waste heat source, during compression of low pressure steam.

<u>5.3.2. Comparison criteria</u>

a) Economic efficiency. Let us recall that the objective of the operation is to transform an energy of low value (or even costless) in an energy with a higher value. The economic efficiency is thus defined as follows :

$$\eta_{eco} \equiv \frac{Q_u v_u}{(Q_e + Q_d)v_s + W_p v_p + W_c v_c} \tag{5.24}$$

where Q_u, Q_e, Q_d are the absorber, evaporator and desorber heat fluxes,
W_p, W_c are the mechanical energies consumed in the solution pumps and in the compressor,
v_u, v_s, v_p, v_c are the values of one energy unit in each case.

The term $(Q_e + Q_d)v_s$ represents an energy of low value (i.e. waste heat) and $(W_p v_p + W_c v_c)$ an energy of high value which are supplied to the system.

We have calculated two types of η_{eco}. The first one (η_{eco1}) was calculated with the following energy values currently utilised in the french market : $v_p = v_c = 0.30$ FF/kWh for electricity ; $v_u = 0.10$ FF/kwh[5] for the heat generated at 180 °C, and also for the steam at 20.5 bar ; v_s = zero for the waste heat at 80 °C.

The second one (η_{eco2}) was calculated assigning to each energy form a value proportional to its Carnot factor (θ), in other words, based on its exergy content ($T_0 = 303$ K). So we have utilized : $v_p = v_c = 0.30$ FF/kwh ($\theta = 1$ for electrical energy) ; $v_u = 0.10$ FF/kwh for heat generated at 180 °C ($\theta = 0.33$) ; $v_s = 0.04$ FF/kwh for the waste heat at 80 °C ($\theta = 0.14$).

It is easy to show that η_{eco2} is exactly equal to the exergy efficiency of the heat transformer :

$$\eta_{ex} \equiv \frac{Q_u \theta_u}{(Q_e + Q_a)\theta_s + W_p + W_c} \tag{5.25}$$

b) Coefficient of performance. When, in the expression of η_{eco}, the same value is given to all forms of energy, one obtains the classical expression of the Coefficient of Performance COP :

$$COP \equiv \frac{Q_u}{Q_e + Q_d + W_p + W_c} \tag{5.26}$$

Such a criterion is of interest here, since the principle of a heat transformer is to increase the value of a low-value waste heat.

---

[5] See footnote 1 page 8.

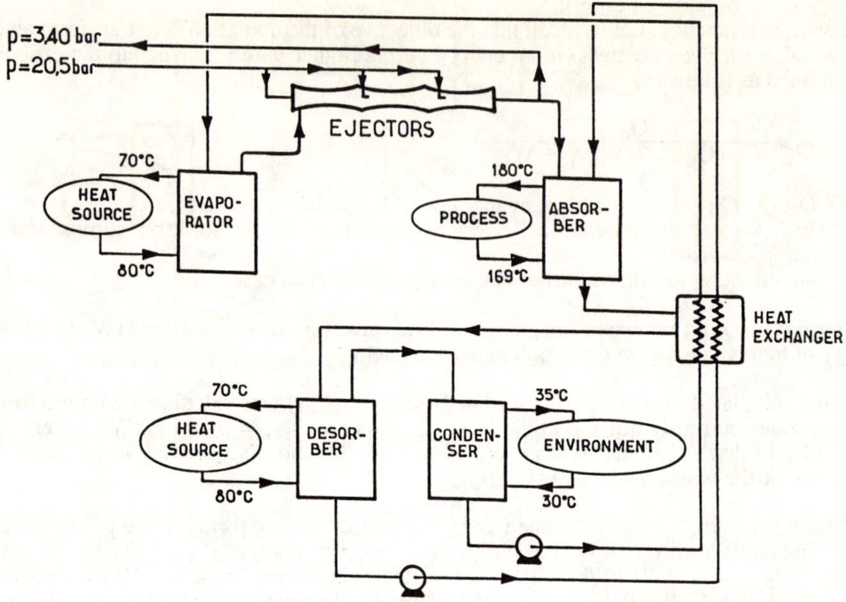

Figure 14. A hybrid absorption-compression heat transformer with steam ejectors.

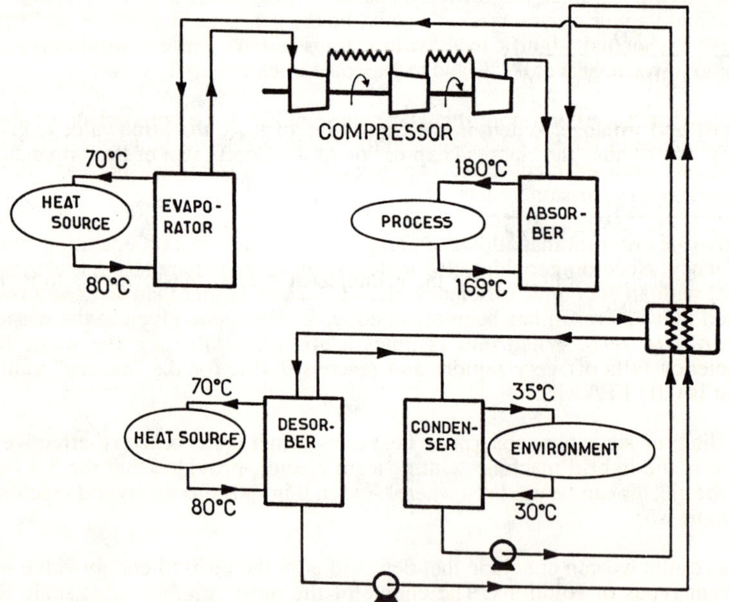

Figure 15. A hybrid absorption-compression heat transformer with
mechanical compressors.

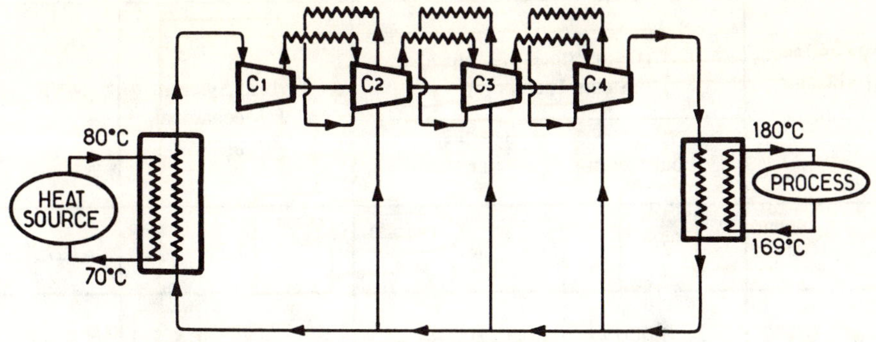

Figure 16. A four-stage mechanical vapour recompression system.

     5.3.3   Results - Discussion. Table 4 gives the results of the calculations for the conditions fixed above.

For the hybrid system with a steam ejector, two cases are considered, whether the low-pressure steam (at 3.4 bar), issued from the ejector, finds an utilization in another machine or is unused and rejected into the environment. In the first case, its value is taken as 0.10 FF/kwh, and in the second case, it is taken as zero.

In the technical literature devoted to heat pumps, one generally uses the COP as the main criterion for comparing the quality of different machines. According to this criterion, the fourth system (mechanical vapour recompression) would be the best one. But, as we said before, the concept of COP, has a small significance for heat transformers, since it would give the same value to one joule of waste heat at 80 °C and to one joule of electricity !

The values of η_{ex} (or η_{eco2}) are more realistic than the COP for a comparison of the energy utilization performance of the HT : according to this value scale, one joule of 80 °C waste heat equals 14 % of one joule of electricity. The greatest η_{ex} is obtained, here again, with the mechanical vapour recompression system.

Finally, the only true criterion that allows to compare the present market operating costs, is the economic efficiency. According to this value scale, the multi-stage pure absorption system is by far the best one, with an economic efficiency of 16.67, much higher than any other one. But it must be recalled that this result has been obtained, when the value given to the waste heat at 80°C was *rigorously zero*. When one assumes a non-zero value for the waste heat, the economic efficiency falls off very rapidly and reaches 0.431, for the "exergy" value of the waste heat, that is 0.04 FF/kwh[6].

In conclusion, the four-stage pure absorption heat transformer is the best cost-effective system. In second place is the hybrid machine with a steam ejector, provided that the 3.4 bar-steam coming out of the ejector can be used elsewhere. And in third place is the hybrid machine with a mechanical compressor.

Based on these results we can conclude that depending on the utilized energy value scale one can find different types of solutions. The choice of the most suitable value scale will be a function of the objectives of the decision-maker.

It is important to note that the present comparison was only based on operating costs. A complete economic analysis would require the evaluation of capital costs and pay-back times.

---

[6] See footnote 1 page 8.

1 Multi-Stage Absorption	2 Hybrid System with Steam Ejector		3 Hybrid System with Mechanical Compressor	4 MVR
	Useful 3.4 bar steam	Useless 3.4 bar steam		
η_{eco} 16.67	2.32	0.04	1.09	0.79
$\eta_{ex}=\eta_{eco2}$ 0.432	0.096	0.036	0.553	0.64
COP 0.176	0.369	0.04	0.404	0.91*

* The usual COP of compression heat pumps, calculated as Q_u/W_e gives here 2.38.

Table 4. Comparison of the four Heat Transformer Systems.

6. THE GENERAL OPTIMIZATION PRINCIPLES FOR AN ENERGY CONVERTER

We have considered four value scales for energy (Le Goff et al., 1990). The first two (based on enthalpy and entropy) lead to objective and rigorously defined values. The third one, based on ecology, depends on many subjective factors such as the estimation of the availability and usefulness of each energy form. The fourth scale, based on economics is also relatively subjective since it includes not only the production costs but also some market terms which are a function of marketing situation.

In any case, it is only after having adopted one or another value scale, and having used it in predictive calculations for the overall cost of an industrial project, that the technical decision-maker could make the optimum choice.

We know in fact that many technical decisions are based on empiricism or direct intuition, without any overall computer optimization had been previously conducted, but only in the own decision-maker's mind, thanks to a more or less conscious synthesis based on experience and knowledge, as well as a more or less conscious choice of a particular value scale.

Anyway, the choice of such a value scale is always necessary. Without any doubt, the universal objective of every technical decision-maker is to minimize this integral cost per useful energy unit, independently of the value scale chosen.

We can now state 9 general optimization principles, based on the equation developed in Part 2, which are applicable to each and every one of the value scales. These principles are :

1. To minimize the specific energy consumption r (to maximize the system's efficiency).
2. To adapt the rejected energy value v_r for potential user's needs.
3. To minimize the feed energy value v_s : to look for the stated objective by using the lowest value energy, for example, low-level heat.
4. To maximize the equipment life-time N, i.e. to make the equipment more corrosion and/or erosion resistant.
5. To minimize the equipment energy content I_e : to design equipment to perform the same function and being lighter and made of cheaper materials.
6. To minimize the value of the energy used for producing the equipment.

7. To have an acceptable level for the residual value of equipment v_f by designing it in such a way that it could be re-used afterwards, at least partially, in other systems.

In practice, these 7 parameters are rarely independent of each other ; they are interrelated with a given number of constraints, for example :

- The residual value v_f of equipment is lower the lower the life-time period N.
- The efficiency of a thermal engine is lower the warmer the rejected water but this water could be more useful : r and v_r are then related.
- The degradation rate of energy r is lower the higher the performance of the engine and consequently its cost.

All of these 7 principles are technical in nature. They must be considered by the engineer whose objective is, for example, the design of a new energy system for a given power E_u of an equally given value v_u (specified by the user).

For the technical decision-maker, E_u and v_u are defined, but for a socio-political decision-maker these parameters may change.
For instance, the fact that a common french man wanted to use his car to travel 10000 km per year, is a given datum for the car's designer engineer. For the sociologist, this fact is but one of the many components of the well-being of the common french man, so it could be neglected or considered.

The sociologist will direct his research to find the effect of the energy consumption E_u on the well-being of every person. It is known that the power of all the "mechanical slaves" made available for each french man is about 4 kw ; for a U.S. citizen it is 12 kw and for the average world human-being on earth only 2 kw. The question is whether these figures are proportional to the corresponding well-being.

Do we have, for example, a real need for eating strawberries in mid-winter, importing them from far countries or heating the place where they can be produced ?

The last two optimization principles are then :

8. To minimize the quantity of energy needed E_u to achieve a given rate of individual or collective satisfaction.
9. To minimize the value assigned to used energy v_u, i.e. to teach human beings for assigning a lower value to material goods (to live with less !).

In the present paper we have only dealt with the technical problems raised by the first 7 principles.

REFERENCES

Gouy, G. (1889). Sur l'énergie utilisable. *Journal de Physique*, 8, 501-518.
Keenan, J.H. (1932). A steam chart for second law analysis. *Mechan. Eng.*, 54, 195-204.
Le Goff, P. (1979a). La valeur de l'énergie a-t-elle une base économique, écologique ou technique ? Critères d'optimisation en énergétique industrielle. *Revue d'Economie Industrielle*, 8, 68-98.
Le Goff, P. (1979b). *Energétique Industrielle*, 1, Analyse thermodynamique et mécanique en économies d'énergie. Lavoisier, Paris.
Le Goff, P. (1980). *Energétique Industrielle*, 2, Analyse économique et optimisation des procédés. Lavoisier, Paris.
Le Goff, P. (1982). *Energétique Industrielle*, 3, Application en génie chimique. Lavoisier, Paris.
Le Goff, P. and M. Giulietti (1982). Is the value of a source of energy based on enthalpy, entropy, economy or ecology ? An example of application : the optimisation of a heat exchanger. *J. Chem. E. Symposium Series*, 78, T1/1-T1/13.

Le Goff, P., L. Le Bec, A. Taieb and M. Bendif (1980). Optimization of industrial processes using two energy sources. *6th International Cost Engineering Congress*, October 20-22, Mexico.
Le Goff, P., S. de Oliveira Jr. and R. Rivero (1990). Industrial process optimization : exergy and economics. *11th International Cost Engineering Congress*, April 22-25, Paris.
Rant, Z. (1956). Exergie, ein Neues Wort für Technische Arbeitsfähigkeit, *Forschung Ing.-Wes.*, 22, 36, 36-37.

LIST OF SYMBOLS

A	coefficient
B	coefficient
a, b	elasticity
CE_k	energy content per kg of energy converter [J/kg]
C_E	energetic operating cost
C_o	net annual operating cost (US \$/year or FF/year)
C_t	total cost (US \$ or FF)
c_p	specific heat (J.kg^{-1}.K^{-1})
COP	coefficient of performance
d	diameter (m)
E	flux of energy (W)
Ex	flux of exergy (W)
F	value of expensive supplies (US \$.year^{-1} or FF.year^{-1})
f	friction factor
I	investment (US \$ or FF)
I_e	energy content of the system (J)
i_e	specific energy content of the system
J	Colburn factor
k_u, k_u'	coefficient
K_m	mass of a technical structure (kg)
L_f	dimensionless number = 2 J/f
M	mass flowrate (kg.s^{-1})
N	number of years
NTU	number of transfer units
P	profit (US \$/year or FF/year)
p	pressure (bar)
Pr	Prandtl number
P_{max}	maximum available power per unit of land area (W.m^{-2})
Q	heat flux (W)
r	specific energy consumption
R	volume of reserves
s	elasticity coefficient
SMOC	specific monetary operating cost (FF.W^{-1})
SOCEx	specific operating consumption of exergy
SOCMEx	specific operating consumption
SOCTEx	specific operating consumption of thermal exergy
T	temperature (°C, K)
T_o	reference temperature (°C, K)
toe	ton of oil equivalent
u	rate of electrical energy
u_m	velocity (m.s^{-1})
U	usefulness
v	energy value
V	flowrate (m^3.s^{-1})
W	mechanical or electrical power (W)
x	dryness rate
Y	price of energy in foreing currencies
α	thermal flux ratio

ß	coefficient
θ	Carnot factor [1 - (T$_o$/T)]
η	efficiency
ρ	mass density (kg.m^{-3})
τ$_e$	energy equivalent period
τ$_m$	monetary equivalent period

<u>Subscripts</u>

comp	compressor
d	destroyed, desorber
e	inlet, evaporator
eco	economic
ex	exergetic
f	final
i	initial
k	capital
ml	log mean temperature difference
min	minimal
o	operating, reference
oe	optimal-energetic
of	optimal-operating cost
od	optimal-foreign currency
opt	optimal
p	primary, pump
q	heat
r	rejected
s	supplied to the system, solution
t	total, thermal
u	useful
w	electricity

FIGURES

Figure 1	Energy balance.
Figure 2	Balance of value.
Figure 3	Exergy losses.
Figure 4	Industrial system as energy converter.
Figure 5	Double-pipe heat exchanger.
Figure 6	The specific operating consumption of exergy and its thermal and mechanical components as a function of the thermal flux ratio α.
Figure 7	Specific monetary operating cost as function of the thermal flux ratio α for three values of the residual warm water.
Figure 8	Drying of beet pulps.
Figure 9	Primary energy and its mechanical and thermal components as a function of the pulp dryness rate x.
Figure 10	Optimization of the drying of beet pulps.
Figure 11	General principle of absorption heat transformers.
Figure 12	A heat transformer as a double exergy converter.
Figure 13	A multi-stage pure absorption heat transformer.
Figure 14	A hybrid absorption-compression heat transformer with steam ejectors.
Figure 15	A hybrid absorption-compression heat transformer with mechanical compressors.
Figure 16	A four stage mechanical vapor recompression system.

Analysis of Cumulative Exergy Consumption and Cumulative Exergy Losses

J. Szargut

Institute of Thermal Engineering
Technical University of Silesia
44-101 Gliwice, Poland

ABSTRACT

Cumulative exergy consumption (CExC) expresses the consumption of exergy of natural resources in all links of a technological manufacturing network leading from raw materials to the final product. When analysing the production of materials and energy carriers, the cumulative degree of thermodynamic perfection (CDP) can be defined as a ratio of exergy to CExC. The calculation method for complex processes, including the major product and by-products, is presented. Illustrative values of CDP are cited.

The difference between CExC and exergy expresses the cumulative exergy loss. These losses have been divided into constituent exergy loss (connected with the fabrication of particular semi-finished products) and partial exergy loss (connected with particular links of the technological network). Analysis of constituent and partial exergy loss provides information about a possible improvements in the technological network. A set of equations can be used for the analysis of cumulative exergy loses. A sequence method of analysis is also presented. An illustrative analysis for the heat delivery from a heat-and-power station has been included.

Exhaustion of unrestorable domestic natural resources can be also determined by means of CExC, but the influence of the imported raw materials and semifinished products should be taken into account. Utilization efficiency of unrestorable domestic natural resources (UDN) can be greater than 100 % if the raw materials do not prevail in the export. Illustrative values of UDN for the Polish economy have been cited.

Ecological cost takes into account not only CExC but also the exergy losses resulting from the deleterious impact of waste by-products on human activity, on human health, on the productivity of the agriculture and forestry, and on the natural resources. An ecological economy can be introduced, based upon a postulate of minimization of the consumption of unrestorable natural resources.

278

INTRODUCTION

All kinds of human activity can be developed thanks to the
consumption of natural resources, which can be divided into unrestorable and restorable ones. The exhaustion of unrestorable
natural resources is very dangerous for the future of mankind and
therefore a measure should be introduced enabling us to evaluate
natural resources and methods should be elaborated for the estimation of the exhaustion of these resources.

The ability of natural resources to drive thermal, chemical
and biological processes results from the deviation of their state and composition, from the thermodynamic equilibrium with the
natural environment. Hence the ability to perform maximum work in
the conditions of natural environment can be accepted as a measure for the evaluation of natural resources. A quantity defined in
such a way has been termed exergy (Szargut, Morris and Steward,
1988).

All production processes form a complicated network whose
links are mutually connected. The consumption of natural resources appears not only in the last step of fabrication of the considered product, but also in precedent links. Therefore the analysis of cumulative exergy consumption, appearing in the technological network, is necessary to determine the total consumption
of natural resources, connected with the fabrication of the considered product. A comparison of exergy with the cumulative exergy consumption for materials or energy carriers informs us about
the thermodynamic imperfection of the utilization of natural resources. The reasons of this imperfection can be investigated by
determining the exergy losses appearing in particular links of
the technological network.

The calculation of cumulative exergy consumption and the allocation of cumulative exergy losses can be made only approximately. The relatively most exact calculation method bases upon the
linear input-output equations. The set of these eqs contains the
consumption coefficients, which can be only approximately determined because of the differences in the attained technological
and operational level in particular plants of the analysed system.
The input-output method requires a simultaneous analysis of all
mutually connected production processes. Fabrication of a particular product in the frame of technological network can be approximately analysed by means of the sequence method, which takes
into account only the most important links of the technological
network leading from the natural resources to the investigated
product.

The exhaustion of natural resources can be analysed in a
global scale or in the scale of particular countries. In the last
case the influence of imported raw materials and semi-finished
products should be taken into account, because such an import can
influence profitably the exhaustion of domestic natural resources.

When analysing the exhaustion of natural resources, the deleterious environmental impact of waste products can be additionally taken into account. So the ecological cost can be defined,

which can be used for the formation of ecological economy aiming
at a minimization of the exhaustion of unrestorable natural resources.

1. EXERGY AND ITS CLASSICAL APPLICATIONS

Exergy expresses simultaneously the quantity and quality of
energy. The quality is characterized by the ability of the considered kind of energy to be transformed into mechanical work in
the conditions determined by the natural environment. Hence exergy can be defined (Riekert 1974) as the shaft work or electrical
energy necessary to produce a material in its specified state
from materials common in the environment, in a reversible way, heat being exchanged only with the environment at its temperature
T_0 .

In contradistinction to energy, exergy is exempt form the
law of conservation. In every real (and hence irreversible) process appears an unrestorable exergy loss, expressed by the law
of Gouy-Stodola

$$\delta B = T_0 \sum \Delta S , \qquad (1.1)$$

where: T_0 - environmental temperature,
$\sum \Delta S$ - sum of entropy increases of all the bodies taking
part in the process.

Internal exergy loss is a result of irreversible processes
occurring within the analysed system. External exergy loss appears if the waste product of the process is discharged to the environment. Irreversible processes in the environment, resulting
from the deviation of thermal parameters and chemical composition
between the waste product and the components of the environment,
lead to the destruction of the exergy of the waste product. Hence
the exergy of a waste product discharged to the environment, expresses immediately the external exergy loss.

The thermal state and chemical composition of the natural
environment represent a reference level for the calculation of
exergy. In the natural environment appear however distinct deviations from thermodynamic equilibrium. Components of the environment differing in their composition or thermal parameters from
prevailing values, can be applied to drive thermal and chemical
processes and therefore should be regarded as natural resources.
Their exergy is positive. Only for the commonly appearing components of the environment a zero value of exergy may be accepted.
The correct assumption of the reference level is essential for
the calculation of external exergy losses. The assumption of the
reference level (if consistently respected) does not influence
the results of the calculation of internal exergy losses.

The most probable chemical interaction between the components of the waste products and the environment occurs with the
participation of the most common components of the environment.
Therefore the correct determination of external exergy losses requires the assumption of reference species commonly appearing in
the environment, for every chemical element. Hence the assumption
of the reference level for the calculation of exergy is to some
extend conventional.

Flow processes usually appear in industry. Therefore the

exergy of the matter flux crossing the immovable system boundary
(termed simply "exergy") is of greatest importance. The main components of this kind of exergy are: kinetic exergy, potential
exergy, physical exergy and chemical exergy. Kinetic and potential exergy are expressed by the kinetic and potential energy
calculated in relation to the environment. Physical exergy results from the deviation of temperature and pressure from environmental values. Chemical exergy is defined for the substance, having environmental temperature and pressure. Its value results
from the deviation of the composition in comparison with the commonly appearing components of the environment.

Exergy balance enables us to represent in a convenient form
the results of exergy analysis. The balance equation must be closed by the term representing the internal exergy loss.

Exergy losses can be exactly distributed according to the
place of occurence. If in the same place some irreversible phenomena appear simultaneously, the partition of exergy losses according to the simultaneously acting causes can be accomplished only
conventionally.

The main task of exergy analysis is to detect and evaluate
the causes increasing the thermodynamic imperfection of analysed
processes. All irreversible phenomena are connected with exergy
losses, which leads to a decrease of useful effects (at a constant consumption of driving exergy) or to an increase of the consumption of driving means (at constant useful effect). Hence exergy analysis indicates the possibilities of improvement of the analysed processes, but cannot decide about the purpose-fulness of
the improvement, because it depends on economic indices.

Exergy is sometimes applied to solve economic problems. Exergy is, however, a thermodynamic notion, not an economic one. The
exact solution of economic problems cannot depend on the applied
method and hence exergy analysis cannot provide new exact results.
It can, however, facilitate some approximate solutions. E.g. the
partition of production costs between the useful products of a
complex process can be done much more reasonably by means of exergy (proportionally to the exergy fluxes) than with energy (proportionally to the enthalpy fluxes). Only the postulate of minimizing the consumption of natural resources may be realized exclusively with exergy, but this postulate leads to a new, ecological
economy, not introduced hitherto.

2. ANALYSIS OF CUMULATIVE EXERGY CONSUMPTION

All useful products of human activity are fabricated in a
complicated network of mutually connected production processes.
This network is supplied by raw materials, fuels and other energy carriers extracted from natural deposits. The quality of natural resources can be expressed by means of exergy. The total consumption of natural resources connected with the fabrication of
the considered product can be expressed by the index of cumulative exergy consumption (CExC). This index can be determined separately for every kind of natural resources

$$r_{kj} = \frac{\sum B_{kj}}{P_j} ,$$

(2.1)

where: P_j - final production of the product j , leaving the
                 investigated system,
$\sum B_{kj}$ - exergy consumption of the natural resource k in
                 the entire technological network, connected with
                 the fabrication of the product j .

Exergy consumption of secondary raw materials should be also taken into account. Secondary raw materials may be treated as restorable natural resources.

Usually the overall CExC is determined, comprising all kinds
of the consumed raw materials and energy carriers

$$r_j = \sum_k r_{kj} \ . \tag{2.2}$$

CExC can be applied not only to evaluate the consumption of
natural resources but also to solve other problems of energy system analysis, e.g:
- to determine the necessary increase of raw materials delivery
  for the planned increase of the production rate of particular
  products,
- to investigate the influence of price changes of raw materials
  on the production costs of particular products,
- to evaluate the cumulative effects of materials and energy saving,
- to compare different production technologies from the point of
  view of thermodynamic imperfection.

The analysis of CExC can be regarded as a further development of the industrial energy analysis considering the cumulative
energy consumption (Chapman 1974, Boustead and Hancock 1979, Bibrowski et all. 1983). Industrial energy analysis can be, however
applied only to determine the exhaustion of fuel resources and
does not determine the degree of the thermodynamic imperfection
of the considered processes.

When analysing the production of materials and energy carriers, CExC can be compared with the exergy of the product. The
ratio of exergy to CExC expresses the cumulative degree of thermodynamic perfection (CDP) of the investigated part of the technological network (Szargut 1985, Szargut and Morris 1987)

$$\eta^* = \frac{b}{r} \ , \tag{2.3}$$

where b, r - specific exergy and CExC of the considered mate-
              rial or energy carrier.

For the fabrication of major products, according to the 2-nd
law of thermodynamics, $\eta^* < 1$. Sometimes, however, for the fabrication of a by-product the value $\eta^* > 1$ can be obtained, if the
exergy of the analysed by-product is greater than the exergy of
the substituted product of a specialized process.

It would not be reasonable to calculate CDP for sophisticated ready products (e.g. cars, tv-receivers) because their usefulness results mainly from their system features, not from the
chemical composition of the components. The calculation of CExC
is however reasonable for all products, because if makes it possible to compare various design variants and fabrication techno-

logies.

CExC can be calculated by means of the system of balance e-
quations, because this quantity fulfils the law of conservation.
The CExC burdening the useful products of the process equals the
sum of CExC of all raw materials and semi-finished products consumed in the considered link of the technological network. The
system of balance eqs can be separately formulated for all kinds
of consumed natural resources. The exergy of secondary raw materials should be also taken into account. For the link j of the
technological network and for the natural resource k , the balance equation takes the form

$$r_{kj}^{(1)} = \sum_i (a_{ij}^{(1)} - f_{ij}^{(1)}) \sum_t x_{it} r_{ki}^{(t)} + B_{kj}^{(1)} , \qquad (2.4)$$

where: k   - serial number of the considered natural resource,
  i, j - serial number of the link of the technological
     network,
  l, t - serial number of the production technology of the
     product i, j ,
  $a_{ij}^{(1)}$ - coefficient of the gross consumption of the semi-
     finished product i per unit of the complex use-
     ful product, containing a unit of the major pro-
     duct j[*],
  $f_{ij}^{(1)}$ - coefficient of the by-production of the useful
     product i ,
  x_{it} - fraction of the technology t of the fabrication
     of the product i in the analysed system,
  $B_{kj}^{(1)}$ - immediate gross exergy consumption of the natural
     resource k per unit of the complex useful pro-
     duct.

According to eq. (2.1) CExC relates to a unit of the product
leaving the considered system. Hence CExC depends distinctly on
the assumption of the system boundary. Usually the system of production processes is analysed, without employees, because in this
case CExC does not depend on the consumption level of the society
and can be applied for the sake of comparison of production processes in various countries. If the employees are included into
the system, CExC becomes much higher than without them, because
the final production leaving the system becomes much smaller.

Eq. (2.4) is formulated for a complex process producing more
than one useful product. In every complex process a <u>major product</u>
can be distinguished, determining the capacity and location of
the process. A useful product substituting the major product of
another specialized process has been termed the <u>by-product</u>. If
more than one useful product of the analysed process does not
substitute the major product of another specialized process, a
complex major product should be accepted, and CExC should be calculated for this complex containing a constant fraction of useful
products. The partition of CExC between the components of a complex major product is optional and can be done e.g. proportionally to their exergy.

---

[*] Coefficients of the net consumption refer separately to the
unit of the major product and to the unit of by-product.

The coefficient of by-production should be expressed in units of the substituted major product, by means of the substitution ratio

$$f_{ij}^{(1)} = f_{uj}^{(1)} \, z_{iu} \, , \qquad (2.5)$$

where: $f_{uj}^{(1)}$ – coefficient of production of the real by-product
            u   per unit of the major product   j ,

       z_{iu} – susbtitution ratio (amount of units of the major
            product   i   substituted by the unit of the by-product   u).

CExC for the fabrication of the real by-product   u   results from the dependence

$$f_{ij}^{(1)} \, r_i = f_{uj}^{(1)} \, r_u \, . \qquad (2.6)$$

Taking into account (2.5), we obtain

$$r_u = r_i \, z_{iu} \, . \qquad (2.7)$$

CDP for the real by-product   u   can be expressed as follows

$$\eta_u^* = \frac{b_u}{r_u} = \frac{b_u}{r_i \, z_{iu}} = \frac{b_u}{b_i \, z_{iu}} \, \eta_i^* = \frac{\eta_i^*}{\eta_{z\,iu}} \, , \qquad (2.8)$$

where: η_i^* – CDP for the major product   i   substituted by the
            product   u ,

      $\eta_{z\,iu}$ – exergetic substitution efficiency,

$$\eta_{z\,iu} = \frac{b_i \, z_{iu}}{b_u} \, . \qquad (2.9)$$

If $\eta_{z\,iu} < \eta_i^*$, the value $\eta_u^* > 1$ results from eq. (2.8). The smaller is the value r_u resulting from the substitution ratio (or substitution efficiency) the greater is CExC burdening the major product   j   of the analysed process.

It is convenient to transform the system of eqs (2.4) to a form containing only two subscripts, because in such a case the matrix notation can be applied. For this purpose different subscripts should be introduced for the same product manufactured by different technologies

$$r_{km} = \sum_{n} (a_{nm} - f_{nm}) \, r_{kn} + B_{km} \, , \qquad (2.10)$$

where

$$a_{nm} = x_{it} \, a_{ij}^{(1)} \, , \quad f_{nm} = f_{ij}^{(1)} \, . \qquad (2.11)$$

To every subscript   j   correspond some subscripts   m , taking into account various technologies   l . Similarly to every subscript   i   correspond some subscripts   n , taking into account the technologies   t .

The set of eqs (2.10) can be formulated exclusively for semi-finished products, consumed in other links of the technological network. Balance eqs for ready products not consumed in production processes are independent of other eqs and therefore can be considered separately after the solution of eqs (2.10).

In matrix notation the system of eqs (2.10) has the form

$$r \ (E - A + F) = B , \qquad (2.12)$$

where: E - diagonal unitary matrix,
     A, F - square matrix MxM , ($M = m_{max}$),
     r, B - rectangular matrix KxM , ($K = k_{max}$) .

From (2.12) it results that

$$r = B(E - A + F)^{-1} = B \ S^* . \qquad (2.13)$$

The elements of the inverse matrix S^* represent the coefficients of cumulative net consumption of intermediate products per unit of the considered major product leaving the considered system

$$r_{kn} = B_{k1} \ S^*_{1n} + B_{k2} \ S^*_{2n} + \dots \qquad (2.14)$$

E.g. the coefficient S^*_{1n} expresses the cumulative net production of the product 1 per unit of the product n .

Some exemplary values of CExC and CDP have been cited in Table 1.

Table 1. CDP for the Production of Some Materials
and Energy Carriers

Material or energy carrier		CDP %	Reference	Notes
Name	Specific exergy			
1	2	3	4	5
Aluminium	32.9 MJ/kg	9.6	(1)	Bayer process and Hall cell 50 % bauxite ore
		13.2	(2)	Electrolytic method from Al_2O_3
Ammonia gas	20.03 MJ/kg	45.4	(1)	Steam reforming of naphta
		64.8	(1)	Steam reforming of natural gas
		41.5	(2)	Semi-combustion of natural gas
Cement	0.635 MJ/kg	10.3	(1)	From raw materials in the ground, dry route
		6.2	(2)	Wet method, medium rotary kiln
Coal (hard)	1.09 C_1	95.3	(1)	
		95.1	(5)	

1	2	3	4	5
Coke	1.06 C_1	86.9 72.5	(1) (5)	Inputs apportionned over outputs proportionally to exergy values Coke-oven gas substitutes the natural gas
Copper	2.11 MJ/kg	3.2 2.6 1.5	(1) (1) (2)	From ore containing Cu_2S, smelting and refining Hydrometallurgical route Electrolytic
Diesel oil	1.07 C_1	83.5	(1)	
Electricity	1 MJ/MJ	24 22.8	(1) (5)	From conventional power plant, fed with bituminous coal, at consumption place
Gas (natural)	1.04 C_1	87.5 99.8	(1) (5)	
Gasoline	1.07 C_1	84	(1)	
Glass	0.174 MJ/kg	0.8 0.5	(1) (2)	From raw material in the ground Panels
Iron (pig)	8.2 MJ/kg	44.0 30.6	(1) (2)	From haematit ore in the ground, liquid
Iron ore sinter	0.73 MJ/kg	27 21	(1) (2)	
Lead	1.20 MJ/kg	3.5 0.62	(1) (3)	From ore in the ground From ore in the ground without utilization of the waste sulphides
Nitric acid	0.69 MJ/kg	8.5 3.6	(1) (2)	From ammonia produced from crude oil
Oxygen gas	2.4 kJ/mol	1.8 1.4	(1) (2)	Non-compressed
Paper	16.5 MJ/kg	18.7 27.5 27.4 74.3	(1) (1) (2) (1)	From standing timber Integrated plant, waste products used as fuels Integrated plant, waste products used as fuels From waste paper
Sodium chloride	0.24 MJ/kg	12.0 76.1	(1) (2)	From mineral in the ground From brine
Sodium hydroxide	1.87 MJ/kg	9.9 8.3	(4) (4)	Mercury cell Diaphragm cell Electrolysis of brine. Inputs apportionned over outputs proportionally to exergy values

1	2	3	4	5
Steel liquid	8.04 MJ/kg	49.9	(1)	Blast furnace, basic oxygen furnace
		47.7	(1)	Open hearth furnace, 70 % scrap
		55.4	(1)	Electric-arc furnace, 100 % scrap
		34.4	(2)	Open hearth furnace, 70 % scrap
		35.4	(2)	Electric furnace, 100 % scrap
		30.1	(2)	Steel converter, 15 % scrap
Steel, cold rolled products	7.04 MJ/kg	14.8	(2)	40 % scrap in raw materials
Steel, hot rolled products	7.04	19.0	(2)	40 % scrap in raw materials
Sulphur	19.01 MJ/kg	74.2	(1)	Frasch process, from underground deposit
		62.7	(2)	
Sulphuric acid	1.666 MJ/kg	18.3	(1)	Frasch process, sulphur combustion
		15.0	(2)	
		32.2	(3)	
Zinc	5.19 MJ/kg	7.6	(1)	From ore, vertical retort
		6.7	(1)	Electrothermic method
		2.7	(2)	From ZnS, metallurgical method
		4.2	(2)	From ZnS, electrolytic method

x) C_1 – lower calorific value
(1) Calculated by means of the data cited by Boustead and Hancock (1979).
(2) Calculated by means of the data cited by Bibrowski and all. (1983).
(3) Szargut and Morris (1990).
(4) Morris (1989).
(5) Szargut and Jełowicki (in press).

3. ANALYSIS OF CUMULATIVE EXERGY LOSSES

The difference of CExC and exergy represents the cumulative exergy loss δb^* appearing in all the parts of the technological network connected with the fabrication of the considered product

$$\delta b^* = r - b . \tag{3.1}$$

The analysis of the components of δb^* provides information about possibilities to improve the technological network.

The difference $(r - b)_n$ characterizing particular semi-finished products delivered to the final link of the production network of the considered product, determines the constituent

exergy loss . It results from the thermodynamic imperfection of
the constituent technological network producing the semi-finished
product n(Szargut 1988, 1990).

In complex processes the raw materials and semi-finished products consumed in the analysed link of the technological network,
are partially used for the fabrication of by-products. Hence the
coefficient of net consumption of semi-finished products and raw
materials per unit of the major product should be determined by
means of the dependence

$$A_{nm} = a_{nm} - \sum_u f_{um} A_{nu} = a_{nm} - \sum_p f_{um} z_{pu} A_{np} \ , \qquad (3.2)$$

where: p – serial number of the major product substituted by
the by-product u ,

z_{pu} – susbtitution ratio of the product p by the product u .

The set of eqs (3.2) enables us to calculate the coefficients
A_{nm} . The coefficient A_{nm} can be negative if the consumption of
the semi-finished product n is greater in the process substituted by the utilization of the by-product, than in the analysed
principal process.

The coefficients A_{nm} serve to calculate the constituent
exergy losses

$$\delta b_{nm} = A_{nm} (r_n - b_n) \ , \quad n \neq m \ , \qquad (3.3)$$

Some of this losses can be negative, which results from the elimination of the constituent exergy losses occurring in the substituted process. Negative exergy losses can appear only in system analysis.

Local gross exergy loss δb_m comprises the sum of internal
and external exergy losses in the analysed link. It can be calculated from the exergy balance, which can be presented for a steady state as follows

$$\sum_n a_{nm} b_n = b_m + \sum_u f_{um} b_u + \delta b_m \ , \qquad (3.4)$$

where the work performed and heat exchanged with external heat
sources or sinks should be treated as one of the useful products
represented by b_m, b_n or b_u . The local gross exergy loss
refers to the complex of useful products containing a unit of the
major product.

The local gross exergy loss burdens partially the major product and partially the by-products of the analysed link. Local
net exergy loss burdening exclusively the major product, results
from the difference

$$\delta b_{mm} = \delta b_m - \sum_u f_{um} \delta b_u \ , \qquad (3.5)$$

where δb_u – local exergy loss burdening the by-product u .
The local exergy loss burdening the by-product results not only
from the local net exergy loss in the substituted process but also from the difference of exergy of the by-product and substitu-

ted major product

$$\delta b_u = z_{pu}\, \delta b_{pp} - (b_u - z_{pu}\, b_p) =$$

$$= z_{pu}\, \delta b_{pp} - (1 - \eta_{z\,pu})\, b_u \,, \qquad\qquad (3.6)$$

where: δb_{pp} - local net exergy loss burdening the major product in the substituted process,

$\eta_{z\,pu}$- exergetic substitution efficiency, eq. (2.9).

Introducing (3.6) into (3.5) we receive

$$\delta b_{mm} = \delta b_m - \sum_p f_{um}\, z_{pu}\, \delta b_{pp} + \sum_u f_{um}\, (1 - \eta_{z\,pu})\, b_u \,, (3.7)$$

The set of eqs (3.7) enables us to calculate the local net exergy losses for all major products.

The correctness of eq. (3.6) can be checked for the extreme case $\eta_z = 0$, $z = 0$. Then from eqs (3.6) and (3.5) it results that

$$\delta b_u = - b_u \,, \qquad \delta b_{mm} = \delta b_m + \sum_u f_{um}\, b_u \,. \qquad\qquad (3.8)$$

In this case the exergy of not utilized by-products represents an additional external exergy loss in the analysed link m .

Sometimes an unfinished by-product requires additional treatment in a separate plant (e.g. SO_2 - containing gases from pyrite roasting processes are utilized in a separate plant for the production of sulphuric acid). CExC of the unfinished by-product results from the CExC-balance of the additional treatment process

$$a_{su}\, r_s + \sum_{i \neq s} a_{iu}\, r_i = \sum_i f_{iu}\, r_i + r_u \,, \qquad\qquad (3.9)$$

where: r_s, r_u - specific CExC of the unfinished and finished by-product,

a_{su} - coefficient of consumption of the unfinished by-products per unit of the finished by-product u .

Eq. (3.9) takes into account the possibility to produce by-products in the additional treatment process [the first term on the right-hand side of eq. (3.9)]. From eq. (3.9) a negative value of r_s can result if the consumption of semi-finished products [the second term on the left-hand side of eq. (3.9)] is large. In such a case the utilization of the unfinished by-product is not advantageous from the thermodynamic point of view, but can be necessary e.g. from the ecological point of view.

If by-products in a unfinished f. m appear, the local net exergy loss burdening the major product should be determined by means of the formula

$$\delta b_{mm} = \delta b_m - \sum_{u \neq s} f_{um}\, \delta b_u - \sum_s f_{sm}\, (r_s - b_s) \,. \qquad\qquad (3.10)$$

The sum of the local net exergy loss and constituent exergy losses equals the cumulative exergy loss

$$\delta b_m^* = \delta b_{mm} + \sum_{n \neq m} \delta b_{nm} \, . \tag{3.11}$$

Partial exergy loss $\overline{\delta b}_{km}$ expresses the local net exergy loss in the link k of the technological network, resulting from the fabrication of the product m

$$\overline{\delta b}_{km} = S_{km}^* \, \delta b_{kk} \, , \tag{3.12}$$

where S_{km}^* results from eq. (2.13).

If the product of the link m is consumed in the precedent links of the technological network, then $S_{mm}^* > 1$ and the partial exergy loss δb_{mm} is greater than the local net exergy loss δb_{mm}.

A partial exergy loss in a complex process can be negative, if the production of the intermediate product k is smaller in the principal network than in the network substituted by the utilization of by-products.

The sum of partial exergy losses equals the cumulative exergy loss

$$\delta b_m^* = \sum_k \overline{\delta b}_{km} \, . \tag{3.13}$$

Relative constituent, local and partial exergy losses can be defined as follows

$$\zeta_{nm} = \frac{\delta b_{nm}}{r_m} \, , \quad \varepsilon_{km} = \frac{\overline{\delta b}_{km}}{r_m} \, , \quad \lambda_{mm} = \frac{\delta b_{mm}}{r_m} \, . \tag{3.14}$$

From eqs (3.12) and (3.14) it results that

$$\eta_m^* + \sum_{n \neq m} \zeta_{nm} + \lambda_{mm} = 1 \, , \tag{3.15}$$

$$\eta_m^* + \sum_k \varepsilon_{km} = 1 \, . \tag{3.16}$$

The analysis of constituent and partial exergy losses provides information about possibilities of improving the technological network:

- If some constituent exergy loss is very large, the possibility of changing the production technology or the substitution of the considered semi-finished product by another more convenient one should be analysed. Usually it may be done for new and future processes.

- If some partial exergy loss is very large, the reduction possibilities of thermodynamic imperfection of the intermediate link of the technological network should be investigated. It may be achieved by a decrease of internal irreversibilities and by a better utilization of waste products or by changing the technology. Also the possibility of decreasing the consumption of some very "expensive" intermediate products should be taken into account.

4. SEQUENCE METHOD OF CExC-ANALYSIS

CExC and cumulative exergy losses can be analysed approximately by means of the sequence method, which consists in the analysis of subsequent links of the technological network, beginning with the final link and ending with the extraction of raw materials and fossil fuels from nature. The sequence method eliminates the necessity of solving the sets of equations, but makes it difficult to take into account the feedbacks appearing in the technological network. Some less important interconnections of the network may be neglected. Sequence analysis may be made by means of a graphical scheme (Szargut and Morris 1987, Szargut [1] 1987, Szargut and Morris 1990). The scheme presented on Fig. 1 concerns an individual link of the technological network. It is a development of the scheme proposed in Guidelines of Energy Analysis (1974).

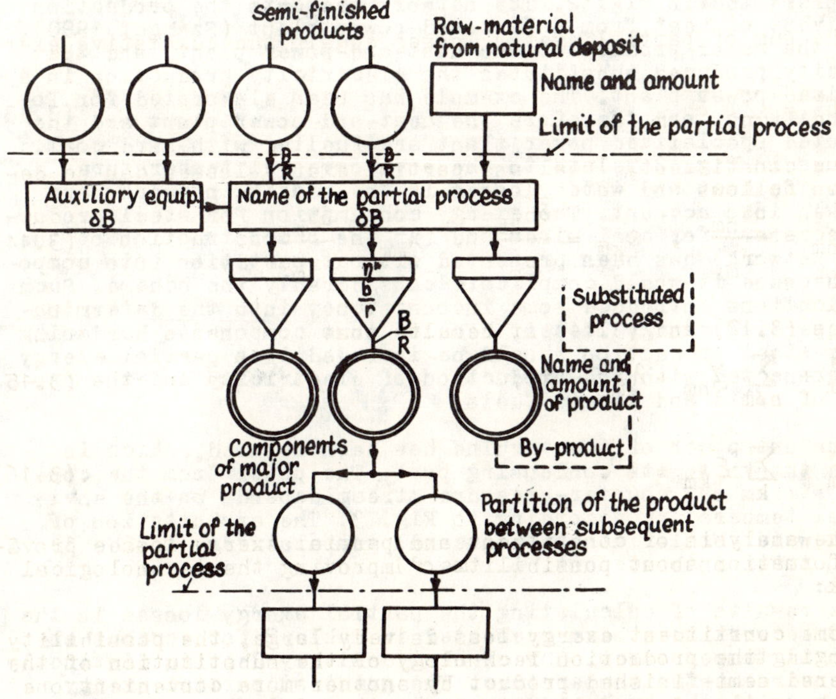

Figure 1. Scheme of the analysis of CExC for an intermediate link of a technological network

The exergy consumption for the construction of machines and installations has been omitted. The exergy losses for transportation have been included into the production process. On the other

hand, the fabrication of by-products and the components of the
complex major product have been shown.

The name of the considered link is given in a great rectangle. The rectangle of auxiliary equipment can be connected with
the principal rectangle. The names and amount of the delivered
semi-finished products are given in circles. The delivery of natural resources extracted immediately from nature is shown in a
small rectangle. Double circles contain the coefficients of cumulative consumption or production of intermediate and final products. In single circles the coefficients of local consumption or
production are shown. The consumption and production coefficients
are related to the unit of the complex final product of the network. Exergy and CExC per unit of the intermediate product are
presented in a triangle. In a small rectangle above the triangle
CDP is given for the intermediate product. Near the connection
lines of the network, the values of exergy and CExC are given per
unit of the final complex major product. The network for the fabrication of the product substituted by the useful by-product can
be presented together with the principal network or separately.

An example of the analysis of CExC and cumulative exergy losses is presented in Fig. 2. The network presents the production
and delivery of heat from a heat-and-power plant (Szargut 1990).
Heat is the major product of the heat-and-power plant, and the
electricity produced substitutes the electricity production in a
specialized power plant. The example has been elaborated for Polish conditions, and therefore the heat-and-power plant and the
substituted specialized power plant are fuelled with hard coal.
The values in Fig. 2 relate to the environmental temperature
+ 2 C . The heat and water losses in the heating network have
been taken into account. The exergy consumption for steel production (necessary for coal mines and for the reconstruction of the
heating network) has been presented without partition into components, because it would complicate considerably the scheme. Such
simplifications introduce some inconsistency into the determination of partial exergy losses, because some components burdening
e.g. the steel production should be included into partial exergy
losses connected with the production of electricity and the production of solid and liquid fuels.

Constant power of the turbine has been assumed, which is
possible thanks to its condensing part. The power from the condensing stream and counter-pressure stream depends on the environmental temperature according to Fig. 3. The exploitation of
the peak-water-boiler at environmental temperatures $t_0 < - 11$ ^{o}C
has been taken into account.

The results of calculating the partial exergy losses in the
analysed network have been presented in Table 2 for various environmental temperatures. CDP of heat delivery depends distinctly
on the environmental temperature. The greatest partial exergy
losses occur in the heat-and-power plant, mainly because of the
irreversibility of the boiler. The exergy losses caused by the
irreversible heat transfer in heat exchangers are important, but
much smaller than in the boiler.

The analysed process is not a good example for the presentation of constituent exergy losses, because no great variety of

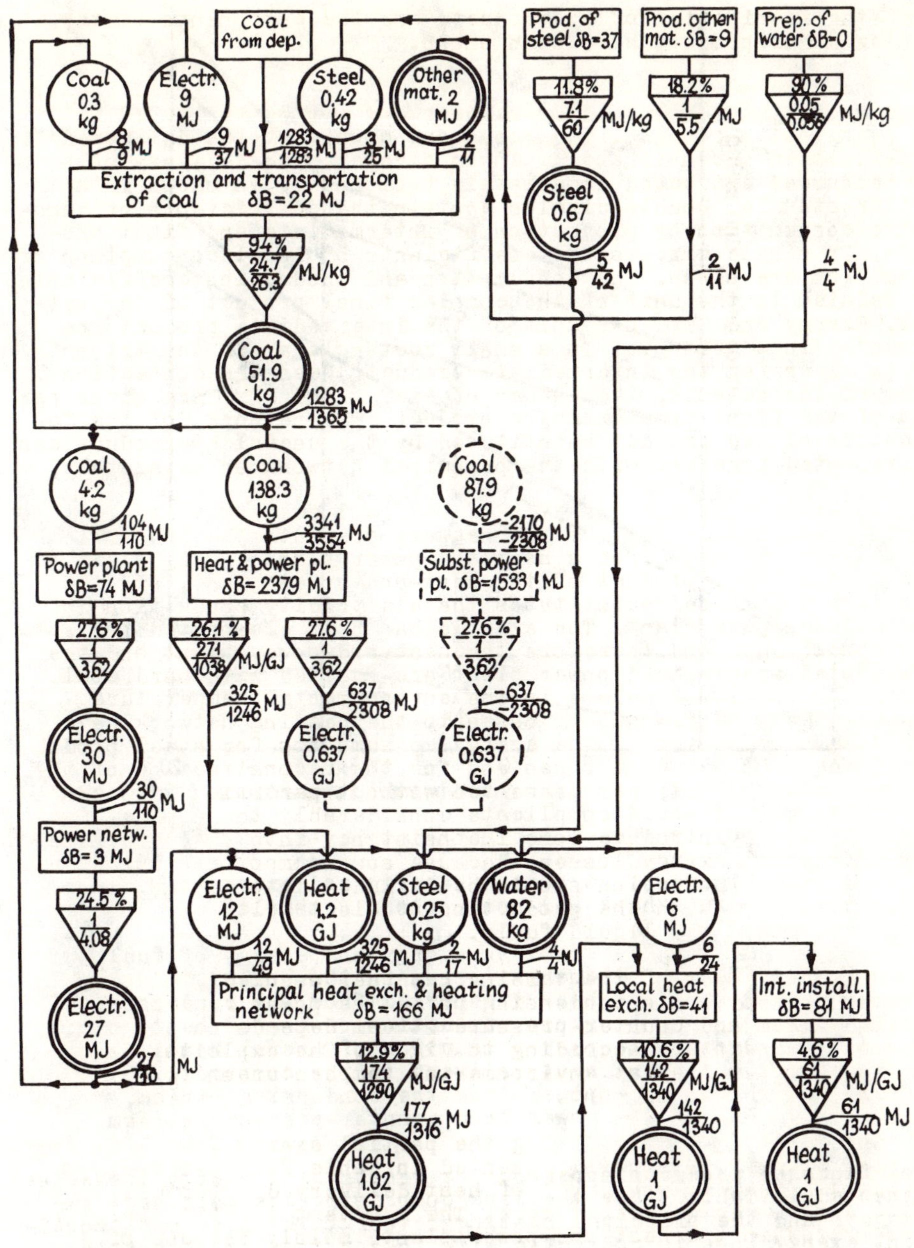

Figure 2. Sequence analysis of CExC for the heat delivery from a heat-and-power plant

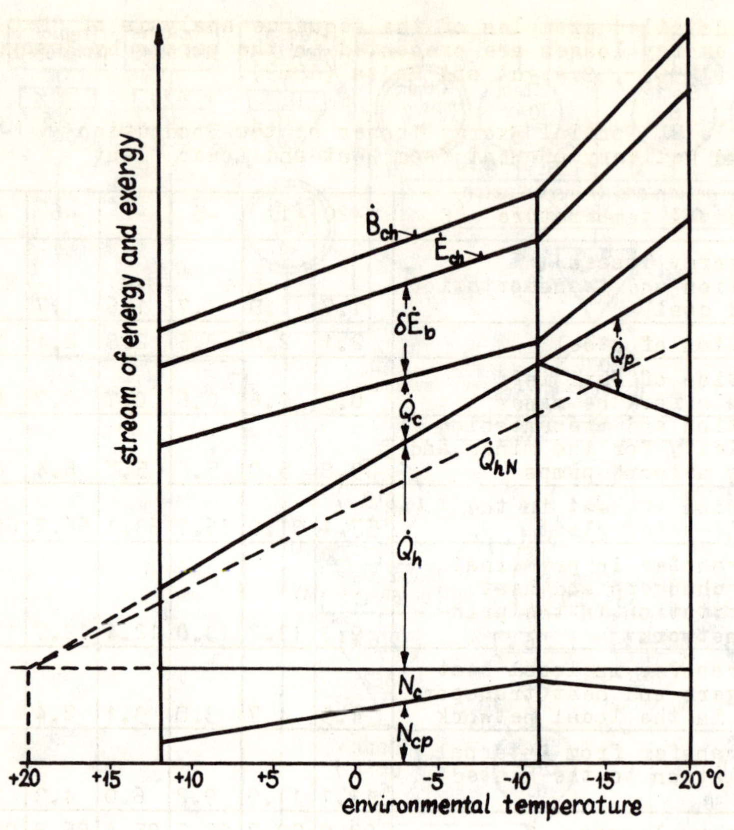

Figure 3. Streams of energy and exergy
in a heat-and-power plant (example)

$\dot{B}_{ch}$, $\dot{E}_{ch}$ – chemical exergy and energy of fuel,
$\delta\dot{E}_b$ – energy losses in the boiler,
$\dot{Q}_c$ – heat discharged from the condenser,
$\dot{Q}_h$ – gross production of useful heat,
$\dot{Q}_p$ – peak production of useful heat,
$\dot{Q}_{hN}$ – heat delivery to the consumers,
N_c – power from the condensing stream,
N_{cp} – power from counter-pressure stream.

semi-finished products appears. The constituent exergy losses are
presented in Table 3 for the link comprising the main heat exchangers and the principal heating network. The greatest constituent exergy loss is connected with the production of counter-
pressure stream.

294

Other detailed examples of the sequence analysis of CExC and
cumulative exergy losses are presented in the papers by Szargut
and Morris (1990) , Szargut and Majza (1989).

Table 2. Partial Exergy Losses at the Production
and Delivery of Heat from Heat-and-Power Plant

Environmental temperature ^{o}C	-20	-11	-5	+2	+6	+12
Partial exergy losses % Extraction and transportation of hard coal	1.7	1.8	1.7	1.6	1.7	1.7
Production of steel	2.1	2.0	2.5	2.8	2.4	3.1
Production of auxiliary materials for the mines	0.5	0.6	0.6	0.7	0.7	0.6
Production and transmission of electricity for the mines and heating network pumps	3.8	5.0	5.2	5.7	6.4	8.4
Production of heat in the heat-and-power plant	57.1	51.0	56.7	63.1	66.2	68.1
Heat transfer in principal heat exchangers and heat transportation in the prin- cipal network	9.3	13.2	13.0	12.4	12.7	13.0
Heat transfer in local heat exchangers and heat transpor- tation in the local network	4.3	4.7	3.8	3.1	2.4	1.6
Heat transfer from internal installation to the heated room	11.1	11.9	9.2	6.0	4.3	2.1
Sum of exergy losses %	89.9	90.2	92.7	95.4	96.8	98.6
CDP %	10.1	9.8	7.3	4.6	3.2	1.4

Table 3. Constituent Exergy Losses for the Link
Comprising the Principal Heat Exchangers and
the Principal Heating Network (at t_0 = +2 ^{o}C)

Component	Constituent exergy loss	
	MJ/GJ	%
Bleed steam	921	70.0
Electricity for driving pumps	37	2.8
Steel	15	1.1
Local exergy loss	166	12.6
Sum	1139	86.5

5. EXHAUSTION OF UNRESTORABLE NATURAL RESOURCES

The exhaustion of unrestorable natural resources is inevitable but very undesirable. Therefore the analysis of the exhaustion of unrestorable natural resources is of great importance. In order to calculate the exhaustion of unrestorable natural resources (ENR) the balance equations (2.10) should be supplemented if domestic unrestorable natural resources are of interest. In this case imported raw-materials, fuels and semi-finished products should be separately taken into account:

$$e_{km} = \sum_n (a_{nm} - f_{um}) e_{kn} + \sum_r a_{rm} e_{kr} + B_{km} \, , \qquad (5.1)$$

where: e_{km}, e_{kn}, e_{kr} — exhaustion of the domestic unrestorable natural resource k per unit of the product m, n and r ,

a_{rm} — coefficient of gross consumption of imported raw material, fuel or semi-finished product r ,

B_{km} — immediate gross consumption of the domestic unrestorable natural resource k in the link m .

The index e_{kr} should be determined according to the assumption, that financial means for the import are gained by the export. The exported products are also burdened with ENR. According to the assumption, that the unit of monetary value of the exported and imported products is burdened with the same ENR, the following relation can be formulated (Szargut [2], 1987):

$$e_{kr} = e_k^{(d)} D_r = \frac{\sum_n X_n e_{kn}}{\sum_n X_n D_n} D_r \, , \qquad (5.2)$$

where: $e_k^{(d)}$ — ENR per unit of the monetary value of exported products,

D_r, D_n — specific monetary value of the imported product r and exported product n ,

X_n — export of the product n (e.g. per year).

After the introduction of (5.2) into (5.1) the set of balance eqs takes the form

$$e_{km} = \sum_n (a_{nm} - f_{nm} + d_{nm}) e_{kn} + B_{km} \, , \qquad (5.3)$$

where

$$d_{nm} = X_n \frac{\sum_r D_r a_{rm}}{\sum_n X_n D_n} \, . \qquad (5.4)$$

In matrix notation the set of eqs (5.3) can be presented as

$$e (E - A + F - D) = B \, , \qquad (5.5)$$

and

$$e = B (E - A + F - D)^{-1} \, , \qquad (5.6)$$

where: A , F , D - square matrices MxM ,
      e - rectangular matrix KxM .

The set of eqs (5.5) is very large, because it comprises all the domestic semi-finished products and all the exported products. For approximate solution it can be assumed that per unit of the monetary value the exported products are burdened with the same ENR as products consumed within the country (Bałandynowicz, 1987). From this assumption it results that

$$e_k^{(d)} = \frac{\sum B_k}{(DN)} , \qquad (5.7)$$

where: $\sum B_k$ - total exhaustion of the domestic natural resour-
               ce k ,
      (DN)   - monetary value of all domestic products consu-
                med within the country together with exported
                products.

The assumption (5.7) is not exact, but it enables us to formulate the set of balance eqs (5.5) exclusively for domestic semi-finished products. ENR for imported products results from the first part of eq. (5.2), and so the second term on the right-hand side of eq. (5.1) is known.

Usually the total exhaustion of domestic unrestorable natural resources is of interest

$$e_m = \sum_k e_{km} . \qquad (5.8)$$

When analysing the production of materials and energy carriers, the <u>utilization efficiency of unrestorable domestic natural resources</u> (UEN) can be defined (Szargut and Jełowicki, 1990)

$$\xi = \frac{b}{e} . \qquad (5.9)$$

Usually $e > b$ and $\xi < 1$, but for secondary raw materials $\xi \gg 1$ and for imported raw materials and fuels, very often $\xi > 1$.

For secondary raw materials ENR contains only the consumption of exergy for the change of the shape and for the transportation. Usually it is much smaller than the exergy of the material under consideration. The inequality $\xi \gg 1$ indicates a great profitability of the utilization of secondary raw materials. These materials substitute the semi-finished products requiring a large consumption of exergy for their production.

The inequality $b > e$ can occur for imported raw materials, fuels and semi-finished products if the exported products are more sophisticated than the imported ones. In countries possessing very limited deposits of unrestorable natural resources (e.g. Switzerland) ENR can be much smaller than the exergy for imported materials.

In Table 4 some values of UEN are cited for the Polish metallurgical and energy industry according to data from 1980 and 1975 (Szargut and Jełowicki, 1990). The values of UEN have been compared with CDP. The interconnections between the two conside-

| Material or energy carrier | | CDP % | UDN % | Notes |
| Name | Specific exergy | | | |
1	2	3	4	5
Calcium oxide	2.26 MJ/kg	28.5 35.7	27.9 35.8	Shaft furnace, coke Maerz furnace
Cement	0.635 MJ/kg	10.6 6.9	13.9 7.0	Dry method Wet method
Coal (hard)	1.09 C_1	95.1	95.2	Transportation losses included
Coal (lignite)	1.17 C_1	94.9	95.0	
Coke	1.06 C_1	72.5	68.7	
Crude oil	1.07 C_1	96.3 –	96.4 239	Domestic Imported
Electricity	1 MJ/MJ	22.8	23.5	Transmission losses included
Gas (blast-furnace)	0.98 C_1	107 116	343 116	Substituting imported natural gas Substituting hard coal
Gas (coke-oven)	1.0 C_1	95.6	305	Substituting imported natural gas
Gas (liquid)	1.07 C_1	84.0	191	Mainly from imported crude oil
Gas (natural)	1.04 C_1	99.8 – 98.2 71.3	99.8 302 98.2 71.4	Domestic, high CH_4 content Imported, high CH_4 content Domestic, contaminated with N_2 Domestic, after removing of N_2
Gasoline	1.07 C_1	83.8	177	Mainly from imported crude oil
Iron (pig)	8.2 MJ/kg	28.8	27.8	Liquid
Oil	1.07 C_1	84.0 83.9	180 178	Fuel, mainly from imported crude oil Diesel, mainly from imported crude oil

1	2	3	4	5
Oxygen (gas)	2.37 MJ/kg	0.70	0.72	
Steel (liquid)	8.04 MJ/kg	26.2	36.6	Open-hearth furnace, 50 % scrap
		36.0	54.4	Electric-arc furnace, 100 % scrap
		26.7	30.8	Oxygen converter, 23 % scrap
Steel sections	7.04 MJ/kg	16.1	21.3	Hot rolled, domestic
		–	21.6	Hot rolled, imported
Steel tubes	7.04 MJ/kg	12.3	16.3	Domestic
		–	8.8	Imported
Steel scrap	7 MJ/kg	96.6	2930	

$^*)C_1$ – lower calorific value

red industry branches and other branches are very small, and therefore the set of eqs (5.1) has been formulated only for the two considered branches. The indices e_{kr} have been determinated from eq. (5.2). The index $e^{(d)}$ of the total exhaustion of domestic unrestorable natural resources per unit of the monetary value of the exported products has been calculated from eq. (5.7) according to the data of Bałandynowicz (1987).

For the products fabricated with a prevailing utilization of the domestic unrestorable natural resources the values UEN < 1 have been calculated.

For the imported fuels (crude oil, natural gas) the values UEN > 1 have been determined. It has been taken into account that the imported natural gas closes the balance of gaseous fuels, because the domestic production of gaseous fuels is smaller than the demand and does not depend on the demand. The difference of the demand and domestic production is covered by the import. The domestic gaseous fuels produced as by-products (coke-oven gas, blast-furnace gas) substitute the imported natural gas and therefore their UEN is also greater than unit[*]. But only the basic part of the production rate of blast-furnace gas substitutes the natural gas; the peak-part of the production rate is very variable. Therefore it is burned in bi-fuel boilers and substitutes the hard coal. CDP of the blast-furnace gas is greater than unit, because its exergetic substitution ratio referred to the natural gas is smaller than CDP of natural gas; see eq. (2.8).

---

[*]It has been assummed that the coke-oven gas and blast-furnace gas substitute the natural gas with the energy efficiency 0.99 and 0.88 respectively. The peak-part of the blast-furnace gas susbtitutes the hard coal with the energy efficiency 0.82.

The relatively large values of UEN characterizing the gaseous by-products (coke-oven gas, blast-furnace gas), cause a decrease of UEN characterizing the major products (coke, pig iron). Therefore for these major products CDP > UEN. On the other hand in processes consuming gaseous fuels CDP < UEN (e.g. for hot rolled steel products).

Liquid fuels are produced mainly from the imported crude oil (the domestic production is very small). Therefore for all the liquid fuels UEN > 1, and for the materials fabricated with a consumption of liquid fuels CDP < UEN (e.g. calcium oxide produced in Maerz-furnace, liquid steel produced in open-hearth furnace).

For electricity production CDP < UEN, because some small part of electricity is produced in hydro-power stations. In countries with great fraction of hydro-power-electricity, this inequality would be much more explicit.

UDN is very large for the steel scrap. It indicates, that the utilization of secondary raw materials secures a very large saving of the domestic natural resources. As a consequence for liquid steel produced from scrap in electric-arc furnaces CDP < UEN.

6. ECOLOGICAL COST

If the ecological impact of production processes should be analysed, it is not sufficient to determine ENR connected with the extraction of raw materials and fuels from natural deposits. Additional destruction of natural resources results from the deleterious influence of waste products discharged to the environment. The waste products accelerate the corrosion of the products of human activity, decrease the productivity of agriculture and forestry, destruct partially the natural resources and influence harmfully the human health. The exhaustion of unrestorable natural resources, taking into account all these factors, has been termed the index of <u>ecological cost</u> (EcC) (Szargut, 1986). If the domestic ecological cost is to be determined, the impact of imported materials and fuels should be respected, according to the method presented in section 5. Thus the following set of eqs determining EcC can be formulated

$$\rho_m = \sum_n (a_{nm} - f_{nm} + d_{nm} + \sum_s{}' B_{sm} x_{ns}) \rho_n +$$

$$+ \sum_s B_{sm} (\sum_k y_{ks} + z_s) + \sum_k{}' B_{km} , \qquad (6.1)$$

where: s – serial number of the deleterious waste product,
     B_{km} – immediate gross consumption of the unrestorable domestic natural resource k per unit of complex useful products containing a unit of the major product m ,
     B_{sm} – exergy of deleterious waste product s ,
     x_{ns} – destruction coefficient of the useful product n per unit of the exergy of waste product s ,
     y_{ks} – destruction coefficient of the unrestorable natural resource k per unit of the exergy of waste product s ,

$$z_s \qquad \text{- multiplier of exergy consumption to eliminate the results of human health deterioration, per unit of exergy of the waste product } s.$$

The destruction coefficients can be defined as

$$x_{ns} = \frac{\delta m_n}{B_s}, \qquad y_{ks} = \frac{\delta B_k}{B_s}, \qquad (6.2)$$

where: δm_n - amount of units of the destructed useful product n,

δB_k - exergy decrease of the damaged natural resource.
The coefficient x_{ns} should also take into account the reduction of agricultural and forestrial production.

Similarly the global EcC can be calculated. In such a case the coefficient d_{nm} in eq. (6.1) should be omitted, and the last term on the right-hand side should take into account all the unrestorable natural resources.

The degree of the harmful impact of the considered process on natural resources can be char terized by means of the <u>ecological efficiency</u>

$$\eta_\varrho = \frac{b}{\overline{\varrho}}. \qquad (6.3)$$

Usually $\eta_\varrho < 1$, but sometimes values $\eta_\varrho > 1$ can appear, if restorable natural resources are applied for the considered process (e.g. the agricultural production of grain).

By means of EcC the ecological economy can be formed, which can be used to minimize the exhaustion of unrestorable natural resources. The desired consumption level should be introduced as a limitation when solving the optimization problem. The ecological economy would differ distinctly from the classical economy. All calculations of ecological economy would be carried out in exergy units.

7. CONCLUSIONS

1. The analysis of cumulative exergy consumption and of cumulative exergy losses provides information about the possibilities of improving the technological networks of the fabrication of particular products.

2. The sequence method of analysis enables us to receive approximate results without analysing the total network of production processes.

3. The analysis of exhaustion of domestic unrestorable natural resources informs us about the profitability of the import of raw materials, fuels and semi-finished products and about the effectiveness of utilization of secondary raw materials.

4. The ecological cost can serve for the introduction of ecological economy aiming at a minimization of the consumption of unrestorable natural resources.

REFERENCES

Bałandynowicz, H. 1987. Structural modell of the fuel-and-energy
    system and investigations of the cumulative transformation
    of the input factors into output factors (in Polish).
    PhD-thesis,Technical University of Silesia, Faculty of Me-
    chanics and Energetics, Gliwice, Poland.
Bibrowski, Z. (Ed.) et all., 1983. Cumulative consumption of ener
    gy (in Polish). Warsaw: PWN.
Boustead, I., Hancock, G.H. 1979. Handbook of industrial energy
    analysis. Chichester: Ellis Harwood.
Chapman, P.F. 1974. Energy costs: a review of methods. Energy
    Policy 2: 91 - 103.
Guidelines of Energy Analysis. 1974. International Federation of
    Institutes of Advanced Study. Stockholm.
Morris, D.R. 1989. Exergy analysis and cumulative exergy consum-
    ption of complex chemical processes. Chem. Engng. Science
    (in press).
Riekert, L. 1974. The efficiency of energy utilization in chemi-
    cal processes. Chem. Engng. Sci. 29: 1613 - 1620.
Szargut, J. 1985. Thermodynamic perfection of the production of
    materials and useful energy (in Polish). Energetyka XXXIX:
    485 - 488.
Szargut, J. 1986. Application of exergy for the calculation of
    ecological cost. Bull. Pol. Acad.: Techn. 34: 475 - 480.
Szargut, J., Morris,D.R. 1987. Cumulative exergy consumption and
    cumulative degree of perfection of chemical processes.
    Energy Research 11: 245 - 261.
[1] Szargut, J. 1987. Analysis of cumulative exergy consumption.
    Energy Research 11: 541 - 547.
[2] Szargut, J. 1987. Influence of the imported goods on the cu-
    mulative energy indices. Bull. Pol. Acad.: Techn. 35: 591 -
    595.
Szargut, J., Morris, D.R., Steward, F. 1988. Exergy analysis of
    thermal, chemical and metallurgical processes. New York:
    Hemisphere Publ. Corp.
Szargut, J. 1988. Exergy losses in the chains of technological
    processes. Bull. Pol. Acad.: Techn. 36: 513 - 521
Szargut, J. 1989. Indices of the cumulative consumption of energy
    and exergy (in Polish). Zeszyty Naukowe Politechniki Ślą-
    skiej, Energetyka No 106: 39 - 55.
Szargut, J., Majza, E. 1989. Thermodynamic imperfection and exer-
    gy losses at the production of pig-iron and steel. Archives
    of Metallurgy 34: 197 - 216.
Szargut, J., Morris, D.R. 1990. Analysis of cumulative exergy
    losses at the production of lead. Energy Research (in press)
Szargut, J. 1990. Analysis of cumulative exergy losses at the
    production and delivery of heat from a heat-and-power plant
    (in Polish). Archiwum Energetyki (in press).
Szargut, J., Jełowicki, A. Exhaustion of unrestorable natural re-
    sources. Archiwum Energetyki (in press).

Glossary of Symbols

(see also more detailed lists of symbols at the end of some papers)

A   area, availability, action, affinity, extremized objective
A_{nm} coefficients of exergy losses
a   accelleration, coefficient
B, b exergy
C   capacity, cost, coefficient
c   price
d   diameter
E   energy, local efficiency
e   unit vector
e   economic value of unit exergy
F   flow, criterion function
F, f enlarged vector rate function and vector rate function,
      respectively
f   driving force, friction factor
G   gas flow, state transformation function
H   hamiltonian function, enthalpy
h   specific enthalpy, heat transfer coefficient, heating function
I   investment, solid enthalpy (dry basis)
i   gas enthalpy (dry basis)
J   total cost, flux
J   flux vector
K   chemical equilibrium constant, coefficient
k   rate constant, coefficient, transfer or exchange property
L   lagrangian, kinetic potential, thermodynamic length,
      phenomenological coefficient
M   molar mass, mass property
N   mole number, total number of stages, power output
n   dimensionality of state vector, stage number
P   pressure, power, profit, production
p   probability
Q   heat, charge
q   heat flux
R   universal gas constant, resistance, recycle flow
r   index
S   entropy, solid flow
T   absolute temperature
t   time or time-type variable

U	internal energy, utility property
u	control vector
V	volume, scalar potential
v	speed, thermodynamic speed
v	velocity vector
W	work, moisture content
X	gas humidity (dry basis)
X, x	enlarged state vector and state vector, respectively
x	liquid phase mole fraction
Y	thermodynamic conjugate of state **X**
y	vapour phase mole fraction
z	multiplier, ratio
Z, z	enlarged adjoint vector and adjoint vector, respectively

α	absorption coefficient, friction coefficient, cost coefficient
β	coefficient of renovations
δ	effective diameter
ε	emissivity, reaction extent, coefficient
η	efficiency
θ	Carnot factor, time increment
λ	Lagrangian multiplier, parameter
μ	chemical potential
ν	stoichiometric coefficient, dynamic viscosity
ρ	density
σ	entropy source, Stefan-Boltzmann constant
τ	time constant, amortization rate, equivalent period
ϕ	radiation flow

Index